Advanced Cybersecurity Services Design

Advanced Cybersecurity Services Design

Editor

Víctor A. Villagrá

Basel • Beijing • Wuhan • Barcelona • Belgrade • Novi Sad • Cluj • Manchester

Editor
Víctor A. Villagrá
Universidad Politécnica de Madrid
Madrid, Spain

Editorial Office
MDPI
St. Alban-Anlage 66
4052 Basel, Switzerland

This is a reprint of articles from the Special Issue published online in the open access journal *Electronics* (ISSN 2079-9292) (available at: https://www.mdpi.com/journal/electronics/special_issues/cybersecurity_services).

For citation purposes, cite each article independently as indicated on the article page online and as indicated below:

Lastname, A.A.; Lastname, B.B. Article Title. *Journal Name* **Year**, *Volume Number*, Page Range.

ISBN 978-3-0365-9240-4 (Hbk)
ISBN 978-3-0365-9241-1 (PDF)
doi.org/10.3390/books978-3-0365-9241-1

© 2023 by the authors. Articles in this book are Open Access and distributed under the Creative Commons Attribution (CC BY) license. The book as a whole is distributed by MDPI under the terms and conditions of the Creative Commons Attribution-NonCommercial-NoDerivs (CC BY-NC-ND) license.

Contents

About the Editor vii

Victor A. Villagrá
Advanced Cybersecurity Services Design
Reprinted from: *Electronics* **2022**, *11*, 2803, doi:10.3390/electronics11182803 1

Andrés Marín-López, Sergio Chica-Manjarrez, David Arroyo, Florina Almenares-Mendoza and Daniel Díaz-Sánchez
Security Information Sharing in Smart Grids: Persisting Security Audits to the Blockchain
Reprinted from: *Electronics* **2020**, *9*, 1865, doi:10.3390/electronics9111865 5

Wenbin Yu, Yiyin Wang and Lei Song
A Two Stage Intrusion Detection System for Industrial Control Networks Based on Ethernet/IP
Reprinted from: *Electronics* **2019**, *8*, 1545, doi:10.3390/electronics8121545 21

Xavier Larriva-Novo, Carmen Sánchez-Zas, Víctor A. Villagrá, Mario Vega-Barbas and Diego Rivera
An Approach for the Application of a Dynamic Multi-Class Classifier for Network Intrusion Detection Systems
Reprinted from: *Electronics* **2020**, *9*, 1759, doi:10.3390/electronics9111759 39

João Henriques, Filipe Caldeira, Tiago Cruz and Paulo Simões
Combining K-Means and XGBoost Models for Anomaly Detection Using Log Datasets
Reprinted from: *Electronics* **2020**, *9*, 1164, doi:10.3390/electronics9071164 57

Hanan Hindy, Robert Atkinson, Christos Tachtatzis Jean-Noël Colin, Ethan Bayne and Xavier Bellekens
Utilising Deep Learning Techniques for Effective Zero-Day Attack Detection
Reprinted from: *Electronics* **2020**, *9*, 1684, doi:10.3390/electronics9101684 73

Youngho Cho
Intelligent On-Off Web Defacement Attacks and Random Monitoring-Based Detection Algorithms
Reprinted from: *Electronics* **2019**, *8*, 1338, doi:10.3390/electronics8111338 89

Hansaka Angel Dias Edirisinghe Kodituwakku, Alex Keller and Jens Gregor
InSight2: A Modular Visual Analysis Platform for Network Situational Awareness in Large-Scale Networks
Reprinted from: *Electronics* **2020**, *9*, 1747, doi:10.3390/electronics9101747 109

John Ibañez Rodriguez, Santiago Rocha Duran, Daniel Díaz-López, Javier Pastor-Galindo and Félix Gómez Mármol
C^3: A Conversational Agent to Detect Online Sex Offenders
Reprinted from: *Electronics* **2020**, *9*, 1779, doi:10.3390/electronics9111779 125

Ferhat Ozgur Catak, Ismail Aydin, Ogerta Elezaj and Sule Yildirim-Yayilgan
Practical Implementation of Privacy Preserving Clustering Methods Using a Partially Homomorphic Encryption Algorithm
Reprinted from: *Electronics* **2020**, *9*, 229, doi:10.3390/electronics9020229 149

Jiyeon Kim and Hyong S. Kim
Intrusion Detection Based on Spatiotemporal Characterization of Cyberattacks
Reprinted from: *Electronics* **2020**, *9*, 460, doi:10.3390/electronics9030460 169

Andrew Ramsdale, Stavros Shiaeles and Nicholas Kolokotronis
A Comparative Analysis of Cyber-Threat Intelligence Sources, Formats and Languages
Reprinted from: *Electronics* **2020**, *9*, 824, doi:10.3390/electronics9050824 . . . . **193**

Ivan Kovačević, Stjepan Groš and Karlo Slovenec
Systematic Review and Quantitative Comparison of Cyberattack Scenario Detection and Projection
Reprinted from: *Electronics* **2020**, *9*, 1722, doi:10.3390/electronics9101722 . . . . **215**

Hyok An, Yoonjong Na, Heejo Lee and Adrian Perrig
Resilience Evaluation of Multi-Path Routing against Network Attacks and Failures
Reprinted from: *Electronics* **2021**, *10*, 1240, doi:10.3390/electronics10111240 . . . . **247**

Tarek Gaber, Yassine El Jazouli, Esraa Eldesouky, and Ahmed Ali
Autonomous Haulage Systems in the Mining Industry: Cybersecurity, Communication and Safety Issues and Challenges
Reprinted from: *Electronics* **2021**, *10*, 1357, doi:10.3390/electronics10111357 . . . . **263**

About the Editor

Víctor A. Villagrá

Víctor A. Villagrá is a professor in the Department of Telematics Engineering at the Telecommunications Engineering School of the Technical University of Madrid (UPM). He gained a PhD in Telecommunications Engineering in 1994 from the UPM, and he has focused on the areas of Cybersecurity and Management of Telecommunications Networks and Services in his teaching and research.

He is currently the Head of Studies at the Telecommunications Engineering School of UPM and the director of the Official Master's Degree in Cybersecurity at the UPM. He has supervised 10 PhD students and published a teaching book about cybersecurity: "Security in Telecommunication Networks". He is an author and co-author of more than 100 contributions to scientific journals and conferences. He has participated and coordinated as main researcher several research projects in the scope of the European Union, international Defence organizations, and Spanish R&D specific nation projects in these areas, as well as various contracts with companies and entities in the area of Cybersecurity.

Editorial

Advanced Cybersecurity Services Design

Victor A. Villagrá

Departamento Ingeniería Telemática (DIT), ETSI Telecomunicación (ETSIT),
Universidad Politécnica de Madrid (UPM), Avda. Complutense 30, 28040 Madrid, Spain;
victor.villagra@upm.es

1. Introduction

Cybersecurity technologies have been researched extensively in the last few years in order to face the current threat landscape, which has shown a continuous growth in the quality and quantity of attacks that are oriented toward any potentially vulnerable items (people, software, firmware, hardware, etc.). Thus, there is a need for more sophisticated cybersecurity services that are able to combine different technologies to cover all the different aspects that such attacks may utilize.

These advanced cybersecurity services must enrich the different areas of cybersecurity, including cyberattacks prevention, detection and response, as well as advanced supporting infrastructures for these services. Nowadays, most of the prevention initiatives rely on increasing users' awareness, in order to prevent social engineering attacks, but there are also some technological areas that can complement a security architecture for the prevention of cyberattacks. In the area of attacks detection, machine learning-based algorithms are one of the most promising techniques for anomaly detection, as well as the use of Cyber Threats Intelligence in order to share knowledge about the attacks. In the area of supporting infrastructure, there is an increasing interest in leveraging traditional cryptographic algorithms with new approaches, such as homomorphic encryption when privacy is concerned, quantum and post-quantum cryptography, and the use of blockchain technologies for different advanced cybersecurity services.

2. The Present Issue

This Special Issue includes 14 contributions that cover these areas. It includes two review contributions: [1] provides a survey with an overview of the state of the art in detecting and projecting cyberattack scenarios, i.e., approaches that automate the analysis of alerts to detect large-scale attacks and predict the attacker's next steps, with a focus on evaluation and the corresponding metrics. On the other hand, Ref. [2] reviews a specific application area, namely the identification and discussion of the relation between the safety of Autonomous Haulage Systems in the mining environment and both cybersecurity and communication; furthermore, the article highlights their challenges and open issues.

The Special Issue also includes 12 research contributions. It includes several contributions related to advanced services for Intrusion Detection Systems. In [3], the use of different machine learning models depending on the specific scenarios and datasets is addressed, as well as the design of an automatic dynamic model selector for anomalies detections scenarios. The article in [4] focuses on the use of deep learning techniques for the detection of zero-day attacks, with an autoencoder implementation that is able to monitor the rate of false-negative detection rates. In [5], an integrated scalable framework is proposed that aims at efficiently detecting anomalous events on large amounts of unlabelled data logs. Detection is supported by clustering and classification methods that take advantage of parallel computing environments using two models: one based on K-Means and the other based on a XGBoost system implementing a gradient tree boosting algorithm. In [6] a spatiotemporal characterization of cyberattacks for detecting cyberattacks is proposed

Citation: Villagrá, V.A. Advanced Cybersecurity Services Design. *Electronics* **2022**, *11*, 2803. https://doi.org/10.3390/electronics11182803

Received: 1 September 2022
Accepted: 3 September 2022
Published: 6 September 2022

Publisher's Note: MDPI stays neutral with regard to jurisdictional claims in published maps and institutional affiliations.

Copyright: © 2022 by the author. Licensee MDPI, Basel, Switzerland. This article is an open access article distributed under the terms and conditions of the Creative Commons Attribution (CC BY) license (https://creativecommons.org/licenses/by/4.0/).

using a stochastic graph model to represent these cyberattacks in time and space. In [7], the area of industrial control networks is addressed with a two-stage intrusion detection system, including a traffic prediction model and an anomaly detection model. A chatbot is proposed in [8] for detecting online sex offenders, based on an Artificial Conversational Entity (ACE) that connects to different online chat services to start a conversation. The ACE was designed using generative and rule-based models in charge of generating the posts and replies that constitute the conversation from the chatbot side. The proposed solution also includes a module to analyse the conversations performed by the chatbot and calculate a set of 25 features that describes the suspect's behaviour. Finally, Ref. [9] introduces Insight2, an open-source platform for manipulating both streaming and archived network flow data in real time in order to understand normal activity and identify abnormal activity.

This Special Issue also contains two protection-related research contributions: [10] proposes a countermeasure for on-off web defacement attacks with a random monitoring strategy, designing and validating two specific strategies for such a purpose. On the other hand, Ref. [11] addressed the evaluation of multi-path routing as a protection feature against network attacks and failures, with the study of the following two different models: first-hop multi-path and multi-hop multi-path routing.

The Special Issue also includes several contributions about related technologies for providing support to the design of advanced cybersecurity services: [12] provides an analysis of a partially homomorphic encryption algorithm for the design of services needing privacy-preserving functionalities. In [13] a system for using blockchain technologies for the accountability of cybersecurity audit results is proposed in order to boost the automation of both digital evidence gathering, auditing, and controlled information exchange. The study in [14] focuses on the area of Cyber-Threat Intelligence Sources, and Formats and Languages, while investigating the landscape of the available formats and languages, along with the publicly available sources of threat feeds, how these are implemented and their suitability for providing rich cyber-threat intelligence.

3. Future Directions

Cybersecurity is nowadays a race between the attackers and the different digital technologies actors (users, manufacturers, IT companies, etc.). A new technology, or a new protection system is usually followed by a new attacker method, so there is a need for new services based on new technologies that might make possible the establishment of a gap with attackers, increasing the confidence on the usage of business and personal services by users. Therefore, there is a need to extend the research on these areas with the adequate support of public and private entities in order to generate new research proposals that might be able to find the different steps to establish a gap with the attackers. The proposals in this Special Issue might provide small steps in pursuit of this aim, but there is a need for many more initiatives to be designed, tested and validated in order to make secure technologies available for everybody.

Funding: This research received no external funding.

Institutional Review Board Statement: Not applicable.

Informed Consent Statement: Not applicable.

Data Availability Statement: Not applicable.

Acknowledgments: First of all, I would like to thank all the researchers who submitted articles to this Special Issue for their excellent contributions. I am also grateful to all the reviewers who helped in the evaluation of the manuscripts and made very valuable suggestions to improve the quality of the contributions. I would like to acknowledge the editorial board of *Electronics*, who invited me to guest edit this Special Issue. I am also grateful to the *Electronics* Editorial Office staff who worked thoroughly to maintain the rigorous peer-review schedule and timely publication.

Conflicts of Interest: The author declares no conflict of interest.

References

1. Kovačević, I.; Groš, S.; Slovenec, K. Systematic Review and Quantitative Comparison of Cyberattack Scenario Detection and Projection. *Electronics* **2020**, *9*, 1722. [CrossRef]
2. Gaber, T.; El Jazouli, Y.; Eldesouky, E.; Ali, A. Autonomous Haulage Systems in the Mining Industry: Cybersecurity, Communication and Safety Issues and Challenges. *Electronics* **2021**, *10*, 1357. [CrossRef]
3. Larriva-Novo, X.; Sánchez-Zas, C.; Villagrá, V.; Vega-Barbas, M.; Rivera, D. An Approach for the Application of a Dynamic Multi-Class Classifier for Network Intrusion Detection Systems. *Electronics* **2020**, *9*, 1759. [CrossRef]
4. Hindy, H.; Atkinson, R.; Tachtatzis, C.; Colin, J.; Bayne, E.; Bellekens, X. Utilising Deep Learning Techniques for Effective Zero-Day Attack Detection. *Electronics* **2020**, *9*, 1684. [CrossRef]
5. Henriques, J.; Caldeira, F.; Cruz, T.; Simões, P. Combining K-Means and XGBoost Models for Anomaly Detection Using Log Datasets. *Electronics* **2020**, *9*, 1164. [CrossRef]
6. Kim, J.; Kim, H. Intrusion Detection Based on Spatiotemporal Characterization of Cyberattacks. *Electronics* **2020**, *9*, 460. [CrossRef]
7. Yu, W.; Wang, Y.; Song, L. A Two Stage Intrusion Detection System for Industrial Control Networks Based on Ethernet/IP. *Electronics* **2019**, *8*, 1545. [CrossRef]
8. Rodríguez, J.; Durán, S.; Díaz-López, D.; Pastor-Galindo, J.; Mármol, F. C3-Sex: A Conversational Agent to Detect Online Sex Offenders. *Electronics* **2020**, *9*, 1779. [CrossRef]
9. Kodituwakku, H.; Keller, A.; Gregor, J. InSight2: A Modular Visual Analysis Platform for Network Situational Awareness in Large-Scale Networks. *Electronics* **2020**, *9*, 1747. [CrossRef]
10. Cho, Y. Intelligent On-Off Web Defacement Attacks and Random Monitoring-Based Detection Algorithms. *Electronics* **2019**, *8*, 1338. [CrossRef]
11. An, H.; Na, Y.; Lee, H.; Perrig, A. Resilience Evaluation of Multi-Path Routing against Network Attacks and Failures. *Electronics* **2021**, *10*, 1240. [CrossRef]
12. Catak, F.; Aydin, I.; Elezaj, O.; Yildirim-Yayilgan, S. Practical Implementation of Privacy Preserving Clustering Methods Using a Partially Homomorphic Encryption Algorithm. *Electronics* **2020**, *9*, 229. [CrossRef]
13. Marín-López, A.; Chica-Manjarrez, S.; Arroyo, D.; Almenares-Mendoza, F.; Díaz-Sánchez, D. Security Information Sharing in Smart Grids: Persisting Security Audits to the Blockchain. *Electronics* **2020**, *9*, 1865. [CrossRef]
14. Ramsdale, A.; Shiaeles, S.; Kolokotronis, N. A Comparative Analysis of Cyber-Threat Intelligence Sources, Formats and Languages. *Electronics* **2020**, *9*, 824. [CrossRef]

Article

Security Information Sharing in Smart Grids: Persisting Security Audits to the Blockchain

Andrés Marín-López [1,*,†], Sergio Chica-Manjarrez [1], David Arroyo [2], Florina Almenares-Mendoza [1] and Daniel Díaz-Sánchez [1]

1. Telematics Engineering Department, Politechnical Engineering School, University Carlos III de Madrid, Avda. de la Universidad, 20, 28911 Leganés, Madrid, Spain; sergio.chica@uc3m.es (S.C.-M.); florina.almenares@uc3m.es (F.A.-M.); daniel.diaz@uc3m.es (D.D.-S.)
2. Institute of Physical and Information Technologies (ITEFI), Spanish National Research Council (CSIC), C/Serrano, 144, 28006 Madrid, Spain; david.arroyo@csic.es
* Correspondence: andres.marin@uc3m.es
† Current address: KIT Campus South, Building 50.34, Am Fasanengarten 5, 76131 Karlsruhe, Germany.

Received: 15 September 2020; Accepted: 4 November 2020; Published: 6 November 2020

Abstract: With the transformation in smart grids, power grid companies are becoming increasingly dependent on data networks. Data networks are used to transport information and commands for optimizing power grid operations: Planning, generation, transportation, and distribution. Performing periodic security audits is one of the required tasks for securing networks, and we proposed in a previous work AUTOAUDITOR, a system to achieve automatic auditing. It was designed according to the specific requirements of power grid companies, such as scaling with the huge number of heterogeneous equipment in power grid companies. Though pentesting and security audits are required for continuous monitoring, collaboration is of utmost importance to fight cyber threats. In this paper we work on the accountability of audit results and explore how the list of audit result records can be included in a blockchain, since blockchains are by design resistant to data modification. Moreover, blockchains endowed with smart contracts functionality boost the automation of both digital evidence gathering, audit, and controlled information exchange. To our knowledge, no such system exists. We perform throughput evaluation to assess the feasibility of the system and show that the system is viable for adaptation to the inventory systems of electrical companies.

Keywords: security auditing; permissioned blockchain; scalability; smart grid security

1. Introduction

A smart grid is an energy delivery system that moves from a centrally controlled system, like the ones we currently have, to a consumer-driven approach, i.e., an iterative system relying on bi-directional communication to adapt and tune the delivery of energy in the real-time market. A smart grid includes a broad range of sophisticated sensors that constantly assess the state of the grid and the electrical power demand and availability, with the aim of optimizing the energy supply. The power grid is evolving into a cyber physical system where smart devices allow advanced monitoring and control. This area is developing fast, as one can guess by seeing the 68 active working groups of the IEEE SA around smart grids. More than half of the EU Member States have reached a 10% installation rate for electricity smart meters, showing the first important step in their large-scale rollout programs. Seven have already reached 80% like Denmark, or even finished their large-scale electricity smart metering roll-out like Estonia (>98% in 2017), Finland (100% by 2013), Italy (95% by 2011), Malta (80–85% by 2014), Spain (100% end of 2018), and Sweden (100% by 2009). Some of them are already proceeding with the second generation rollout, like Italy, or are planning to [1]. Other devices are also expected to be incorporated into the so-called smart grid, such as domestic power micro-generators, smart protection and storage devices, and even electric vehicles. Electrical data provided by such devices will be collected not only by customer applications but also by other stakeholders, most notably power distribution and power transportation companies. Besides enhanced control, such data will allow new intelligent features such as self-healing, resilience, sustainability, and will improve the efficiency of energy critical infrastructure. Near-real time communication between the power grid subsystems (from the concentrators to the legacy systems) with smart sensors and devices will be required to achieve optimization, automation and control of the smart grid. These new features will bring obvious benefits for electrical companies, but also customers are expected to benefit from smart grids. Specifically, from smart meters they can have three important benefits: (1) A better customer service comes from (a) fewer and shorter duration of outages as smart meters report instantaneously and (b) from a faster service (remote operations); (2) viewing energy usage in near real-time and comparing with current tariffs, which can lead to adjusted monthly bills; and (3) manual periodic metering no longer being required.

Besides the mentioned benefits of smart grids, the risks are obvious if the system fails. In this work we address the protection of smart grids as critical infrastructures against cyber attacks, and we specifically address the security auditing of devices and networks that comprise smart networks. The security auditing we are referring to in this article is also known as pentesting or vulnerability testing.

Critical infrastructures require automated security control assessment, as described in NISTIR-8011 Volume 3 and 4 [2,3]. This can be achieved by the integration of auditing with the inventory system, so as to have a realistic view of the company assets together with their associated risks in near real time. We foresee this integration as automatic or semi-automatic auditing procedures fired from the inventory system. The auditing processes incorporate security metadata to the inventoried assets of the company: Releases and patch status history, known vulnerabilities and their respective severity information. The risk metrics incorporate the severity of exposures, and facilitate the selection of the vulnerabilities that have to be mitigated, according to the risk appetite of the company.

Cybersecurity information sharing and collaboration between organizations can decrease the time of threat detection and increase the accuracy of detection. Several researches have pursued privacy in cybersecurity information [4,5]. We aim to contribute in the accountability of this information sharing.

In [6] we described AUTOAUDITOR, a system for automatic or semi-automatic security auditing to be integrated with power grid companies inventory systems. In this article we address the accountability of the system via the integration of auditing results records in a permissioned blockchain. Certainly, in the energy sector there exist a vast set of requirements and needs than can be fulfilled by an adequate blockchain architecture [7]. In our case, the decentralized nature of the blockchain is very appealing as the backbone of a least privilege strategy along the chain of custody of digital evidences. Furthermore, the tamper-resistant nature of blockchain makes it a very good candidate for protecting the integrity of audit trails. Finally, a permissioned blockchain enables the design of an adequate access control for the overall process of auditing information systems. Indeed, without a proper separation of duties this task cannot be satisfied.

This paper is organized as follows. Section 2 examines related automatic auditing systems, some of them using machine learning, and related works with a different usage of distributed ledgers in energy systems. Section 3 presents the previous version of AUTOAUDITOR [6], and discusses the requirements of an automatic auditing system in the field of smart grids. Section 4 outlines the benefits of introducing blockchains in the system, analyzes and justifies our selection of distributed ledger technology. Section 5 gives an overview of the implementation details, explaining the execution flow, identified roles, and the smart contract implemented. Finally, Section 6 describes performance results and Section 7 formulates our conclusions.

2. Related Work

2.1. Security Assessment. Works in Automatic Auditing

Security assessment is defined as "a circular process of assessing assets for their security requirements, based on probable risks of attack, liability related to successful attacks, and costs for ameliorating the risks and liabilities", according to the standard developed for handling the security of power systems and associated information exchange, IEC TS 62351 [8].

The U.S. National Institute of Standards and Technology (NIST) and the Department of Homeland Security (DHS) are working in defining capabilities of automation support for ongoing security control assessments such as software asset management (SWAM) [2] and software vulnerability management (VULN) [3]. Security control assessment is a crucial part to manage information security risks across a company. Risk management is a complex and multifaceted activity, which starts establishing a realistic "risk frame" where assumptions about threats, vulnerabilities, consequences/impact, and likelihoods are identified, followed by information security assessment and monitoring [9]:

- *Software Asset Management* capability is defined as part of Continuous Diagnostics and Mitigation (CDM) process. Its main purpose is to control risk created by unmanaged or unauthorized software installed on a supervised network. The authorized software installed on every device is inventoried and can be as small as a line of source code or as large as a software suite made up of multiple products, thousands of individual executables, and countless lines of code, e.g., firmware, BIOS, operating systems, applications, services, and malware. Thus, a software asset is usable and automated, being described in terms of Common Platform Enumeration (CPE) names.
CPE is a standardized method of describing and identifying classes of applications, operating systems, and hardware devices present among an enterprise's computing assets. For that, Software Identification (SWID) tags for identifying software installed is included. A CPE name includes at least four unique attributes: Part, vendor, product, and version, and four additional attributes: Language, sw_edition, target_hw, and update. These second group of attributes are used to identify where software vulnerabilities may be found.
SWAM supports vulnerability management and configuration settings management. Likewise, it directly supports Hardware Asset Management (HWAM) because checking software asset requires knowing where it was or should be installed.
- *Software Vulnerability Management* addresses defects present in software on the network. Thus, once software and hardware are part of the inventory, VULN capability provides visibility into the vulnerabilities in software authorized to operate or access to the company's network(s), in order to manage and patch them in an appropriate manner. Vulnerable software is software in use on a system that has a vulnerability, but has not yet been patched or otherwise mitigated, being a key target of attackers in order to initiate an attack [3]. VULN manages and assesses directly two kinds of software flaws: Common Vulnerabilities and Exposures (CVEs), whose program works with software providers, vulnerability coordinators, bug bounty programs, and vulnerability researchers to provide a list of publicly disclosed vulnerabilities, and Common Weakness Enumeration (CWE), which provides identifiers for weaknesses that result from poor coding practices and have the potential result in software vulnerabilities.

Researchers, developers, and industry have developed tools for automating these capabilities. The project Software Assurance Marketplace (SWAMP) offers a service to provide continuous software assurance capabilities to researchers and developers. They also offer a self-contained, standalone version of SWAMP [10]. Besides this tool, there are many other available, some of them as part of the Black Hat Arsenal or/and Kali, which try to automate mainly vulnerability assessment such as:

- *DeepExploit* [11] is a fully automated penetration test tool linked with Metasploit. It identifies the status of all opened ports on the target server and executes an automated attack. Among the execution possibilities, it may be launched in a self-learning mode (using reinforcement learning);

- *VAPT framework* [12] is an automated Vulnerability Assessment and Penetration Testing tool that identifies vulnerabilities, retrieves exploits from open databases, e.g., ExploitDB, and performs penetration tests. The results are stored in graph-based database, Neo4j, at each stage;
- *APT2* [13] is a console-based Automated Penetration Testing Toolkit, whose results are stored locally and used to launch exploits and enumeration modules. This performs a nmap scan or import the results of other scanners;
- *Archery* [14] uses open source web and network vulnerability scanners (e.g., zap, nmap, openvas, selenium, etc.) to create a vulnerability assessment and management tool;
- *Lynis* [15] is a shell-script that runs on *NIX-based operating systems. It is an extensible security audit and vulnerability analysis tool, which performs security tests to check configuration errors, software vulnerabilities, or weaknesses, in order to perform vulnerability assessments and penetration tests;
- *CROZONO framework* [16,17] allows gathering information about possible attack vectors and performing automated penetration tests from autonomous devices (e.g., drones, robots, etc.) that could ease the access to the logical infrastructure of an industrial facility [16]. This framework has a key feature because it generates reports about gathered information identifying weak points and exposure levels;
- *Faraday platform* [18] reuses the available tools in the community to perform penetration-tests. This introduces the concept of Integrated Penetration-Test Environment (IPE), which automates distribution, indexation, and analysis of the data generated during a security audit;
- *Intrigue Core* [19] discovers assets (i.e., applications and infrastructure) and vulnerabilities utilizing APIs and OSINT techniques to discover an attack surface. This can be used from a docker image or web interface;
- *Leviathan framework* [20] is a python-based audit toolkit which includes service discovery (using Shodan and Censys), brute force, SQL injection detection, and running custom exploit capabilities. This tool allows to do massive scans (using masscan) on several systems at once;
- *Trommel* [21] is a python tool that sifts embedded device files to identify potential vulnerabilities, such as, protocol key files, email addresses, shell scripts, etc. It integrates vFeed for in-depth vulnerability analysis.

A comparison can be found in Table 1, where X indicates that functionality is satisfied and—means the opposite. We can see that none of the tools support all the analyzed characteristics. In particular, none are designed for exchanging or sharing logs with security information. Likewise, there is no a ledger that allows tracking risks over time. As mentioned before, these tools have been designed to assess vulnerabilities, which is the first step in the life cycle of vulnerabilities management, but these do not implement or facilitate the remaining steps. Our approach supports the smart grid companies to design, perform, and integrate their tests in their continuous auditing processes, and the use of secure procedures to grant access to security logs for the sake of advancing security countermeasures. This is the base to report, remedy, and verify such threats. Besides we aim at using containers, library objects, and well known components in pentesting, and to use common network infrastructure to provide autonomy to the auditing companies.

A practical risk assessment method applied to Austrian smart grid is presented [22]. The method follows a twofold approach: A conceptual and implementation-based assessment. For the conceptual assessment it uses a reference architecture based on the Smart Grid Architecture Model (SGAM) [23], in order to analyze the deployed architectures or, to be deployed in the near future, for mapping them to SGAM. The outcome of this first phase is a risk matrix and mitigation strategies. Then, the implementation-based assessment consists in evaluating details of systems, such as poor configurations and potential software implementation vulnerabilities. This second phase deals with existing systems that allow a security audit to assess the security with respect to the potential attack vectors and vulnerabilities, resulting in a set of possible exploits.

Table 1. Comparison of tools to automate software asset management (SWAM) and/or vulnerability management (VULN).

	Asset Mgmnt	VULN	Pen. Tests	Shared Logs	Full Automation
DeepExploit	-	-	X	-	-
VAPT	X	X	X	-	-
APT2	-	-	X	-	-
Archery	-	X	X	-	-
Lynis	X	X	X	-	X (docker)
CROZONO	-	-	X	-	X (dron & robots)
Faraday	-	-	X	-	-
Intrigue Core	X	X	-	-	X
Leviathan	X	-	X	-	X
Trommel	-	X	-	-	-

Though in this paper we restrict to network and computer security audits, recent works related to a broader concept of auditing, and the role of Supreme Audit Institutions, explore the dependency between auditing and trust [24]. Lack of trust in public organizations can contribute to the necessity of frequent auditing. Audits can enable auditors to correct errors and irregularities, strengthening the audited entities, and thus enhancing public trust. In [24], conclusions include that enforcing the informative function of audit institutions can strengthen SAI's trustworthiness for their customers. Blockchain-based architectures can be conceived not as a replacement of auditors but to achieve a less expensive and more effective auditing of information systems [25].

2.2. Distributed Ledger Technologies in Smart Grids

Researchers have proposed some solutions focused on smart grids and IoT (Internet of Things). In [26], an ISO/IEC 15408-2 compliant security auditing system based on a blockchain network as the underlying communication architecture is proposed for IoT. Certainly, the inner characteristics of blockchains and, in general, Distributed Ledger Technologies (DLT) paves the way to construct transparent and traceable procedures of major interest for the energy sector [7]. The heterogeneity of this ecosystem makes cumbersome to deploy management solutions and governance schemes to ponder security, reliability, but also regulatory compliance. Although blockchain was initially interpreted as channel to guide arising functional needs in the energy sector (e.g., P2P energy trading), there exists an increasing trend to handle cybersecurity and cyber safety goals in this ecosystem by means of blockchain procedures. Continuous cybersecurity management can be perfected with the guidance of blockchain logics [27], governance, and interoperability matters can be articulated in a more nuanced way with the assistance of on-chain and off-chain blockchain protocols. In this sense, it is worth noting current efforts in organizations as the International Association of Trusted Blockchain [28] or the European Telecommunications Standards Institute (ETSI) [29] to align efforts in pursuing standard data models and procedures for a broad set of application contexts, including the energy sector. Regarding incident management, standardization is crucial to ease the sharing, traceability, and trust evaluation of cyber threat intelligence sources [30]. Provenance evaluation is also critical for the identification of threats associated with inadequate software and firmware updates in the ecosystems of smart and micro grids. Blockchain could be used to leverage the root of trust of physical devices and get genuine information about critical security updates across the endpoints of the ecosystem [31].

There are other researches in the interconnection strategy of federated smart grids (power networks, control systems, market, customer premises). Ref. [32] proposes a three layer interconnection architecture, being L3 the distributed ledger layer which handles transactions of type <object><resource><action>. L3 acts thus as a distributed database across all partners of the federation. The paper argues that: "the own nature of blockchain in L3 already addresses itself" (the need to manage provenance measures for traceability). As for auditors, denoted as SECAUD in IEC-62351-8, they are in charge of verifying the performance of the infrastructure, even ensuring the correct application of the authorization policies with the verification of access registers (stored as transactions in L3).

This work is a step towards the deployment of an auditing system aligned with [32]. We address to improve the interoperability and collaboration between the agents involved in forensic research of attacks and failures in smart grids, potentially federated.

3. AUTOAUDITOR System Description

Power grid companies hire third parties for performing security analysis. Those third-party companies have to perform security audits of the critical infrastructures deployed. Our proposal was to design a system that can be controlled by the power grid companies automatically, using a tailored and preconfigured system provided by the security company. This system has two main benefits: (1) The security company is not required to do the bulk work of auditing the whole installation, and (2) the power grid company does not have to grant access to the critical infrastructures for periodic auditing. AUTOAUDITOR [6] is designed to test elements of different types: Smart meters, smart meter concentrators, other smart grid equipment (power chargers, etc.), networking elements (switches, routers, proxies, gateways), and servers in the core of the company's network(s).

In AUTOAUDITOR, the security company role is to do an initial fingerprinting of a sample device or network equipment to test followed by the identification of potential vulnerabilities to test and the selection of the most suited auditing modules. The fingerprinting process output can be used by the company to update the company inventory. This will bring two additional benefits: Increase in the accuracy of the fingerprinting process and improve the automatic inventory with an additional online source. Inaccuracies in the automatic fingerprinting may be detected manually or in the reconciliation with the inventory system. This procedure can help the company to confirm the inventory with respect to production elements at different points of their network. For instance, imagine we test a smart meter X and end with a fingerprinting test procedure. The company can run this test to confirm that a given population of smart meters of type X, according to the inventory are indeed of type X, and have not been replaced by another element type.

Zero-day vulnerabilities and exploits are out of the scope of AUTOAUDITOR. The proposed approach towards zero-day(s) is to have the devices subject to continuous monitoring processes and further inspected by behavior analysis. Such a behavior analysis may trigger alerts and subsequent manual inspection, or even subject to preventive block or quarantine. We expect also to improve the defenses through the security information sharing and collaboration with other companies.

At this step AUTOAUDITOR offers the possibility of packing the collected information in an attack plan persisted to a JSON file. The attack plan is encapsulated in a container. The reason why we use containers is to better scale with the number of potential devices which will be delivered to the power company. It is up to the power company: (1) To configure the addresses of the devices to test, and (2) to decide upon the resources devoted for the testing. The result is the test plan which can include parameters, either manually defined or template based. Those parameters are useful for defining configurable connections including different network parameters for running the different tests. AUTOAUDITOR presently uses VPN connections, though other network connection configurations can be added. The main reason for this is that the equipment of a power grid operator is scattered along the geography of the region or country supplied by the operator. The operator internal routers and firewalls ensure the separation of the interconnecting networks. The system must include the capability of defining configurable connections including different network parameters for running different tests.

Finally, the encapsulated test plan can be manually or automatically executed, according to the inventory system needs to update the data regarding the included assets and their related security information (vulnerability meta-information). The test will fire the instantiation and execution of the tests. That may require setting up connection elements according to the defined network configuration and the concrete parameters of the test. The execution of the system will output evidences of the vulnerabilities tested. AUTOAUDITOR outputs the tested CVEs, whether successful or not, and other information obtained in the testing. A link to a supporting video is included as Supplementary Material.

This information is part of the security assessment to better evaluate the vulnerabilities and incorporate the success likelihood for the tested attacks. This is the information we propose to improve its accountability, and to persist and share with other actors involved in the smart grid. The next section justifies the selection of distributed ledger technologies for such purposes.

4. Evolving AUTOAUDITOR with Blockchains

AUTOAUDITOR was designed with the requirements identified in the previous section for the protection of critical infrastructures such as power grids. In a first iteration, AUTOAUDITOR was conceived to automate security audits in power grids [6]. In this second iteration, AUTOAUDITOR is extended by integrating a blockchain protocol to enable collaboration in cyber threat intelligence. By forcing a strict AAAA (Access, Authorization, Audit, Accountability) policy on the basis of a blockchain, a set of functions will be implemented using smart contracts. These functions are targeted at persisting security trails on a blockchain and to provide access to such evidences according to a concrete policy. As an append-only log, and taking into account the tamper-proof nature of blockchain, AUTOAUDITOR is thus enhanced to foster accountability along the life cycle of the continuous security auditing of power grids.

Among the different types of blockchain, the AAAA goal demands functionality to restrict the set of users/entities with permissions to write in the blockchain, and to establish the information to share with other users/entities according to specific agreements and objectives in the construction of collaborative consortia in the sphere of cyber threat intelligence. In other words, audit information is not going to be publicly accessible and access is granted on the basis of explicit agreements among the concerned parties. Therefore, we are going to adopt a permissioned blockchain to extend AUTOAUDITOR in order to boost the automation of audits in power grids, and to establish accountable channels for sharing outcomes and insights in concrete investigations about security and safety incidents in this domain.

HyperLedger Fabric (HLF) provides one of the most adopted permissioned blockchains. As part of the Hyperledger initiative, HLF entails functionality to create and manage a Public Key Infrastructure by means of a so-called Membership Service Provider (MSP). The architecture of HLF is deployed through a meaningful set of nodes, which can have different roles according to their implication in the underlying execute-order-validate protocol [33]. This separation of duties is very relevant in a context of collaborative frameworks where information disclosure must be conducted according to data minimization criteria. Moreover, the throughput, latency, and scalability of HLF is adequate for the application context of AUTOAUDITOR (https://www.hyperledger.org/learn/publications/blockchain-performance-metrics [last accessed on 13 September 2020]). Finally, HLF comes with a mature Software Development Kit that paves the way for prototyping. All in all, HLF has been adopted in AUTOAUDITOR to extend the first version of the tool and to deploy a platform to favor information exchange among the different agents and entities involved in the investigation of security threats and incidents in power grids.

5. Implementation Details

5.1. Scenario and General Workflow

The scenario we address is a failure or an attack, where the affected company can help the forensic investigation with the details of audit results prior to the incident, i.e. which security audits have been performed in the organization including all the details about devices and vulnerabilities tested.

This can narrow the investigation and reveal sooner the details of the attack, so that the systems can be easily protected from similar attacks. However the most important gain of this accountability is obtained from sharing some of the audit results with other power grid companies and relevant players, therefore, learning from the attacks on other companies so as to be better protected.

AUTOAUDITOR workflow is illustrated in Figure 1. It starts with an auxiliary tool CVEScanner [34] and the list of metadata about potential vulnerabilities in the database, including available metasploit modules. The list is saved in the JSON format and the expert supervises and configures the attack plan, which is then encapsulated for execution. The configuration also comprises the VPN connection including addresses and credentials. AUTOAUDITOR outputs the composition of the containers including the connection and audit application. The composition is then manually instantiated the first time so that the expert can verify its correctness. After potential corrections are made, the company launches the automatic deployment and execution of as many instances as required for the equipment under test (EUT). Finally, the results are collected, processed, and stored in the blockchain.

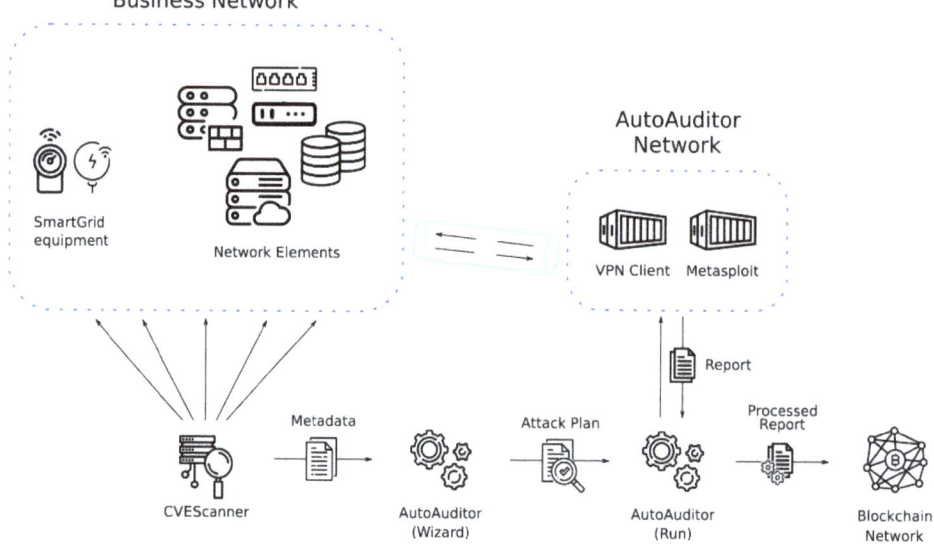

Figure 1. AUTOAUDITOR workflow.

5.2. Auditing Objects and Roles

As a first step in the implementation of the blockchain-based AAAA system upon AUTOAUDITOR, sensitive information has to be protected by defining and implementing a suitable access control policy [29]. We have identified the following information objects that are subject to access control:

- Network topology information including concrete network addresses;
- Fingerprint of the equipment under test (EUT):
    - Hardware architecture;
    - Firmware manufacturer and version;
    - Operating system version;
    - Network address;
    - Identified subcomponents.
- Details on the auditing tests execution results:

- Vulnerabilities tested and their severity;
- Test timestamps;
- Used modules;
- Successful modules;
- Hosts proved vulnerable;
- Information retrieved from the attack (possibly including credentials).

On the ground of the previous classification, we have identified three different roles for access control to AUTOAUDITOR audit results:

- Role A: Total access to the whole records. This access corresponds to people in charge of security as the CSO or CISO, and similar staff of the power grid company. It will also be granted to forensic investigators in case of attack;
- Role B: Partial access to records. This access is envisaged for Critical Infrastructure regional or national managers. They require some details of the audits to have a clear picture of the level of risk accepted by the companies, typically including last audit timestamp, periodicity of audits, vulnerabilities tested, and general information on the equipment (number of devices). Network topology, network addresses and some details of the devices are hidden;
- Role C: Limited access to records. Power grid companies should share information with this level of access. Some of the quantitative information granted for the B level like the number of devices will be hidden, but qualitative information used and successful modules will be available.

5.3. Smart Contract and Distributed Ledger Workflow

AUTOAUDITOR implements storage of reports in a HLF network. The version of HLF of this prototype is 2.1.1. We have developed a smart contract (chaincode as coined in HLF) compatible with AUTOAUDITOR, allowing audits reports storage and query by some given identities. This different access is programmed by defining two different collections, which makes possible to protect private data and to keep the secrecy of specific types of audit reports. In the blockchain-based AAAA system of AUTOAUDITOR, these two collections have been defined:

- Collection A: Stores very basic information, i.e. year and month of report and number of machines affected per vulnerability;
- Collection B: Stores more detailed information, i.e. accurate timestamp of report, metasploit modules tested, and network address of affected machines.

In addition, each collection stores common information, namely, number of vulnerabilities, tested vulnerabilities, and vulnerability score.

Developed smart contract exposes multiple functions for storage, querying, and deletion:

- **NewReport():** Main functionality of the smart contract. Enables AUTOAUDITOR to store reports in the blockchain. Makes use of transient map in order to not track input data in a transaction record;
- **GetReportById():** To query a report by a given unique identifier. Returns a report in JSON format;
- **GetReportsByOrganization():** To query all reports from a given organization. Returns a list of reports in JSON.
- **GetReportsByDate():** to query all reports from a given year and month. Returns a list of reports in JSON;
- **GetReportHash():** Allows data integrity check. Returns data hash;
- **DeleteReport():** Sets the status of the register as deleted so that it does not appear in queries. No evidence is ever deleted from the blockchain.

A general overview of the HLF workflow is presented in Figure 2.

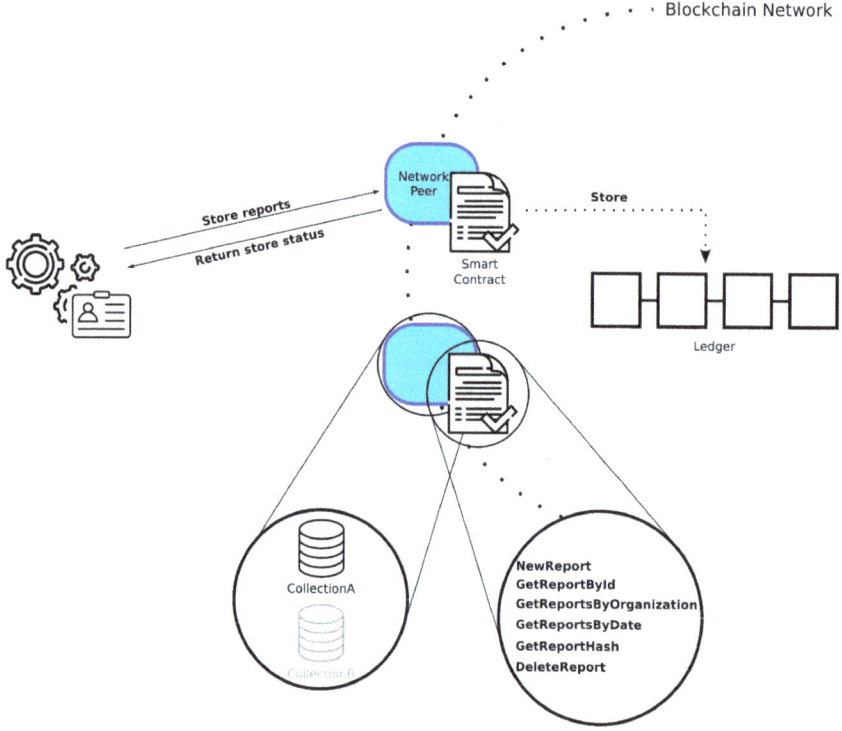

Figure 2. Hyperledger Fabric workflow.

6. Results

To test the correct behavior of the system, we have developed a closed environment including the EUTs, VPN server, the application performing the audit, and the HLF network. It is orchestrated as a python application that sets up the configuration and composition, instantiates and connects the components, and launches the auditing. In addition, it collects the results, and after the analysis, uploads the relevant information to the blockchain. Such results should be similar to the ones obtained in a company, without the time required to instantiate the EUTs, which have been modeled as a set of vulnerable containers.

The auditing process analyzes a list of 10 vulnerabilities. The list contains different vulnerabilities of software products that can be used in clientes, servers, IoT devices, and other devices in the smart grid. They are just examples of vulnerabilities with publicly available exploitation modules we have selected: CVE-2010-2961 correspond to a privilege escalation in Unix filesystems; CVE-2012-2122 is an authentication bypass in mySQL; CVE-2014-0160 is an OpenSSL implementation bug which leaks data, even the server private key; CVE-2014-6271 is a bash bug that allows attackers to execute arbitrary commands; CVE-2017-5638 allows attackers to execute remote commands in Apache Struts; CVE-2017-11610 can be used to issue remote commands by XML-RPC; CVE-2017-12635 is a Apache couchDB vulnerability which allows attackers to execute commands without being authenticated; CVE-2018-10933 exploits may lead to unauthorized OpenSSH server access; CVE-2018-15473 allows attackers to enumerate the users of a OpenSSH server; and CVE-2019-5418 may disclose files in RubyOnRails.

The test environment comprises a HLF network with two organizations and each organization has its own Certification Authority (CA) and both are connected to the same channel. In addition, there is an *orderer* with its own CA.

The study was executed in a processor Intel i7-8565U (8) @ 4.600GHz and was divided in three experiments. The first one studied the performance with 100 reports stored in the blockchain, second and third ones had 500 and 1000 reports stored, respectively.

Figure 3 presents the time needed for storing the reports in 500 executions of the audit. The majority of the measurements are depicted in blue, except for the first execution that took much longer since it required AUTOAUDITOR to gather vulnerabilities information from external sources, i.e., vulnerability score and related metasploit modules. As soon as AUTOAUDITOR collects vulnerabilities metadata, a SQLite database is generated and populated with that information, acting as a cache in subsequent executions. Due to extremely different times between first and following executions, we opted to add another y-axis on the right side for the first measurement.

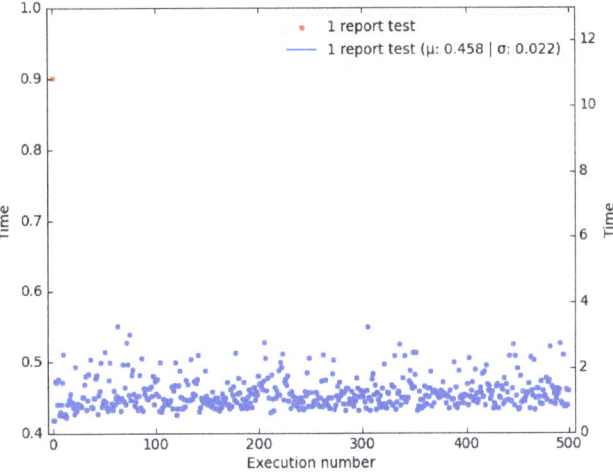

Figure 3. Time required to store a report.

Storing a report spent 0.458 s in average with a standard deviation of 0.022. Similar results were obtained when experimenting with executions storing 100 and 1000 reports, with average times of 0.442 and 0.466 s and standard deviations of 0.05 and 0.028 respectively. That slight penalty during upload shows that the system is usable for storing the audit reports of even large organizations. This performance can cope with changes in the number of devices, networks, maintenance operations, and applicable vulnerabilities published by CERTs, even if continuous monitoring is required.

Figure 4 shows the time spent querying a single report with GetReportById() in multiple executions. Querying a single report spent 0.665 s in average with a standard deviation of 0.069.

Figure 5 shows the time spent doing a bulk query, calling GetReportsByOrganization() in multiple executions. We observe a significant improvement in time compared to single report queries, taking almost the same time for 100 reports.

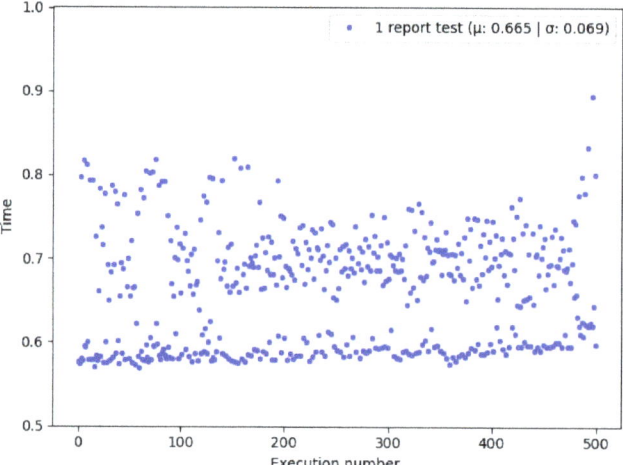

Figure 4. Time required to query a single report.

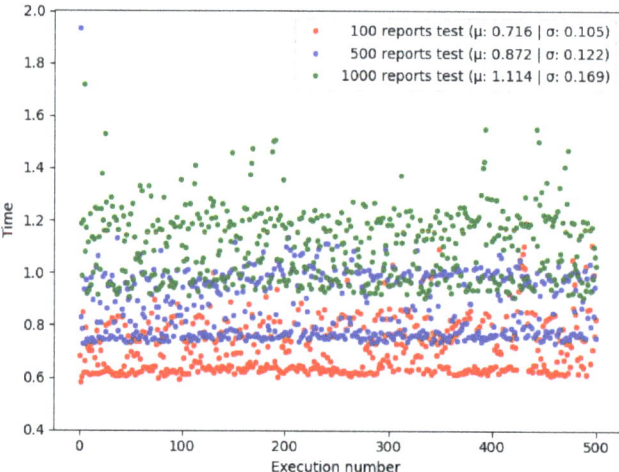

Figure 5. Comparison of time required to query multiple reports at once: Organization.

Figure 6 shows the time spent doing a bulk query, calling `GetReportsByDate()` in multiple executions. It can be seen that batch queries by date take slightly less time than queries by organization.

Table 2 compares the mean and standard deviation between the time required to complete a query in the experiments. As was expected, doing a batch query took significantly less time than querying reports individually due to the extra time added processing the transactions. We can see that HLF allows the querying of many reports at the same time without being notoriously affected, keeping the times low.

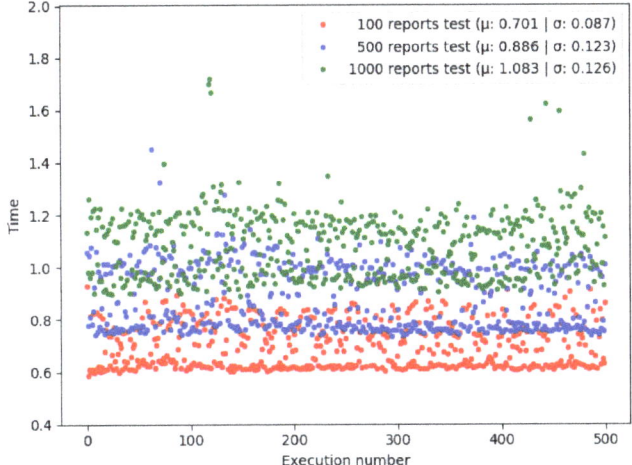

Figure 6. Comparison of time required to query multiple reports at once: Date.

Table 2. Comparison of mean and standard deviation in every experiment.

Query	Value	Reports			
		1	100	500	1000
GetReportById	μ	0.665	-	-	-
	σ	0.069	-	-	-
GetReportsByOrganization	μ	-	0.716	0.872	1.114
	σ	-	0.105	0.122	0.169
GetReportsByDate	μ	-	0.701	0.886	1.083
	σ	-	0.087	0.123	0.126

The apparent improvement in time of date-query can be attributed to the use of integer in `GetReportsByDate()` composite keys, allowing faster query indexing, unlike its counterpart `GetReportsByOrganization()` that uses string in composite keys.

7. Conclusions

Similar to other countries, the policies defined by the Spanish National Commission for the Protection of Infrastructures and Cybersecurity (abbreviated in Spanish as CNPIC) thrust two driving forces:Information sharing and public-private partnership. Sharing information of security audits results can bring benefits to the overall system. Our aim is not to substitute other tools that are already being used for security information exchange. In the case of Spain, we do not want to substitute neither the PI3 (Platform of Infrastructures-based Information Exchange), nor HERMES. The main contribution of this paper is enhancing AUTOAUDITOR with a blockchain-based AAAA scheme to gather audit information and to share in an accountable way cyberthreat intelligence. We provide AUTOAUDITOR as a new tool that can be integrated as a new source of security information and as platform to foster collaboration within the community of cyberthreat intelligence.

The tool helps to achieve continuous monitoring by integrating the audit system in a semi-automated way in the inventory control system of electrical grid companies. The audit result records are persisted in a permissioned blockchain, since blockchains are by design resistant to data modification. The results of the performance tests show that the system can be adapted to the inventory systems of electrical companies.

Future work will be intended to perfect the tokenization of audit trails and to define more fine-grained access control policies by leveraging HLF channels and defining more precise access control lists in the chaincode associated with AUTOAUDITOR. In this way, the set of sharing policies would be improved, which eventually paves the way for a more fluid habit of collaborative work in investigations of security incidents.

Supplementary Materials: A supporting video is available at https://www.youtube.com/watch?v=iAJVCirZFCg.

Author Contributions: Conceptualization, D.D.-S. and F.A.-M.; methodology, A.M.-L. and S.C.-M.; software, S.C.-M.; validation, S.C.-M. and D.A.; investigation, A.M.-L., S.C.-M., D.A., D.D.-S. and F.A.-M.; writing–original draft preparation, A.M.-L., S.C.-M. and D.A.; project administration, D.D.-S. and D.A. All authors have read and agreed to the published version of the manuscript.

Funding: This work has been supported by National R&D Projects TEC2017-84197-C4-1-R, TIN2017-84844-C2-1-R, by the Comunidad de Madrid project CYNAMON P2018/TCS-4566 and co-financed by European Structural Funds (ESF and FEDER), and by the Consejo Superior de Investigaciones Científicas (CSIC) under the project LINKA20216 ("Advancing in cybersecurity technologies", i-LINK+ program).

Acknowledgments: We want to thank the anonymous reviewers for their helpful comments in improving this manuscript.

Conflicts of Interest: The authors declare no conflict of interest.

References

1. Alaton, C.; Tounquet, F. *Benchmarking Smart Metering Deployment in the EU-28*; Final Report, Technical Report, Directorate-General for Energy (European Commission); Tractebel Impact: Brussels, Belgium, 2020. [CrossRef]
2. Dempsey, K.; Goren, N.; Eavy, P.; Moore, G. *Software Asset Management*; Technical Report NISTIR 8011; NIST: Gaithersburg, MD, USA, 2018; Volume 3.
3. Dempsey, K.; Takamura, E.; Eavy, P.; Moore, G. *Software Vulnerability Management*; Technical Report NISTIR 8011 (Draft); NIST: Gaithersburg, MD, USA, 2019; Volume 4.
4. Vakilinia, I.; Tosh, D.K.; Sengupta, S. Privacy-preserving cybersecurity information exchange mechanism. In Proceedings of the International Symposium on Performance Evaluation of Computer & Telecommunication Systems SPECTS, Seattle, WA, USA, 9–12 July 2017; pp. 1–7.
5. de Fuentes, J.M.; González-Manzano, L.; Tapiador, J.; Peris-Lopez, P. PRACIS: Privacy-preserving and aggregatable cybersecurity information sharing. *Comput. Secur.* **2017**, *69*, 127–141. [CrossRef]
6. Chica-Manjarrez, S.; Marín-López, A.; Díaz-Sánchez, D.; Almenares-Mendoza, F. *On the Automation of Auditing in Power Grid Companies*; Ambient Intelligence and Smart Environments; IOS Press EBooks: Amsterdam, The Netherlands, 2020; Volume 28, pp. 331–340. [CrossRef]
7. Andoni, M.; Robu, V.; Flynn, D.; Abram, S.; Geach, D.; Jenkins, D.; McCallum, P.; Peacock, A. Blockchain technology in the energy sector: A systematic review of challenges and opportunities. *Renew. Sust. Energy Rev.* **2019**, *100*, 143–174. [CrossRef]
8. Power Systems Management and Associated Information Exchange—Data and Communications Security—Part 1: Communication Network and System Security—Introduction to Security Issues. Available online: https://webstore.iec.ch/publication/6903 (accessed on 1 April 2020).
9. Initiative, J.T.F.T. *Managing Information Security Risk: Organization, Mission, and Information System View*; Technical Report; NIST: Gaithersburg, MD, USA, 2011.
10. Aydemir, B.; Stienen, C. SWAMP-in-a-Box v1.34.5. Available online: https://github.com/mirswamp/deployment/ (accessed on 1 April 2020).
11. Takaesu, I. DeepExploit: Fully Automatic Penetration Test Tool Using Machine Learning. Available online: https://github.com/13o-bbr-bbq/machine_learning_security/tree/master/DeepExploit (accessed on 1 April 2020).

12. Veracode Vulnerability Assessment Software. Available online: https://www.veracode.com/security/vulnerability-assessment-software (accessed on 1 April 2020).
13. Compton, A.; Lane, A. APT2: An Automated Penetration Testing Toolkit. Available online: https://tools.kali.org/information-gathering/apt2 (accessed on 1 April 2020).
14. Tiwari, A. ArcherySec: Centralize Vulnerability Assessment and Management for DevSecOps Team. Available online: https://archerysec.github.io/archerysec/ (accessed on 1 April 2020).
15. Boelen, M. Auditing, System Hardening, Compliance Testing. Available online: https://cisofy.com/lynis/ (accessed on 1 April 2020).
16. Berta, S.A.; Villanueva, N.S.; Romanos, P.; Benítez, D.; Pepe, M. Crozono: Leveraging Autonomous Devices as an Attack Vector on Industrial Networks. Available online: https://www.blackhat.com/eu-16/arsenal.html (accessed on 1 April 2020).
17. Romanos, P.; Berta, S. A Framework to Test Your Security Perimeter with Drones & Robots. Available online: https://github.com/johnjohnsp1/CROZONO (accessed on 1 April 2020).
18. Riera, G.; Medina, M.A.R. Python-Faraday: A Multiuser Penetration Test IDE. Available online: https://tools.kali.org/information-gathering/faraday (accessed on 1 April 2020).
19. Cran, J.; Kaiser, T.; Bensalah, A. Intrigue Core: Discover Your Attack Surface. Available online: https://core.intrigue.io/ (accessed on 1 April 2020).
20. Jopling, B. Leviathan: Wide Range Mass Audit Toolkit. Available online: https://github.com/utkusen/leviathan (accessed on 1 April 2020).
21. Sift Through Embedded Device Files to Identify Potential Vulnerable Indicators. Available online: https://github.com/CERTCC/trommel (accessed on 1 April 2020).
22. Langer, L.; Skopik, F.; Smith, P.; Kammerstetter, M. From old to new: Assessing cybersecurity risks for an evolving smart grid. *Comput. Secur.* **2016**, *62*, 165–176. [CrossRef]
23. Smart Grid Coordination Group. Smart Grid Reference Architecture (SGAM). Available online: https://ec.europa.eu/energy/sites/ener/files/documents/xpert_group1_reference_architecture.pdf (accessed on 1 April 2020).
24. Dobrowolski, Z.; Sulkowski, L. Supreme Audit Institutions and importance of their trustworthiness. In Proceedings of the 35th International Business Information Management Association Conference (35th IBIMA Conference), Seville, Spain, 1–2 April 2020.
25. Wang, K.; Zhang, Y.; Chang, E. A Conceptual Model for Blockchain-Based Auditing Information System. In Proceedings of the 2020 2nd International Electronics Communication Conference, Singapore, 8–10 July 2020; pp. 101–107. [CrossRef]
26. Cha, S.; Yeh, K. An ISO/IEC 15408-2 Compliant Security Auditing System with Blockchain Technology. In Proceedings of The 6th IEEE Conference on Communications and Network Security (CNS 2018), Beijing, China, 30 May–1 June 2018. [CrossRef]
27. White, J.; Daniels, C. Continuous Cybersecurity Management Through Blockchain Technology. In Proceedings of 2019 IEEE Technology Engineering Management Conference (TEMSCON), Atlanta, GA, USA, 12–14 June 2019; pp. 1–5.
28. International Association for Trusted Blockchain Applications. Available online: https://inatba.org/wp-content/uploads/2020/06/Co-Chairs_presentations_GA_10June.pdf (accessed on 1 September 2020).
29. ETSI GR PDL 001: Permissioned Distributed Ledger (PDL); Landscape of Standards and Technologies. Available online: https://standards.iteh.ai/catalog/standards/etsi/1dea1899-1b85-4e6c-974f-78a6546f037d/etsi-gr-pdl-001-v1.1.1-2020-03 (accessed on 1 April 2020).
30. Cha, J.; Singh, S.K.; Pan, Y.; Park, J.H. Blockchain-Based Cyber Threat Intelligence System Architecture for Sustainable Computing. *Sustainability* **2020**, *12*, 6401. [CrossRef]
31. Ángel Prada-Delgado, M.; Baturone, I.; Dittmann, G.; Jelitto, J.; Kind, A. PUF-derived IoT identities in a zero-knowledge protocol for blockchain. *Internet Things* **2020**, *9*, 100057. [CrossRef]
32. Alcaraz, C.; Rubio, J.E.; Lopez, J. Blockchain-assisted access for federated Smart Grid domains: Coupling and features. *J. Parallel. Distr. Com.* **2020**. [CrossRef]

33. Androulaki, E.; Barger, A.; Bortnikov, V.; Cachin, C.; Christidis, K.; Caro, A.D.; Enyeart, D.; Ferris, C.; Laventman, G.; Manevich, Y.; et al. Hyperledger Fabric: A Distributed Operating System for Permissioned Blockchains. In Proceedings of the Thirteenth EuroSys Conference, EuroSys 2018, Porto, Portugal, 23–26 April 2018. [CrossRef]
34. Nmap Security Tool Used to Discover Potentially CVEs that Affects Services in Detected Open Ports. Available online: https://github.com/alegr3/CVEscanner (accessed on 1 April 2020).

Publisher's Note: MDPI stays neutral with regard to jurisdictional claims in published maps and institutional affiliations.

© 2020 by the authors. Licensee MDPI, Basel, Switzerland. This article is an open access article distributed under the terms and conditions of the Creative Commons Attribution (CC BY) license (http://creativecommons.org/licenses/by/4.0/).

Article

A Two Stage Intrusion Detection System for Industrial Control Networks Based on Ethernet/IP

Wenbin Yu, Yiyin Wang and Lei Song*

Department of Automation, Key Laboratory of System Control and Information Processing, Ministry of Education of China, Shanghai Jiao Tong University, Shanghai 200240, China; yuwenbin@sjtu.edu.cn (W.Y.); yiyingwang@sjtu.edu.cn (Y.W.)
* Correspondence: songlei_24@sjtu.edu.cn; Tel.: +86-21-34204022

Received: 25 October 2019; Accepted: 12 December 2019; Published: 15 December 2019

Abstract: Standard Ethernet (IEEE 802.3 and the TCP/IP protocol suite) is gradually applied in industrial control system (ICS) with the development of information technology. It breaks the natural isolation of ICS, but contains no security mechanisms. An improved intrusion detection system (IDS), which is strongly correlated to specific industrial scenarios, is necessary for modern ICS. On one hand, this paper outlines three kinds of attack models, including infiltration attacks, creative forging attacks, and false data injection attacks. On the other hand, a two stage IDS is proposed, which contains a traffic prediction model and an anomaly detection model. The traffic prediction model, which is based on the autoregressive integrated moving average (ARIMA), can forecast the traffic of the ICS network in the short term and detect infiltration attacks precisely according to the abnormal changes in traffic patterns. Furthermore, the anomaly detection model, using a one class support vector machine (OCSVM), is able to detect malicious control instructions by analyzing the key field in Ethernet/IP packets. The confusion matrix is selected to testify to the effectiveness of the proposed method, and two other innovative IDSs are used for comparison. The experiment results show that the proposed two stage IDS in this paper has an outstanding performance in detecting infiltration attacks, forging attacks, and false data injection attacks compared with other IDSs.

Keywords: intrusion detection; Ethernet/IP; industrial control networks

1. Introduction

An industrial control system (ICS) is composed of various automatic control components and real-time data acquisition components together. The main purpose of the ICS is to monitor and control industrial manufacturing to ensure the normal operation of industrial equipment. The core components of the ICS include the supervisory control and data acquisition system (SCADA) [1,2], distributed control system (DCS), programmable logic controller (PLC), remote terminal unit (RTU), human–machine interface (HMI), and a variety of communication interface technologies [3–5]. The ICS has been widely applied in the energy industry, transportation, metallurgy, electric power systems, etc.

In the traditional ICS, experienced engineers mainly focus on the physical safety of the production and ignore the information security because of the natural isolation of industrial networks, which makes it impossible for malicious hackers to interact with the traditional ICS [6–8]. With the rapid development of information technology (IT), standard Ethernet (IEEE 802.3 and the TCP/IP protocol suite) has been gradually implemented in the ICS communication interface. As a consequence,many automation companies designed the standardized industrial field bus and Ethernet. Schneider Electric also released Modbus/TCP [9] to replace the previous Modbus/RTU [10], which used the original RS-485 as a communication interface. At the same time, Profinet [11], which achieved Profibus over industrial Ethernet, was defined by PROFINET International. Rockwell Automation proposed Ethernet/IP [12], which supports data communications over industrial Ethernet, which enlargen the bandwidth of the communication. Ethernet/IP was first introduced in 2001 and now is the most developed, mature, and complete industrial Ethernet solution available for manufacturing automation, with rapid growth as users are eager to take advantage of the open technologies and Internet. Ethernet/IP implements the common industrial protocol (CIP) over standard IEEE 802.3 and the TCP/IP protocol suite.

Implementing standard Ethernet in the modern ICS improves the interoperability of the ICS and greatly reduces the cost of application developments. However, it also breaks the natural isolation of industrial networks. The modern ICS are facing more advanced threats from the Internet outside the factory. However, the original ICS security mechanisms, such as the industrial firewall and white list, cannot handle these threats effectively enough. On the one hand, an industrial firewall cannot dissect industrial communication protocols (e.g., Ethernet/IP), which makes it impossible to inspect the application layer payloads in packets or automatically generate proper filter rules according to the specific industrial scenarios [6]. On the other hand, the white list can only function as an access control list, and it is easy to forge as a result of many brilliant penetration testing tools, such as Metasploit [13,14].

As a second line of defense, the intrusion detection system (IDS) is an effective approach to detect malicious intruders, who are trying to disrupt the ICS networks from the Internet. By analyzing the information collected from the key points in the network, the IDS can find out whether there is a violation of the security policies and decrease the probability of attack occurrence. According to the analysis and inspection of the problems, the IDS takes appropriate countermeasures, such as raising an alarm or blocking the suspicious connections [3].

Without a doubt, there has been also many innovative works in designing the IDS for industrial networks. The work in [15] designed a telemetry based IDS by measuring the statistical data about client server sessions from the traffic flow. Although it was practical to detect anomalies in the ICS networks, it had a strong precondition that the existence of time delays was introduced by spoofing. Apart from this, the data related to the network protocols, which were more valuable, were not utilized to design the IDS. The work in [16] constructed an anomaly based IDS according to the normal behaviors of function control and process data. The behavior extraction algorithm was attractive to researchers because it considered the information entropy of the function code used by the Modbus/TCP protocol. However, the IDS used the function code sequence in the time interval as the input. If a packet were fabricated, which contained the same function code in the same time interval, but at the wrong time point, it would be impossible to detect the fake packets, which could be a serious threat to the ICS. The work in [17] built a model for normal system behavior to distinguish the normal and abnormal system operations. None these methods considered the characteristics of the data traffic and the normality modeling. In order to overcome the deficiencies of previous works, it is required to construct an intrusion detection system that could reflect the behavior characteristics in the ICS networks, strongly correlated to the ICS protocols, and be able to cope with vulnerabilities. Furthermore, it should have a satisfactory overall accuracy and false alarm rate.

Above all, it is better to think like a hacker before stopping a hacker. This paper firstly considers the infiltration attacks, forging attacks, and false injection attacks. In addition to this, an Ethernet/IP structure fabricates explicit messaging that uses the TCP as the transmission protocol. To prevent the attacks mentioned above, this paper designs a two stage intrusion detection system for the ICS based on the Ethernet/IP. The two stage IDS has the ability to dissect the Ethernet/IP protocol and mainly contains a traffic prediction model and an anomaly detection model. The traffic prediction model based on the autoregressive integrated moving average (ARIMA) can protect the ICS networks from infiltration attacks. The anomaly detection model based on one class support vector machine (OCSVM) is able to detect the elegant fabricated Ethernet/IP packets and protect against the forging attacks and false injection attacks. Compared with other creative IDSs [15,16], the proposed method gives satisfactory results in terms of overall accuracy and false alarm rate.

The rest of this paper is organized as follows: Section 2 introduces the related works. Section 3 describes the simulated industrial scenario and gives a brief introduction about the Ethernet/IP protocol. Section 4 outlines the attack models, especially fabricating malicious Ethernet/IP packets. Section 5 elaborates on the two stage intrusion detection system, which consists of a traffic prediction model and an anomaly detection model. Section 6 is the simulations and analysis. Finally, Section 7 gives the conclusions of this paper.

2. Related Works

As a hot research topic, many researchers are focusing on the two stage intrusion detection approaches. A classical chemical process and went over all the attack vectors were studied in [18]. A data driven approach to detect anomalies was designed for early indicators of malicious activity. The model of the attack process was built to profile some kinds of fault disturbances. A machine learning based approach was proposed in [19], which could reduce the amount of manual customization required for different ICS networks. Furthermore, a series of features for the machine learning input was selected, and the structure was used for a real industrial process control network. A SCADA based IDS was designed in [20], which could deal with the denial of service attacks. The network was investigated to resist response injection and denial of service attacks. A periodicity characteristic analyzing algorithm was designed for SCADA networks, and this feature was used for intrusion detection. Similarly, these papers all considered detecting the intrusion by modeling the normal operation features, which is also the main idea of this paper. Differently, this paper focused on the traffic of the ICS network based on the control and data stream. Furthermore, the application scenario in this paper was basic data transmission for the industrial network.

After the normal transmission traffic is modeled, a classifier needs to be designed to judge whether the receiving data are normal or attack data. Various machine learning algorithms were compared to reduce the false alarm rate and maintain high accuracy for industrial intrusion detection [21]. The work introduced A machine learning based classifier was introduced to reduce the false alarms and guarantee the precision for intrusion detection in the industrial area [22]. The work in [23] focused on the IDS and compared different techniques for the Industrial Internet of Things including machine learning and non-machine learning methods. The machine learning structure is a widely used method for industrial intrusion detection. However, as is known to all, the training of the structure requires a large quantity of labeled data. In this paper, an SVM based structure was designed, and an optimization problem was formulated to build the classifier, which could make full use of the labeled data.

3. Industrial Scenario

In order to make the proposed approach more effective and practical in the real ICS, an ICS demo platform was established based on Ethernet/IP. As shown in Figure 1, the platform consisted of three layers: the process monitor layer, the process control layer, and the plant-floor layer. The process monitor layer contained the operation station, where engineers could monitor the entire manufacturing procedure and modify the program or parameters in the PLC. This layer also contained the malicious attacker and the intrusion detection system (IDS). The process control layer in the middle constituted a set of devices that served the production, including the PLC, industrial router, and variable frequency drive (VFD). The PLC was the Allen-Bradley Micro 850 series, and the VFD was the PowerFlex 525 from Rockwell. The plant-floor layer consisted of a whole auto-guided vehicle (AGV) system, which included an optical-electricity encoder and three phase asynchronous motor encoders.

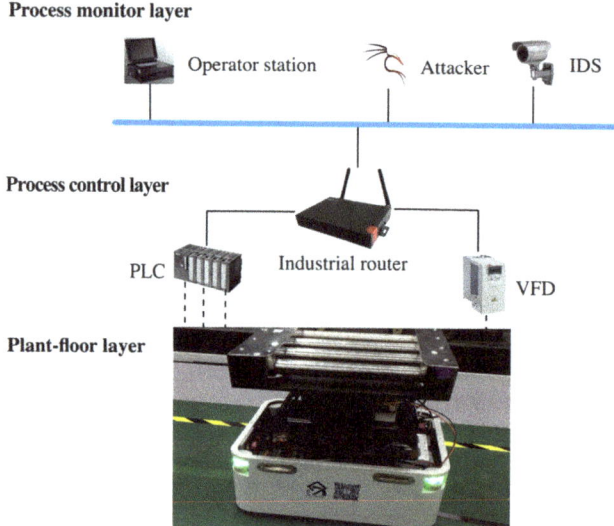

Figure 1. Simulated industrial control system (ICS) based on Ethernet/IP. PLC, programmable logic controller; VFD, variable frequency drive.

In this ICS, the AGV system was regarded as the controlled object, and the controller was the AB Micro 850 PLC. The feedback from the optical-electricity encoder was used to determine the current position of the encoder and the speed of the motors. The motions of the AGV system were controlled by the motors, which were driven by the PowerFlex 525 VFD. The entire process can be summarized as follows: according to the feedback from the optical-electricity encoder, the PLC made decisions and sent Ethernet/IP control packets to the VFD, and the VFD adjusted the motors speed and position by running the motors. This was a precision linear motion mechanism that converted the motor rotation into linear motion, and it is widely used in various linear motion applications.

The utilized communication protocol was Ethernet/IP. The physical layer and data link layer were based on Ethernet, while the transport layer and network layer were based on the TCP/IP protocol suite including the transmission control protocol (TCP), user datagram protocol (UDP), Internet protocol (IP), address resolution protocol (ARP), etc. The CIP was used as the application layer, and it defined two primary types of communications: implicit and explicit messaging. Implicit messaging is often used to transfer real-time control data from a remote I/O device with UDP, while explicit messaging, which is mainly discussed in this paper, is utilized with the TCP for request/reply transactions [24].

4. Attack Models

In this section, some kinds of infiltration attacks will be modeled. Besides that, a technique to fabricate an Ethernet/IP packet containing explicit messaging with the help of scapy, which is a powerful interactive packet manipulation program, is creatively proposed. The attack platform was based on Kali Linux, which is an advanced penetration testing Linux distribution. All the Ethernet/IP packets were parsed out by using the Python scripts. However, this paper mainly focuses on cyber-attacks and safety protection after entering the network.

4.1. Infiltration Attacks

Port scanning is usually used to identify some services and systems in the traditional network. By sending a TCP SYN packet to establish a connection at each port, it can be recognized that the port was opened according to the response. Through the open situation of the port, it was possible to understand what services and operating system were running. Utilizing port scanning in the target ICS system, the results showed that the TCP port number (e.g., port 44818) was open, which indicated that the Ethernet/IP service was running on this ICS. However, the information collected by port scanning was incomplete, and device enumeration was needed to identify the ICS devices.

Device enumeration was used to identify the device information in the target ICS, and it was achieved by sending an Ethernet/IP packet to the remote device that had some TCP port number open. The packet was a request, using the Ethernet/IP list identity command, whose function code was, e.g., 0x63. Once a response was received, the information was parsed out, including the vendor ID, device type, product name, device IP, etc. Having achieved device enumeration in the ICS, the PLC, VFD, and the corresponding address could be identified.

According to the results of the device enumeration, the ARP spoof can be implemented to hijack the Ethernet/IP session between the PLC and VFD. The ARP was used to convert an IP address (network layer address) to the MAC address (link layer address). By sending a malicious ARP reply, attackers can be disguised as a VFD in front of a PLC, as shown in Figure 2, and spoof the VFD in the same way. With the ARP spoof, attackers can hijack the Ethernet/IP session and monitor the data flow between the PLC and VFD. The communication packets between the PLC and VFD were based on explicit messaging in Ethernet/IP. It mainly included two parts: periodic maintenance and control instructions. The periodic maintenance packets refer to the read and write messaging for device parameters, which used the class code (e.g., 0x93) and the whole period including the request. The control instructions contained rotate clockwise with function code (e.g., 0x2A), rotate anticlockwise with, e.g., 0x29, stall clockwise with, e.g., 0x01, and stall anticlockwise with, e.g., 0x19. The specific rotate speed was appended after the function code.

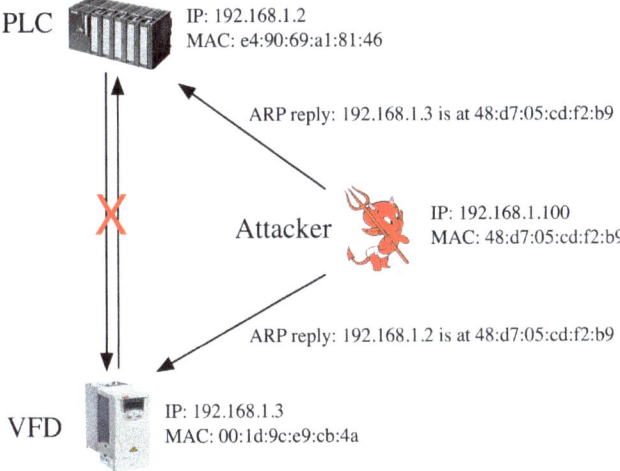

Figure 2. ARP spoof hijack Ethernet/IP session.

In order to avoid raising an alarm and disturbing the production process, it was better to select the forwarding periodic maintenance packets and discard the control instruction packets at the same time. Selective forwarding maintenance messages could ensure that the interface of the VFD had Ethernet/IP traffic and it would not report errors. Dropping the control messages would prevent the PLC from controlling the VFD and destroy the production process.

4.2. Forging Attack: Fabricate Ethernet/IP Explicit Message

As mentioned in Section 3, Ethernet/IP defines two kinds of communications: implicit and explicit messaging. The communication between the PLC and VFD in the ICS was based on explicit messaging because it used TCP as the transport layer protocol. The TCP provided reliable, ordered, and error checked delivery of a stream of octets between the PLC and VFD, and it was impossible to inject fake data [25–27].

However, this paper proposes a technique to inject fabricated Ethernet/IP explicit messaging. Algorithm 1 shows the principle of the proposed mechanism. The algorithm was realized by writing a tailored Python script, and the third party library used here was scapy, which is an elegant packet manipulation tool for networks. The inputs contained the IP addresses of the PLC and VFD and specific control instruction, which were acquired by the infiltration attacks mentioned above.

The forging attack consisted of three steps. First was capturing any Ethernet/IP packets to get the session handle in the transaction between the PLC and VFD, where the session handle could be obtained by dissecting the session handle field in the encapsulation header of Ethernet/IP. Second was capturing the TCP ACK packet sent by the PLC to obtain some key fields (seq, ack, TCP source port, IP identification). Third was to forge and send the Ethernet/IP packet according to the information previously obtained. The forged packet contained certain control instructions (e.g., 0x2A00DC05), which indicated rotating clockwise at a specific rotation speed.

Algorithm 1 Fabricate Ethernet/IP packet using scapy.

1: **function** FORGE ENIP PACKET(*plc ip, vfd ip, control*)
2:   *enip_pkt* ← sniff(filter: TCP port 44818)
3:   *session_handle* ← *enip_pkt*[session]
4:   *ack_pkt* ← sniff(filter: ip src *plc_ip* and dst port 44818 and length 64)
5:   *seq* ← *ack_pkt*[TCP][seq]
6:   *ack* ← *ack_pkt*[TCP][ack]
7:   *sport* ← *ack_pkt*[TCP][sport]
8:   *ip_id* ← *ack_pkt*[IP][id] + 1
9:   *enip_data* ← unhexlify(*control, session handle*)
10: *forged_pkt* ← IP(src: *plc_ip*, dst: *vfd_ip*, id: *ip_id*) + TCP(sport: *sport*, dport: 44818, seq: *seq*, ack: *ack*, flags: PSH and ACK) + *enip_data*
11: send(*forged_pkt*)
12: **end function**

4.3. False Data Injection Attack

As mentioned in Section 3, the Ethernet/IP based network structure would transmit several kinds of data including an optical-electricity encoder and three phase asynchronous motor encoders. The false data injection attacks tried to damage the sampled data and change the control instructions by injecting false data. The attack vector was defined as $a = [a_1, a_2, ..., a_m]^T$, and the sampled data would be changed as:

$$s^a = s + a = Hx + e + a \tag{1}$$

When the false data injection attack was operating, the attack vector a was non-zero and the state error vector was c, while the sampled data value would be changed to $s + a$. The proposed attack model focused on the data collecting and the motion control instructions. When the attack vector satisfied $a = Hc$, the attack model could bypass the classical attack detection approaches [28]. The attack model is defined as:

$$\begin{aligned} r_a &= ||s^a - H(\hat{x} + c)|| \\ &= ||s - H\hat{x} + a - Hc|| \\ &\leq ||s - H\hat{x}|| + ||a - Hc|| \\ &= r + \tau_a, \end{aligned} \qquad (2)$$

in which r_a indicates the error sampled after the false data injection attack worked. τ_a indicates the error change caused by the attack vector. When $\tau_a = 0$, the attack could not be detected because no change would be caused on the real sample data value. Furthermore, the attack strategy could perform the attack at the data collecting control by changing the sensor ID or even the sensor type.

5. Two Stage IDS

The attack models proposed in Section 4 were feasible and practical because Ethernet/IP does not define any explicit or implicit security mechanisms. In order to protect the ICS from these threats, a two stage IDS for the ICS networks based on Ethernet/IP is proposed. The two stage IDS was located in the process monitor layer, as shown in Figure 1, and its workflow is shown in Figure 3. The inputs were captured packets in one time slot, which could be modified by users. The two stage IDS had the ability to capture and analyze all communication packets from the monitoring port on the industrial router. Besides that, it used the libpcap technique to capture packets and dissected all the Ethernet/IP packets by running our Python scripts. In the Python scripts, the Ethernet/IP packets layer was dissected by layer and obtained necessary information according to the key fields, such as the control instructions in Section 4.

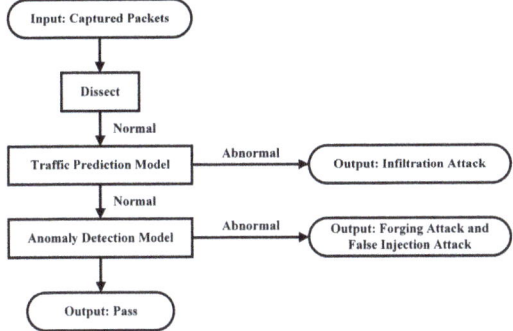

Figure 3. The workflow of two stage intrusion detection system (IDS).

The two stage IDS mainly consisted of two parts: the traffic prediction model and anomaly detection model. The traffic prediction model based on the ARIMA model was designed to estimate the number of packets in next time slot according to the captured packet flow previously. It was feasible to detect the infiltration attacks listed in Section 4 because the infiltration attacks would lead to an abnormal increase or decrease of the ICS network traffics at a specific time slot. If abnormal instances did not occur, the packets would be delivered to the anomaly detection model. That was because it was still possible for a forging attack, which did not lead to traffic changes. The anomaly detection model based on OCSVM was practical to detect the proposed forged Ethernet/IP packet because it could acquire essential control instruction by analyzing the Ethernet/IP packets in depth and detect abnormal behaviors by comparing with the normal pattern.

5.1. Traffic Prediction Model

A well designed traffic prediction model can precisely reflect the traffic characteristics of the ICS network. Since infiltration attacks lead to abnormal traffic fluctuations, it is possible to be detected with the traffic prediction model. The traffic flow data are a kind of time series, and the ARIMA model is the most commonly utilized model for time series prediction. The ARIMA model is used for short term forecasting, and the fundamental patterns of the time series should not be changed, which means the ICS networks should be immutable and the production process stable.

The raw input time series of the ARIMA, which is the count of traffic packets in one time slot, is not stable and fluctuates periodically. In order to use the ARIMA model, it is required to preprocess the raw input by using logarithmic transformation and differentiating. Logarithmic transformation is mainly done to reduce the vibration amplitude of the sequence, making the linear rule more obvious. Differentiating is able to make the series stable, and the difference periods are the periods of the ICS. The augmented Dickey–Fuller test is used to test the stationarity of the time series. Furthermore, the Ljung–Box test for autocorrelation is essential to ensure that the time series is not white noise. After preprocessing and testing, a stable and non-white noise time series $\{x_t\}_{t=1}^n$ is obtained, and it is suitable for the ARIMA model.

The prediction function of ARIMA can be depicted as:

$$\hat{x}_t = \psi_1 x_{t-1} + \psi_2 x_{t-2} + ... + \psi_p x_{t-p} + \epsilon_t + \theta_1 \epsilon_{t-1} + ... + \theta_q \epsilon_{t-q}, \tag{3}$$

where x_t is the stable and non-white noise time series and p and q are the order of the autoregressive (AR) model and moving average (MA) model. ψ_i and θ_i are the parameters of the ARIMA model. The prediction error must be uncorrelated and obey a normal distribution $\epsilon_t \sim N(0, \sigma^2)$.

The orders p and q determine the accuracy of the ARIMA model, and they can be estimated by calculating the autocorrelation function (ACF) and partial autocorrelation function (PACF) [29].

In addition to the ACF and PACF, enumerating many ARIMA models with different orders and using some criterion can also determine the proper model, i.e., the optimal parameters p and q. The criterion contains the Akaike information criterion (AIC), Bayes information criterion (BIC), and Hannan–Quinn information criterion (HQIC). The statistical ideas of these criteria are the same, that is they consider the fitting of the residuals and imposing punishments related to the number of variables at the same time. After calculating the predicted value x_t, the traffic prediction values are recovered by the inverse difference and exponentiation. The results also need to be rounded to the nearest integer.

5.2. Anomaly Detection Model

The anomaly detection model acts as a second line of defense after the traffic prediction model. Malicious attackers may drop the original Ethernet/IP control packet and replace it with the fabricated one that contains the wrong control instructions. The forging attack and false data injection attack cannot be detected by the traffic prediction model because it has little impact on traffic flow. Therefore, it is necessary to establish an anomaly detection model for the forging attack and false data injection attack.

The anomaly detection model firstly filtered out the Ethernet/IP control packets according to the field of service. The service name of the control packets was set single attribute, whose code was, e.g., 0x10. After obtaining the control packets, specific control instructions should be extracted from the packets. The control instructions had four features, i.e., relative time, action, direction, and speed. The relative time refers to the packet time stamp relative to the control period, which was a control cycle for the application. The action refers to rotate, whose value was, e.g., 0x02, or stall (0x01), and the direction refers to clockwise (0x0A) or counterclockwise (0x09). The speed simply refers to the rotation speed. After feature selection, the feature samples (control instructions) were obtained $\{\mathbf{x}_i\}_{i=1}^N$ and each sample $\mathbf{x}_i$ with the four features.

In order to detect the forged packets with the wrong control instructions, an OCSVM was constructed with the collected samples, which were all normal data. The OCSVM was a modified algorithm based on the SVM and has been widely used for one class classification problems, such as anomaly detection. According to [30–32], the OCSVM firstly mapped a sample x_i from the input space to the feature space F using the kernel function. The feature space had a higher dimension, and the separation may be easier in the feature space. Secondly, the OCSVM considered the origin as abnormal and the training samples as normal and constructed an optimal hyperplane between normal and abnormal by maximizing the margin.

The OCSVM mainly resolved the following quadratic programming optimization problem:

$$\min_{w \in F} \quad \frac{1}{2}||\mathbf{w}||^2 + \frac{1}{vn}\sum_{i=1}^{n}\xi_i - \rho \tag{4}$$
$$s.t. \quad \langle \mathbf{w}, \phi(\mathbf{x}_i) \rangle \geq \rho - \xi_i, \xi_i > 0, i = 1...n,$$

in which $\mathbf{w}$ is the normal vector in the feature space; n is the number of training samples; ξ_i is the slack variable to handle outliers; and $\phi(\cdot)$ is the mapping function mentioned above. ρ is the compensation parameter, and v defines the upper bound on the fraction of training errors and a lower bound of the fraction of support vectors.

By utilizing the Lagrangian method, the dual formulation and transformation of the original problem were obtained calculate the Lagrangian operator $\{\alpha_i\}_{i=1}^{n}$. The Lagrangian operator could be resolved by sequential minimal optimization (SMO) [33–35]. Finally, the decision function could be obtained by using the kernel method:

$$f(\mathbf{x}) = \langle \mathbf{w}, \phi(\mathbf{x}_i) \rangle - \rho = \sum_{i=1}^{n}\alpha_i \Phi(\mathbf{x}_i, \mathbf{x}) - \rho, \tag{5}$$

where $\Phi(\mathbf{x}_i, \mathbf{x}_j) == \langle \phi(\mathbf{x}_i), \phi(\mathbf{x}_j) \rangle$ is the kernel function. For any new input sample $\mathbf{x}$, if $f(\mathbf{x}) \geq 0$, then $\mathbf{x}$ was labeled as normal. Otherwise, if anomaly instances were detected, this indicated that the input control instruction was different from the normal behavior and that it was likely to be a fake packet. The anomaly detection model was implemented based on OCSVM using Python. The kernel function selected was the Gaussian kernel, as shown in the following formulation:

$$\Phi(\mathbf{x}_i, \mathbf{x}_j) = exp\left(-\frac{||\mathbf{x}_i - \mathbf{x}_j||^2}{\gamma}\right). \tag{6}$$

6. Simulations

6.1. Scenarios

The two stage IDS was tested in the ICS network based on Ethernet/IP. As mentioned in Section 3, the ICS platform was built for experiments to testify to the effectiveness of the proposed approach. The platform consisted of three layers: process monitor layer, process control layer, and plant-floor layer. The process monitor layer contained the malicious attacker and IDS. The process control layer was constituted by the PLC, industrial router, and VFD. The PLC used was the Allen-Bradley Micro 850 series, and the VFD used was the PowerFlex 525 from Rockwell. Moreover, the network structure was used to transmit the AGV localization data including an optical-electricity encoder and three phase asynchronous motor encoders. In this paper, according to the technological requirements, the time slot was selected to be 1 s, and the ICS ran for one day including 86,400 time slots.

The packets extracted from each time slot were delivered to the IDS for real-time inspection, which lasted for less than 10^{-4} A. One-thousand infiltration attacks, 1000 forging attacks, and 1000 false data injection attacks were randomly launched during the day. Furthermore, the corresponding 3000 normal instances were used for the simulation. The simulation results were displayed using a confusion matrix, which is shown in Table 1. The confusion matrix was designed for a two class classifier and was useful to evaluate the performance of the IDS [36].

Table 1. Confusion matrix for the IDS system.

Actual Class	Predicted Class	
Class	Normal	Attack
Normal	True negative (TN)	False Positive (FP)
Attack	False Negative (FN)	True positive (TP)

6.2. Simulation Results

6.2.1. Metrics

After calculating the predicted value x_t, the traffic prediction values were recovered by inverse difference and exponentiation. The results also needed to be rounded to the nearest integer. The traffic prediction model in the real ICS network was implemented, and the results of the rolling prediction are depicted in Figure 4. The root mean squared error (RMSE), which is calculated by:

$$RMSE = \sqrt{\frac{\sum_{t=1}^{n}(x_t - \hat{x}_t)^2}{n}}, \tag{7}$$

was used to determine the threshold between normal and abnormal status. At time slot t, if the difference $d_t = x_t - \hat{x}^t$ was larger than $1.5 * RMSE$, the ICS network was suspected to be infiltration attacked, and the two stage IDS would notify the engineer to check the traffic log. If no anomaly was detected, the packets would be delivered to the anomaly detection model for the next step of inspection. The kernel coefficient was set to be 0.1 according to the validation. The training samples (control instructions) were extracted from the ICS network in normal operation, and the smallest training error was guaranteed.

To evaluate the performance of the proposed method, four metrics were selected, the overall accuracy decision rates, false positive rate, false negative rate, and precision rate [36,37].

The overall accuracy is defined as:

$$Overall\,Accuracy = \frac{TP + TN}{TP + TN + FP + FN}, \tag{8}$$

which demonstrates the accuracy of the behavior of the attack detection. Literally, the overall accuracy denotes all the correct classifications against all the classifications.

The false positive rate (FPR) is defined by:

$$FPR = \frac{FP}{TN + FP}, \tag{9}$$

which describes the rate of the wrong predictions for the normal instances. The FPR can also be denoted as the fallout rate [37].

Conversely, the false negative rate (FNR) is defined as:

$$FNR = \frac{FN}{TP + FN}, \tag{10}$$

which describes the rate of the wrong predictions for the normal instances. The FNR can also be denoted as the miss rate [37].

Furthermore, the precision rate (PR) is defined as:

$$PR = \frac{TP}{TP + FP'}\tag{11}$$

which describes the precision of the prediction when an attack prediction is made [37].
All the above metrics were used for simulations comparing other IDS methods.

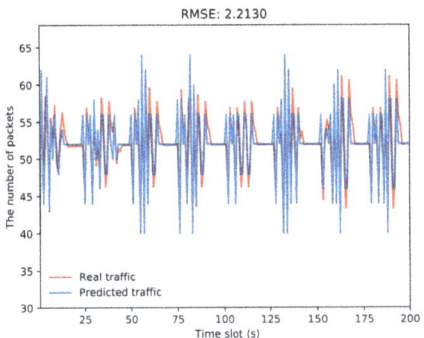

Figure 4. The result of the traffic prediction model based on ARIMA.

6.2.2. Performance of the Traffic Prediction Model

The orders p and q determined the accuracy of the ARIMA model, and they could be estimated by calculating the autocorrelation function (ACF) and the partial autocorrelation function (PACF) [29]. According to the results, both the ACF and PACF had trailing characteristics, and they both had obvious first order correlations. Therefore, we set $p = 1$ and $q = 1$.

The ARIMA model using Python and the fit model by the exact maximum likelihood via Kalman filter was simulated. After comparing the models by the criteria, the selected model was ARIMA(3,1,1), where $p = 3$, $q = 1$, and $d = 1$, which indicated the first difference. The given orders p and q were different from the orders observed by the ACF and PACF. The results of the observation were partly influenced by the real operation data. Furthermore, the ARIMA model needed to be updated periodically, and it was more convenient to select the proper model by using the criteria than observing the ACF and PACF. The criteria of the selected model were small enough and satisfactory.

$$AIC = -867.42, BIC = -887.15, HQIC = -875.41$$

The final parameters were as follows:

$$\psi_1 = -0.8561, \psi_2 = 0.6794, \psi_3 = -0.446, \theta_1 = -0.7998$$

Then, the prediction function can be calculated by the following equation.

$$\hat{x}_t = -0.8561 x_{t-1} + 0.6794 x_{t-2} - 0.446 x_{t-3} + \epsilon_t - 0.7998 \epsilon_{t-1},\tag{12}$$

6.2.3. Performance of the Anomaly Detection Model

In this section, the performance of the OCSVM is tested. As the proposed OCSVM approach is a kind of classification algorithm, the other two machine learning based methods were selected for the comparisons, which were called the semi-supervised machine learning method [22] and the boosting based machine learning method [28].

The size of the training data seriously affects the performance of each classification approach. In order to simulate the methods equally, three data sizes were selected to be the labeled training data, 877 (380 min), 3323 (24 h), and 6646 (48 h).

The simulation results are shown in Figure 5. It is depicted in Figure 5 that almost all the algorithms were improving as the training dataset size increased. When the data size was large, the overall accuracy and precision rate of all the methods could obtain an acceptable performance; both the overall accuracy and precision rate were more than 85%, which is shown in Figure 5c. However, when the data size was small, the proposed method performed better than the others, which is shown in Figure 5a. Furthermore, when the data size increased by about six times, the FPR of the two machine learning methods dropped from the highest of more than 20% to below 5%. Furthermore, the FNR dropped from more than 35% to 6%. This means that as the data size increased, the machine learning based performed better, and the smallest data size would be selected as 6646. Comparatively, the proposed method in this paper performed much better when the data size was small. As the data size increased, the performance even became less effective, which may be because as the training data size was enlarged, the classification model was overfitted.

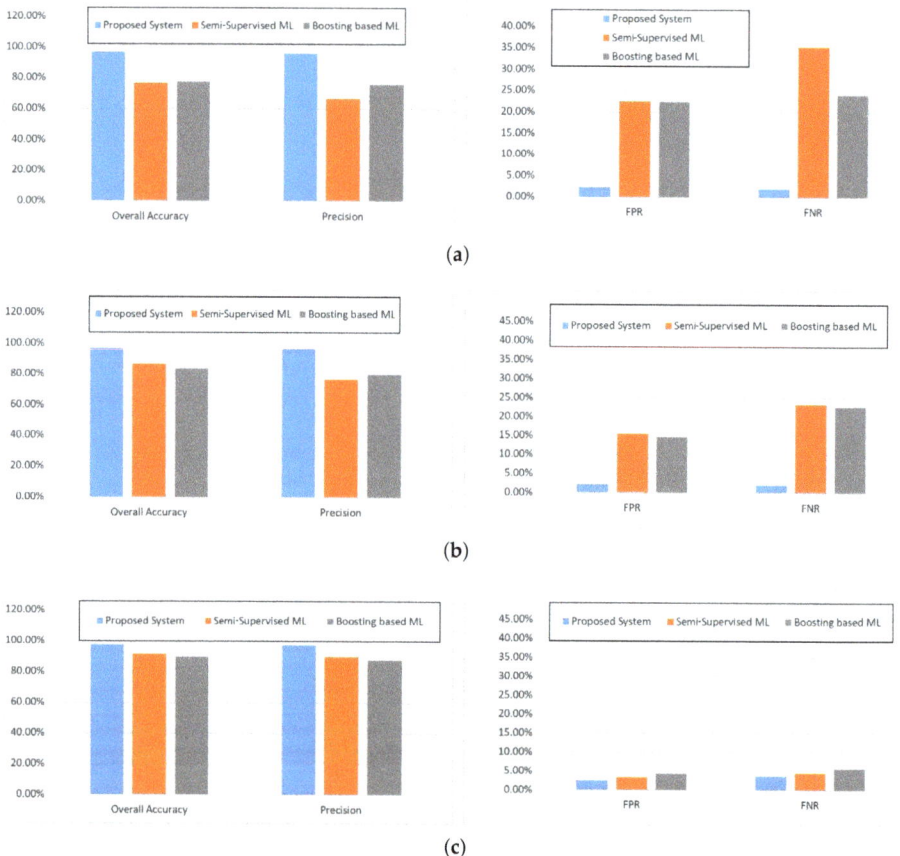

Figure 5. The classification experiment results between OCSVM and machine learning methods. (**a**) Train data size = 877. (**b**) Train data size = 3323. (**c**) Train data size = 6646.

6.2.4. Performance of the Proposed Two Stage IDS

A comparison was made among our two stage IDS, telemetry based IDS [15] and double behavior characteristics IDS [16]. The simulation results of the confusion matrices are shown in Table 2.

Table 2. Confusion matrix results for the IDS system.

Attacks	Infiltration				Forging				False Data Injection			
Confusion Matrix Parameters	TN	TP	FN	FP	TN	TP	FN	FP	TN	TP	FN	FP
Proposed two stage IDS	977	980	20	23	967	957	43	33	911	879	121	89
Telemetry based IDS	918	779	221	82	927	789	211	73	768	878	122	232
Double behavior characteristics IDS	982	821	179	18	964	866	134	36	837	855	145	163

Table 2 illustrates that the performance of the proposed method was better than the other two algorithms. For the infiltration attack, the proposed method obtained the highest precision rate (higher value of TP), and the double behavior characteristics IDS had the highest TN, which reflected the overall accuracy; however, the low TP also undermined the overall accuracy. For both forging and false data injection attacks, the proposed two stage IDS resulted in having the best performances. Although the defense to the false data injection attack may not have be as good as the other two kinds of attacks, it still worked much better than the other IDSs. A more detailed analysis is as follows, and the experimental results are illustrated in Figures 6–8, which was simulated under the proposed attack model mentioned in Section 4.

As shown in Figure 6, when the simulation was carried out under infiltration attacks, the proposed two stage IDS approach performed much better than the other two IDS methods, which was because the proposed method could detect the infiltration attacks by the traffic detection model. For the overall accuracy, the three selected methods all performed well, and the accuracy of the proposed method was better than 95%. Correspondingly, the precision rate performance of the proposed method was about 20% better than the other two methods, which means the proposed method could predict the attack instance more precisely, as shown in Figure 6a. Furthermore, the double behavior characteristics IDS method could perform well in FPR metrics, and not very well for FNR, which was because this method could classify more normal data into the attack group, as shown in Figure 6b. According to the simulation results, the proposed method could detect the infiltration attacks precisely.

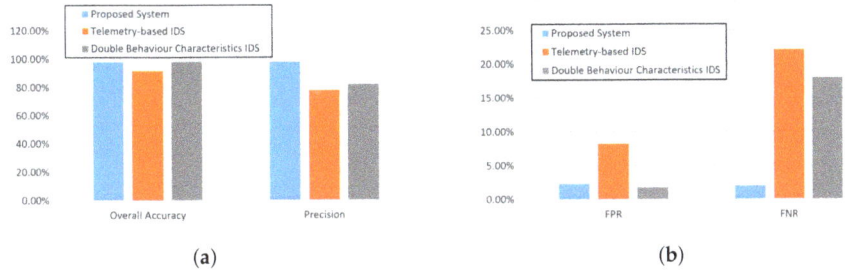

Figure 6. Experimental results under infiltration attacks.

As shown in Figure 7, when the simulation was carried out under forging attacks, the proposed two stage IDS approach performed much better than the other two IDS methods, which was because the proposed method could detect the forging attacks by the traffic detection model. The three selected methods all had good performance for the overall accuracy. The proposed method performed as well as the double behavior characteristics IDS, around 90%, which is shown in Figure 7a. It is to be mentioned that the proposed method could maintain a high precision rate also, because the proposed OCSVM methods could precisely extract the features of the forging attack. On the other hand, double behavior characteristics IDS had a competitive performance in terms of FPR, but the FNR was much more than that of our two stage IDS because it was impossible to detect the forging attack that was launched at a malicious time point. The proposed method had an outstanding performance in detecting the infiltration attacks and forging attack because of its models being strongly related to the ICS scenario, which is shown in Figure 7b. The two stage IDS processed the data collected from the ICS protocol and was capable of precisely reflecting the behavior characteristics in the ICS network.

As shown in Figure 8, when the simulation was carried out under false data injection attacks, it was hard to conclude that the proposed two stage IDS approach performed much better than the other two IDS methods, which was because the false data injection attacks could conceal themselves by changing their parameters, and the proposed approach, the traffic detection model, and the other two methods could not precisely detect the attacks. The proposed method could maintain the overall accuracy and precision rate over 80%, as shown in Figure 8a. Both the FPR and the FNR were a little bit higher than the former two kinds of attacks, which is shown in Figure 8b. To discuss this more, the two stage IDS processed the data collected from ICS protocol and was capable of precisely reflecting the behavior characteristics in the ICS network. If the attack models performed differently from the normal traffic, the proposed method would perform better. Otherwise, the method would not detect all the attack data or misjudge the normal data.

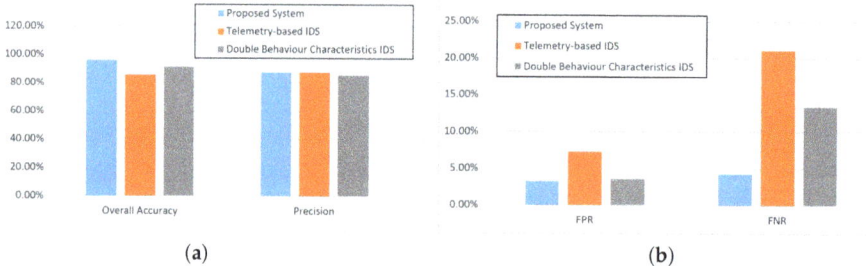

Figure 7. Experimental results under forging attacks.

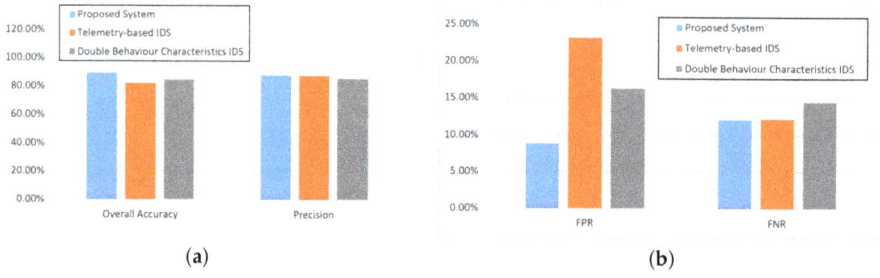

Figure 8. Experimental results under false data injection attacks.

7. Conclusions

This paper proposed a two stage IDS for the ICS based on Ethernet/IP. The two stage IDS contained a traffic prediction model and an anomaly detection model. Compared with machine learning methods, the proposed method could distinguish normal data and attack data with much fewer data. Furthermore, compared with telemetry based IDS and double behavior characteristics IDS, it offered excellent performance in detecting infiltration attacks and forging attack. It is to be mentioned that the performance under false data injection attack was not as good as the above two attack models. Furthermore, the proposed approach could not cope with the situation of asynchronous protocols such as the IEC 60870-5 series, which is mainly used in energy distribution networks and others because the transmission time must be known before two stage IDS is ready to work. Future work can be done in this specific area.

Future work can be divided into two parts: on one hand, evaluating the two stage IDS performance in a complex ICS scenario, which contains more controllers and actuators; on the other hand, refining the two stage IDS to defend I/O data transfers based on Ethernet/IP.

Author Contributions: Conceptualization, W.Y., and L.S.; methodology, Y.W. and W.Y.; system, W.Y.; validation, Y.W. and L.S.; experiment, Y.W.; analysis, Y.W. and W.Y.; writing, original draft preparation, Y.W. and W.Y.; writing, review and editing, L.S.

Funding: The research is sponsored by the National Key Research and Development Program of China (2018YFB1308304, 2017YFB1301103), the National Natural Science Foundation of China (61803261), and the Shanghai Natural Science Foundation of China (18ZR1421100).

Acknowledgments: The authors would like to thank and appreciate the support of all the scholars for helping us with this piece of work and the problems encountered.

Conflicts of Interest: The authors declare no conflict of interest.

References

1. Almalawi, A.; Tari, Z.; Fahad, A.; Khalil, I. A Framework for Improving the Accuracy of Unsupervised Intrusion Detection for SCADA Systems. In Proceedings of the 2013 12th IEEE International Conference on Trust, Security and Privacy in Computing and Communications, Melbourne, Australia, 16–18 July 2013; pp. 292–301. [CrossRef]
2. Oliver, E.; Philipp, K.; Tavolato, P. Identifying S7comm Protocol Data Injection Attacks in Cyber-Physical Systems. In Proceedings of the 2018 Proceedings of the 5th International Symposium for ICS & SCADA Cyber Security Research, Hamburg, Germany, 29–30 August 2018.
3. Kargl, F.; van der Heijden, R.W.; König, H.; Valdes, A.; Dacier, M.C. Insights on the Security and Dependability of Industrial Control Systems. *IEEE Secur. Priv.* **2014**, *12*, 75–78. [CrossRef]
4. Berhe, A.B.; Kim, K.; Tizazu, G.A. Industrial control system security framework for ethiopia. In Proceedings of the 2017 Ninth International Conference on Ubiquitous and Future Networks (ICUFN), Milan, Italy, 4–7 July 2017; pp. 814–817. [CrossRef]
5. Paridari, K.; O'Mahony, N.; El-Din Mady, A.; Chabukswar, R.; Boubekeur, M.; Sandberg, H. A Framework for Attack-Resilient Industrial Control Systems: Attack Detection and Controller Reconfiguration. *Proc. IEEE* **2018**, *106*, 113–128. [CrossRef]
6. Cheminod, M.; Durante, L.; Valenzano, A. Review of Security Issues in Industrial Networks. *IEEE Trans. Ind. Inform.* **2013**, *9*, 277–293. [CrossRef]
7. George, G.; Thampi, S.M. A Graph-Based Security Framework for Securing Industrial IoT Networks From Vulnerability Exploitations. *IEEE Access* **2018**, *6*, 43586–43601. [CrossRef]
8. Fan, X.; Fan, K.; Wang, Y.; Zhou, R. Overview of cyber-security of industrial control system. In Proceedings of the 2015 International Conference on Cyber Security of Smart Cities, Industrial Control System and Communications (SSIC), Shanghai, China, 5–7 August 2015; pp. 1–7. [CrossRef]
9. Meza, G.; Carpio, C.; Vinces, N.; Klusmann, M. Control of a three-axis CNC machine using PLC S7 1200 with the Mach3 software adapted to a Modbus TCP/IP network. In Proceedings of the 2018 IEEE XXV International Conference on Electronics, Electrical Engineering and Computing (INTERCON), Lima, Peru, 8–10 August 2018; pp. 1–4. [CrossRef]
10. Hittanagi, K.N.; Ramesh, M.; Kumar, K.N.R.; Mahadeva, S.K. PLC based DC drive control using Modbus RTU communication for selected applications of sugar mill. In Proceedings of the 2017 International Conference on Circuits, Controls, and Communications (CCUBE), Bangalore, India, 15–16 December 2017; pp. 80–85. [CrossRef]
11. Dias, A.L.; Sestito, G.S.; Turcato, A.C.; Brandão, D. Panorama, challenges and opportunities in PROFINET protocol research. In Proceedings of the 2018 13th IEEE International Conference on Industry Applications (INDUSCON), Sao Paulo, Brazil, 11–14 November 2018; pp. 186–193. [CrossRef]

12. Davies, S. Industrial ethernet—The fundamentals of ethernet/IP - Ethernet/IP has reached the million-node landmark, but what is making this protocol so attractive to industrial control engineers? *Comput. Control Eng. J.* **2007**, *18*, 42–45. [CrossRef]
13. Denis, M.; Zena, C.; Hayajneh, T. Penetration testing: Concepts, attack methods, and defense strategies. In Proceedings of the 2016 IEEE Long Island Systems, Applications and Technology Conference (LISAT), Farmingdale, NY, USA, 29 April 2016; pp. 1–6. [CrossRef]
14. Shebli, H.M.Z.A.; Beheshti, B.D. A study on penetration testing process and tools. In Proceedings of the 2018 IEEE Long Island Systems, Applications and Technology Conference (LISAT), Farmingdale, NY, USA, 4 May 2018; pp. 1–7. [CrossRef]
15. Ponomarev, S.; Atkison, T. Industrial Control System Network Intrusion Detection by Telemetry Analysis. *IEEE Trans. Dependable Secur. Comput.* **2016**, *13*, 252–260. [CrossRef]
16. Wan, M.; Shang, W.; Zeng, P. Double Behavior Characteristics for One-Class Classification Anomaly Detection in Networked Control Systems. *IEEE Trans. Inf. Forensics Secur.* **2017**, *12*, 3011–3023. [CrossRef]
17. Oliver, E.; Philipp, K.; Tavolato, P. Attacks on Industrial Control Systems—Modeling and Anomaly Detection. In Proceedings of the 2018 4th International Conference on Information Systems Security and Privacy, Madeira, Portugal, 22–24 January 2018.
18. Keliris, A.; Salehghaffari, H.; Cairl, B.; Krishnamurthy, P.; Maniatakos, M.; Khorrami, F. Machine learning based defense against process-aware attacks on Industrial Control Systems. In Proceedings of the 2016 IEEE International Test Conference (ITC), Fort Worth, TX, USA, 15–17 November 2016; pp. 1–10. [CrossRef]
19. Mantere, M.; Sailio, M.; Noponen, S. Network Traffic Features for Anomaly Detection in Specific Industrial Control System Network. *Future Internet* **2013**, *5*, 460–473. [CrossRef]
20. Zhang, J.; Gan, S.; Liu, X.; Zhu, P. Intrusion detection in SCADA systems by traffic periodicity and telemetry analysis. In Proceedings of the 2016 IEEE Symposium on Computers and Communication (ISCC), Messina, Italy, 27–30 June 2016; pp. 318–325. [CrossRef]
21. Haripriya, L.; Jabbar, M.A. Role of Machine Learning in Intrusion Detection System: Review. In Proceedings of the 2018 Second International Conference on Electronics, Communication and Aerospace Technology (ICECA), Coimbatore, India, 29–31 March 2018; pp. 925–929. [CrossRef]
22. Wagh, S.K.; Kolhe, S.R. Effective intrusion detection system using semi-supervised learning. In Proceedings of the 2014 International Conference on Data Mining and Intelligent Computing (ICDMIC), Delhi, India, 5–6 September 2014; pp. 1–5. [CrossRef]
23. Chaabouni, N.; Mosbah, M.; Zemmari, A.; Sauvignac, C.; Faruki, P. Network Intrusion Detection for IoT Security Based on Learning Techniques. *IEEE Commun. Surv. Tutor.* **2019**, *21*, 2671–2701. [CrossRef]
24. Mathur, A.P.; Tippenhauer, N.O. SWaT: A water treatment testbed for research and training on ICS security. In Proceedings of the 2016 International Workshop on Cyber-physical Systems for Smart Water Networks (CySWater), Vienna, Austria, 11 April 2016; pp. 31–36. [CrossRef]
25. Shah, M.; Soni, V.; Shah, H.; Desai, M. TCP/IP network protocols—Security threats, flaws and defense methods. In Proceedings of the 2016 3rd International Conference on Computing for Sustainable Global Development (INDIACom), New Delhi, India, 16–18 March 2016; pp. 2693–2699.
26. Bobade, S.; Goudar, R. Secure Data Communication Using Protocol Steganography in IPv6. In Proceedings of the 2015 International Conference on Computing Communication Control and Automation, Pune, India, 26–27 February 2015; pp. 275–279. [CrossRef]
27. Ponmaniraj, S.; Rashmi, R.; Anand, M.V. IDS Based Network Security Architecture with TCP/IP Parameters using Machine Learning. In Proceedings of the 2018 International Conference on Computing, Power and Communication Technologies (GUCON), Greater Noida, India, 28–29 September 2018; pp. 111–114. [CrossRef]
28. Wei, L.; Gao, D.; Luo, C. False Data Injection Attacks Detection with Deep Belief Networks in Smart Grid. In Proceedings of the 2018 Chinese Automation Congress (CAC), Xi'an, China, 23–25 November 2018; pp. 2621–2625. [CrossRef]
29. Wei, M.; Kim, K. Intrusion detection scheme using traffic prediction for wireless industrial networks. *J. Commun. Netw.* **2012**, *14*, 310–318. [CrossRef]
30. Xiao, Y.; Wang, H.; Xu, W. Parameter Selection of Gaussian Kernel for One-Class SVM. *IEEE Trans. Cybern.* **2015**, *45*, 941–953. [CrossRef] [PubMed]

31. Maglaras, L.A.; Jiang, J.; Cruz, T. Integrated OCSVM mechanism for intrusion detection in SCADA systems. *Electron. Lett.* **2014**, *50*, 1935–1936. [CrossRef]
32. Li, Y.; Zhang, T.; Ma, Y.Y.; Zhou, C. Anomaly Detection of User Behavior for Database Security Audit Based on OCSVM. In Proceedings of the 2016 3rd International Conference on Information Science and Control Engineering (ICISCE), Beijing, China, 8–10 July 2016; pp. 214–219. [CrossRef]
33. Keerthi, S.S.; Shevade, S.K.; Bhattacharyya, C.; Murthy, K.R.K. Improvements to Platt's SMO Algorithm for SVM Classifier Design. *Neural Comput.* **2001**, *13*, 637–649. [CrossRef]
34. Toyoda, K.; Okamoto, T.; Koakutsu, S. An optimal routing search method on the network routing problem using the sequential minimal optimization. In Proceedings of the 2017 56th Annual Conference of the Society of Instrument and Control Engineers of Japan (SICE), Kanazawa, Japan, 19–22 September 2017; pp. 805–810. [CrossRef]
35. Sheenu; Joshi, G.; Vig, R. A multi-class hand gesture recognition in complex background using Sequential minimal Optimization. In Proceedings of the 2015 International Conference on Signal Processing, Computing and Control (ISPCC), Solan, India, 24–26 September 2015; pp. 92–96. [CrossRef]
36. Khraisat, A.; Gondal, I.; Vamplew, P.; Kamruzzaman, J. Survey of intrusion detection systems: Techniques, datasets and challenges. *Cybersecurity* **2019**, *2*, 1–22. [CrossRef]
37. Hindy, H.; Brosset, D.; Bayne, E.; Seeam, A.; Tachtatzis, C.; Atkinson, R.C.; Bellekens, X.J.A. A Taxonomy and Survey of Intrusion Detection System Design Techniques, Network Threats and Datasets. *arXiv* **2018**, arXiv:1806.03517.

© 2019 by the authors. Licensee MDPI, Basel, Switzerland. This article is an open access article distributed under the terms and conditions of the Creative Commons Attribution (CC BY) license (http://creativecommons.org/licenses/by/4.0/).

Article

An Approach for the Application of a Dynamic Multi-Class Classifier for Network Intrusion Detection Systems

Xavier Larriva-Novo, Carmen Sánchez-Zas, Víctor A. Villagrá *, Mario Vega-Barbas and Diego Rivera

ETSI Telecomunicación, Universidad Politécnica de Madrid (UPM), Avda. Complutense 30, 28040 Madrid, Spain; xavier.larriva.novo@upm.es (X.L.-N.); carmen.szas@alumnos.upm.es (C.S.-Z.); mario.vega@upm.es (M.V.-B.); diego.rivera@upm.es (D.R.)
* Correspondence: victor.villagra@upm.es

Received: 3 September 2020; Accepted: 20 October 2020; Published: 23 October 2020

Abstract: Currently, the use of machine learning models for developing intrusion detection systems is a technology trend which improvement has been proven. These intelligent systems are trained with labeled datasets, including different types of attacks and the normal behavior of the network. Most of the studies use a unique machine learning model, identifying anomalies related to possible attacks. In other cases, machine learning algorithms are used to identify certain type of attacks. However, recent studies show that certain models are more accurate identifying certain classes of attacks than others. Thus, this study tries to identify which model fits better with each kind of attack in order to define a set of reasoner modules. In addition, this research work proposes to organize these modules to feed a selection system, that is, a dynamic classifier. Finally, the study shows that when using the proposed dynamic classifier model, the detection range increases, improving the detection by each individual model in terms of accuracy.

Keywords: intrusion detection system; dynamic classifier; ensemble machine learning; multiclass; cybersecurity

---

1. Introduction

Intrusion detection systems (IDS) are computer systems designed to monitor network traffic. These systems are capable to find atypical records and attack patterns based on the behavior of the networks. Thus, the aim of IDS is the early detection or prediction of possible real harm to the network, host, or cloud caused by a security issue. To perform this, IDS, as a software application, analyze possible anomalies detected at the network layer. This process is, traditionally, static and linked to the rules or algorithms used for detecting cyberattacks. Nevertheless, this static process is difficult to adapt to the detection of new types of attacks because it implies updating it with new rules in the cases of signature-based IDS [1], or the re-training of the detection model in the case of anomaly-based IDS [2]. Specifically, anomaly-based IDS are related directly to the application of machine learning (ML) techniques. These techniques, depending on the underlaying classification model, are capable of detecting anomalies by means of binary classifiers, or different types of attacks by means of multiclass classifiers. In general, the aim of ML-based IDS is to increase their ability of attack detection by reducing the quantity of possible false positives [2]. The reduction of possible false positives is a crucial issue in the design of IDSs, as the continuous development of automated malware forces the IDSs to be as accurate as possible, trying to be one step ahead of attackers. Nevertheless, the accuracy of this software still has important limitations, and therefore, there is a margin for their improvement.

In recent studies, it is possible to observe this use of ML techniques for detecting possible cyberattacks in a more efficient way [3,4]. Most of these studies are based on a binary classification, but some investigations have improved the prediction by establishing a multiclass classification [5]. The main drawback of the systems presented in these studies is that they offer low levels of general precision [6,7] although they do offer a high level of efficiency in very specific attacks. Thus, it could be possible to identify and characterize certain types of models that allow a better detection of specific classes of attacks.

However, IDS performance is impaired and, in combination with the fact that the same algorithm is applied to a probably complete dataset, a problem to identify certain types of attacks has been posed. When a model is trained, it recognizes better patterns in some features, usually related to a type of intrusion. This is the main reason for the need of an adaptable system, to take advantage from the inclination of each algorithm to detect better some types of attacks over others.

In order to mitigate this disadvantage of current models, this research proposes a dynamic classification model that can obtain the main contributions in terms of rate detection of each individual ML model with the aim of generating reliable predictions of attacks from them. To perform this, the system introduces the separate training, validation, and metrics of different algorithms such as classical ML models and neural networks, and then join their predictions into a module, which aims to compare the different outcomes and extract the most suitable one.

This paper presents the related work in the state of the art in Section 2. Furthermore, in Sections 3–5, we explain the problem statement, the proposed methodology, and experimentation, defining the new architecture for the dynamic classifier introduced in this research. These sections also include several tests for different ML learning techniques, including data preprocessing and feature selection. Section 6 provides the obtained results for the individuals ML models proposed and the application of the dynamic classifier proposed using the dataset UNSW-NB15, concluding with a comparison between results obtained by the different tests done and related works. Finally, Section 6 includes the discussion of the main conclusions and future lines for this article.

2. Related Works

Nowadays, the research related to the selection of the most suitable IDS is addressed from different perspectives based on the description of the input problem and the analysis of the performance of underlaying algorithms. In general, the aim of these research works is to study the algorithm selection problem in order to determine the most proper one.

Some studies are based on basic ML techniques, selecting an algorithm, and training it with a complete dataset. Such as the case of [8], where the application of a multi-layer perceptron (MLP) network is applied to detect large scale datasets and predict malicious attacks. In this case, the selection of the attributes is based on the co-variance, standard deviation, or correlation. The authors of the paper obtained an accuracy of 0.9935 selecting the attributes with a near perfect correlation. In [2], Larriva-Novo et al. propose an analysis based on a categorization of a cybersecurity dataset, where they used a static algorithm based on MLP. The analysis was done in order to select the best hyperparameters related to the best performance in terms of accuracy. Furthermore, the different characteristics selected were based on: basic connection, content, traffic statistical and direction characteristics. The model proposed below presented an accuracy near to 99%, based on anomaly detection.

A typical approach for the design of IDS based on anomalies is the application of the support vector machine (SVM) algorithm, to predict if the input data represents an anomaly or not. The authors of [9] improved the results of their system using a non-linear scaling method for data preprocessing. The classifications conducted were binary and multi class, measured by accuracy (AC), detection rate (DR) and false positive rate (FPR). The accuracy obtained was 85.99% for binary classification and 75.77% for multi-class classification. In [10], they tried to go deeper in the classification based on SVM algorithms. The authors developed an efficient IDS, with an improved performance in terms of accuracy due to the binary gravitational search. The results obtained were 86.62% in terms of accuracy without the application of feature selection and 94.4% with its application, improving the results obtained in [9] for the binary classification.

Another approach based on deep learning techniques such as feed-forward neural networks is presented in [11]. This method consisted on selecting the optimal activation and features. The authors conducted three experiments: selecting the best activation function, deploying a feature selection and applying the obtained results to new data. The optimal results exposed an accuracy of 99.5%. Despite these results, basic algorithms for anomaly detection are not effective enough to perform the classification problem in IDS based on ML algorithms because they must be optimized, related to its hyperparameters and the quality of the data [12].

On the other hand, a large variety of algorithms have been used with the objective of finding a better performance in IDS multiclass attack detection. This is the case of the extreme gradient boosting (XGBoost) [13], which was used to deploy better results. The authors experimented with 39 numerical features, excluding those like IP Address or Protocol Type. The best results were achieved with an accuracy of 88% for the testing dataset [14]. Furthermore, the research conducted to include a data preprocessing phase, where the optimal features were reduced from 39 to 23. The same models were applied, obtaining again the best accuracy by the XGBoost model. However, in this case, the accuracy rate fell to a 76%. It should be noted that in this case the category "attack analysis" presented difficulties to be detected.

Recent works proved the effectiveness of different techniques to make predictions in IDS. However, just a few of them used time-series information and categorical information. This is the case in [15], where the authors used a long short-term memory (LSTM) network with feature embedding. The model makes a multi class classification based on chronologically order information. The research presented multiples combinations in order to deploy the most accurate model: feature embedding selection, transformation of categorical data, and projection of the features as vectors values in the space. The results in terms of accuracy was of 83% for a multiclass classification.

Another standpoint to confront the algorithm selection problem is based on the use of a Bayesian approach, because these models are able to reason under uncertainly. In [16], a static automatic selection is proposed, designing an expert system that presents knowledge representation, learning, and inference ability. The methodology presented in that work was based on the identification of representative characteristics and a list of suitable candidates to generate a random training dataset. They then use the dataset to analyze the performance of each algorithm. This method obtained an accuracy of 76.08% in its prediction, which was the best result in comparison with other proposed candidate algorithms. This approach was also used to select other hyperparameters among a wide space of choices as kernel functions [17].

Alternatively, a method called HYDRA is presented in [18], which represents a new approach to ML models setup that combines automated algorithm configuration and portfolio-based algorithm selection techniques [19,20]. The base of this approach is to join the advantages of techniques, i.e., less domain knowledge required, mechanization from automated algorithm configuration, and variety of candidate algorithms. HYDRA accepts five inputs: a parametrized solver, a set of training problem instances, an algorithms configuration procedure, a performance metric to optimize, and portfolio-based algorithm selection. During the conducted test, it was proved that HYDRA outperformed in almost every case, the accuracy of the proposed algorithms.

Over time, completely automatic ML models were developed which tries to outperform the best accuracy of selected algorithms over a determined output. An example of this is Auto-WEKA [21], designed to choose the most appropriate algorithm and its optimal hyperparameters automatically, between a large number of possible combinations (39 WEKA's algorithms are available) using a Bayesian optimizer. Since its original release in 2013 some updates were included in Auto-WEKA 2.0 [22].

In [23], the authors discuss the need for a dynamic ML system. The objective they propose is to avoid the loss of efficiency, characteristic of a static model. This derives from the choice of the selected features, inferring that a single model is not sufficient. The authors discuss the need of a dynamic selection of a model. At first, Ensemble ML is proposed as a solution, but the drawback of the need for continuous training makes it an inefficient option. The author of the research mentioned above, introduced a new methodology based on a cloud device operations architecture. The research proposed different containers to train a model, so each of them made a prediction to be sent to a model selector. The model selector was able to choose the most suitable one in terms of precision. Analyzing obtained results based on supervised options, an accuracy of 61% was obtained. This value increased substantially when the system was integrated with label categorization. Taking into account these results, unsupervised learning by clustering was only recommended for those attacks with lowest accuracy. The objective was to select the most accurate model according to real-time data received. The authors mentioned that it is imperative to automate this task, as there were at least 70 models to choose from. Furthermore, the dataset used for the researches introduced above was the benchmark dataset UNSW-NB15 [14]. Finally, [24] introduced a multiclassification approach which combines the outputs from related classifiers in the state of the art, such as the one proposed in [25], for mobile and encrypted traffic classification based on hard/soft combiner enhancing up to 9.5% of recall.

As a conclusion, this section has identified and detailed several related works focusing on the study of statics models for the application of IDSs based on ML algorithms. These studies conducted several analysis based on the performance of different ML models in terms of accuracy. The models and ML techniques applied obtained, in the best cases, an overall accuracy of 99% for a binary classification. Furthermore, some of these researches included the study and application of multiclass classification evaluation, experimenting a low rate of detection in terms of accuracy. In addition, other works such as [23], lead to the need of dynamic auto selection ML models with the objective to perform the rate detection by a dynamic selection of a model. All these works are based on several ML techniques such as correlation, feature selection, multiclass classification evaluation and dynamic model selection. Therefore, most of these techniques are covered and evaluated in this proposal.

3. Background

3.1. Cybersecurity Datasets

Nowadays, there exist different cybersecurity datasets that can be used for IDS based ML experimentation, i.e., UNB-ISCX-1012 [26], CTU-13 [27], MACCDC [28], UGR-16 [29], CICDS [30], KDD-99, NSL-KDD [31], or UNSW-NB15 [32]. Some of them have been widely used, like for instance the dataset KDD-99, which has been stablished as the main benchmark dataset for the different studies cases in the application of ML-based IDS. In [6], there is a representation of the most used datasets in the last decade for anomaly detection based on ML algorithms. This study points out that the NSL-KDD and the KDD-99 are the most used, with a 11.6% and 63.8% of use respectively.

In this study, however, the chosen dataset was UNSW-NB15 [14], because it has been considered as a benchmark dataset for the evaluation of IDS based on ML models thanks to the variety of the current cybersecurity attacks to date, now being widely used in cybersecurity [32].

3.1.1. Attack Categories in UNSW-NB15

The following are the nine different attack categories considered in the UNSW-NB15 dataset used in our study:

Fuzzers: Injection of an invalid random data into a program to cause crashes and exceptions.
Analysis: Involves network monitoring i.e., port scanning, spam, or penetration through HTML files.
Backdoor: Action of avoid security mechanisms of a system to access to stored data.
Denial of Service (DoS): Attempt to make a system temporarily unavailable to its users.
Exploits: Take the advantage of a security flaw to gain access of it.
Generic: Technique that works against all blockages, regardless of the encryption used.

Reconnaissance: Malicious actions that involves the collection of information.

Shellcode: Code used as a payload against an identified vulnerability in a system.

Worms: Replication of an attacker to spread an intrusion to other systems through the network.

3.1.2. UNSW-NB15 Description

This dataset contains information about multiple connections being labeled to differentiate those data that represent an attack and the normal behavior of the network, making suitable for supervised learning such as the case of this research.

The UNSW-NB15 includes 47 different features. In addition, it contains two fields that represent if the traffic flow is an attack or not, and the attack category, being one of those mentioned in Section 3.1.1. The features of the dataset are classified into flow features, basic features, time features, content features, additional features, and labeled features [14,32,33].

The dataset is divided in 10 categories including the nine presented before and the normal behavior among them, as presented in Table 1, where the number of records of each category and the percentage of records from the total are also shown.

Table 1. UNSW-NB15 distribution by types of attacks.

Type	% Records	No. Records
Normal	36.092	2,218,761
Fuzzer	9.409	24,246
Analysis	1.039	2677
Backdoors	0.904	2329
DoS	6.346	16,353
Exploits	17.279	44,525
Generic	22.847	215,481
Reconnaissance	5.428	13,987
Shellcode	0.586	1511
Worms	0.067	174

3.2. Machine Learning Applied to Intrution Detection Systems

Nowadays, ML has been incorporated in cybersecurity, specially to the design of IDS. It has been widely used, involving different levels of complexity, from the choice of an algorithm, to complex systems with the ability to autoconfigure themselves [23,34]. It is a field which focuses on the estimation of different mathematical functions with the objective of extract and represent behavioral generalizations of the data [3]. In addition, ML can be defined as a subset of the artificial intelligence with the ability to produce desired outputs referred to an input data without being programed.

ML algorithms can be classified into different categories: supervised learning, unsupervised learning, and semi-supervised learning. In this paper, we focus specifically in supervised learning methods, as the others are currently out of the scope of our research.

Supervised learning algorithms are based on what they are capable to learn about a set of features with an explicit label. Therefore, if the function is capable to adequate the input data to a desired output label, the algorithm is able to predict outputs given inputs in a similar scenario. After the training of the model, the algorithm can be executed until it reaches an acceptable level of performance [35].

3.3. Machine Learning Algorithms under Study

In the context of supervised learning, ML offers different types of algorithms because each one presents a different response to a unique input [23]. We have evaluated the results of the works presented in Section 2, analyzing the algorithms evaluated in those researches, in order to improve their results with our proposal. Additionally, other algorithms with similar procedures were taken into account, for those whose working methods fit the requirements of this research. After that evaluation, the models selected are the following:

1. K-nearest neighbors (KNN): It is an algorithm that calculates and orders the distance from new data to the existing one, classifying this input according to the frequency of the labels of the K-nearest ones. This distance is usually measured by the Euclidian norm presented in Equation (1). For the correct adjustment of this method [36], a correct value of the number of neighbors considered is essential.

$$d(x,y) = \sqrt{\sum_{i=1}^{k}(x_i - y_i)^2} \qquad (1)$$

2. Decision trees (DT): In the DT algorithm each node represents a test over an attribute, and each branch is a possible outcome from it. It uses a decision support to show the prediction lead by features splits [37]. In this case, among the hyperparameters, we can list maximum depth of the diagram, minimum number of samples required to split an internal node or minimum number of samples required to be at a leaf node.
3. Random forest (RF): RF is composed by many DT [34], each one created from a subset of features to be considered. Each one votes and the algorithms compute all of them to choose the best prediction. The main advantage of this method is the prevention of the over-fitting [38].
4. Support vector machines (SVM): SVM allows find the optimal classification, maximizing the margin between classes, whose border is defined by support vectors. In case there is not a lineal separation, the kernel trick is applied to redefine inner products [39].
5. XGBoost: It is an algorithm that starts with weak classifiers over a set of data with the objective to enhance their results by means of a sequential processing with loss function to minimize the error with every iteration, obtaining a strong model at the end [40].
6. Multi-layer perceptron neural network (MLPNN): It is composed by a group of linear classifiers denominated perceptron. The perceptron itself contains a group of layers (i.e., input layer, output layer and hidden layer). MLPNN is a type of feed forward neural network (FFNN), where the layers have a non-linear activation function. MLPNN is trained by a back-propagation model which stablish the relation between the input features and the output features, with the objective to minimize the error [2]. The different hyperparameters (i.e., number of hidden neurons, layers, iterations, activation function, etc.) must be optimized in order to achieve better results.
7. Long short-term memory neural network (LSTMNN): LSTMNN are recurrent neural networks (RNN). These units are capable of connect previous information to the input data, learning those long-term dependencies. This model is configured to maintain the back-propagation error constant through the time between the different layers [2]. In this case, the hyperparameters to be optimized are batch size and number of hidden neurons.

4. Proposal

The dynamic classifier proposed in this research is designed to achieve the objective described throughout this document, a system capable of obtaining the best prediction results from various ML algorithms based on a multiclass classification. To develop the dynamic classifier, previously optimized models are required [41]. The proposed system is presented in Figure 1. As it can be seen in the figure, we propose an architecture composed of different modules: a series of static ML algorithms manually preconfigured by means of a study of hyperparameter selection and feature selection, and finally by a dynamic classifier.

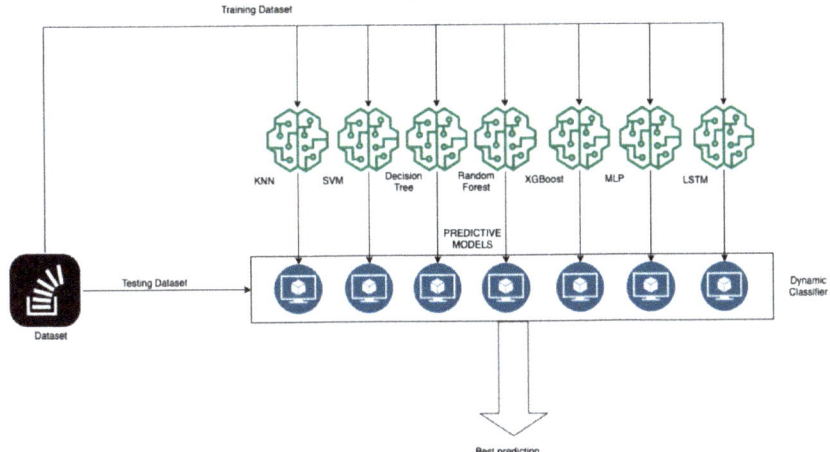

Figure 1. System proposed, composed by Multi ML (machine learning) algorithms, a real time streaming data and a dynamic classifier.

In our research work we include the development of that series of static ML models, capable of predicting possible attacks. Each model will count on its own configuration of hyperparameters and a specific feature selection. The related work study from Section 2 indicates that not all models are able to achieve good prediction accuracy for every type of attack. Instead, each ML model seems to be more suitable for certain attack categories. In the following sections we confirm this, by carrying out tests using different models to predict different attacks.

The dynamic classifier works by evaluating the incoming information by each one of these static ML models. The output of each one, consisting on a prediction on the possible attack (based on the attack categories defined by the UNSW-NB15 dataset, shown in Section 3.1.1 of this paper) is then used as input data for the actual classifier. The classifier is then able to determine the best prediction from those generated by the ML models, and select it, regardless of the attack category. This method improves the accuracy of attack prediction.

5. Methodology and Development

5.1. Dataset Preparation

For this research, the dataset selected was the UNSW-NB15 [14], as was mentioned before in Section 3.1. In [2], the importance of selecting the most important features between all the features available is established, as it has an enormous impact in the algorithm's performance. The less important features do not bring performance in terms of accuracy while consuming computer resources. For this purpose, the features were classified according to its type—i.e., numerical values were transformed into z-score [42] values—and categorical features were transformed into numerical values [2]. As was presented in Table 1, some attack categories expose a low distribution, what may not produce accordance results in terms of accuracy. To prevent this, some researches have used SMOTE [43] to balance the dataset UNSW-NB15, obtaining good results [44]. This algorithm was applied in this research with the objective to compare the results between a balanced dataset by SMOTE, and an unbalanced dataset as the original dataset.

Additionally, we have performed a feature selection based on the correlation Kendall coefficient [45], which has improved the results of the selection task. We improved the tests with both a balanced and unbalanced dataset. In Figure 2, the features correlation matrices for the UNSW-NB15 are presented.

For this research, the best results were achieved by removing the features with a correlation higher than 0.8 for both balanced and unbalanced dataset. Table 2 presents the features selected for both datasets.

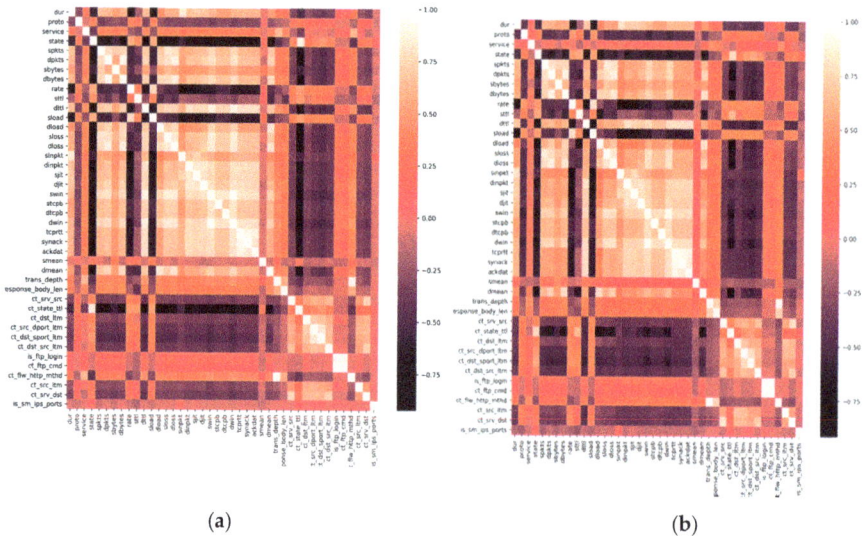

Figure 2. Features correlation matrix UNSW-NB15: (**a**) balanced dataset; (**b**) unbalanced dataset.

Table 2. UNSW-NB15 selected features after correlation Kendall coefficient feature selection.

Balanced Dataset	Unbalanced Dataset
'sloss', 'dinpkt', 'rate', 'dwin', 'synack', 'dmean', 'dload', 'sjit', 'ct state ttl', 'stcpb', 'ct w http mthd', 'djit', 'dloss', 'ct ftp cmd', 'ct dst sport ltm', 'ackdat', 'dtcpb', 'tcprtt', 'swin', 'dbytes', 'sinpkt', 'sload', 'dpkts', 'dttl'.	'ackdat', 'synack', 'dpkts', 'ct srv dst', 'dloss', 'dwin', 'sload', 'sinpkt', 'ct w http mthd', 'ct dst sport ltm', 'sloss', 'stcpb', 'tcprtt', 'dinpkt', 'ct ftp cmd', 'dtcpb', 'sjit', 'dbytes', 'ct state ttl', 'rate', 'djit', 'swin', 'dmean'.

After the selection process, the dataset was divided into training (75%) and testing (25%) data. The training data was used for the generation of models in each algorithm selected. The testing data was used for the validation of each algorithm and for testing the dynamic classifier proposed.

In addition, some algorithms present a better performance using less features [45]. To achieve better results over the correlated feature selection, K-Best [46] function was applied. The final feature selection is presented in the following section.

5.2. Machine Learning Models Applied

All the considered algorithms were trained both for the balanced and Unbalanced dataset presented in Table 2, to analyze the performance in terms of rate detection differences of each method.

1. KNN: This model was trained using the six features obtained with K-Best. From the balanced dataset 'sbytes', 'dbytes', 'sload', 'smean', 'dmean', 'ct srv dst' were extracted, meanwhile for the unbalanced case, the selected features were 'dur', 'proto', 'service', 'sbytes', 'dttl', 'smean'. The best result was obtained for 12 neighbors (balanced dataset) and 27 neighbors (unbalanced dataset).
2. SVM: Due to the large increase of data from the balanced dataset, the training of this model for this dataset could not be completed, due to memory restrictions in the experiment environment. The model was trained for the unbalanced dataset using features from K-Best analysis ('dur',

'proto', 'service', 'sbytes', 'dttl', 'smean'). The best results were obtained with the hyperparameters: kernel = 'rbf', gamma = 'scale', C = 50 and max_iter = 50,000.
3. DT: The best results were achieved by using the entire dataset. The best hyperparameters obtained were max depth = 28 and random_state = 4 for both balanced and unbalanced datasets.
4. RF: The best results were achieved by using the entire dataset. The best hyperparameters obtained were criterion = 'entropy', max_depth = 50 and n_estimators = 50 for both balanced and unbalanced datasets.
5. XGBoost: The best results were achieved by using the entire dataset. The best hyperparameters obtained were n_estimators = 500.
6. MLPNN and LSTMNN: The best results were carried by using all the features of the dataset. The model configuration applied was with the model introduced in [2].

In the case of the KNN, SVM, DT, RF, and XGBoost algorithms the cross validation function with GridSearch [47] was applied in order to find the optimal hyperparameters.

5.3. Dynamic Classifier

The dynamic classifier, as presented in Figure 3, was designed to aggregate the predictions from the individual ML models, and make an automatic selection of the optimal prediction obtained from each one for a single sample, while the models are executed in parallel to make predictions over the testing dataset.

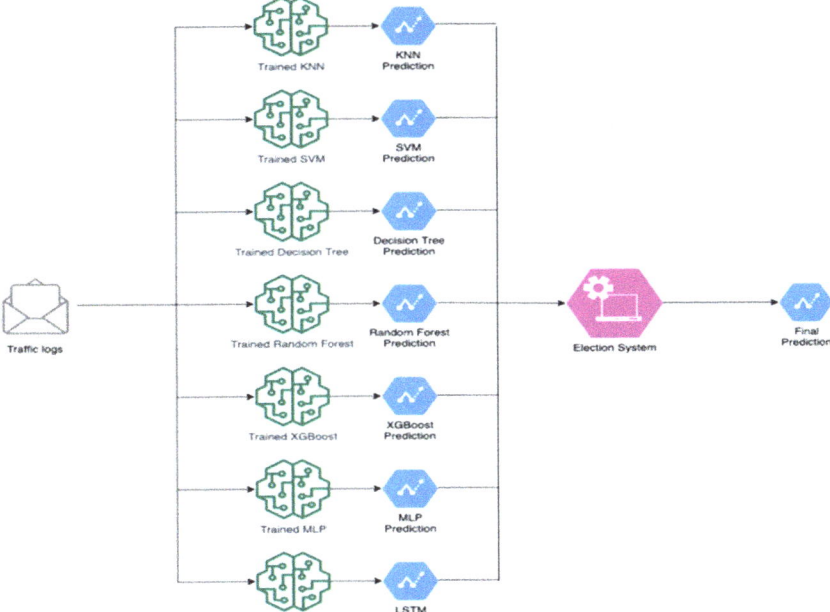

Figure 3. Dynamic classifier system.

Diverse tests were done in order to design the dynamic classifier system i.e., probability analysis, voter model [48], and ensemble ML model [12,23]. After several tests, the dynamic classifier was designed by an ensemble ML model based on XGBoost. The ensemble model was more accurate than the voter model by 3% of difference in comparison with the balanced dataset. In relation with the probability analysis the results were not relevant, so this method was excluded. Also, the use of a

ML algorithm permits increasing the detection rate by optimizing its hyperparameters. Taking into account these results, the ensemble model based on XGBoost was selected for the dynamic classifier.

The goal of the dynamic classifier proposed is to combine opinions of the base experts (individual ML models) in an ensemble, which is shown in the Algorithm 1. Each expert assigns a dynamic coefficient based on its own prediction, over the decision taken from the input data. This weight coefficient (0–1 scale) depends on the general accuracy by each individual model. For this purpose, an alternative dataset was created with the accuracy of each model presented in Section 5.2. The fitting was done selecting as input features, the predictions from the individual models mentioned below, and as outcome the desired classification attack. The proposed system classifies the predictions from the individual models, and the output is based on the relations that the ensemble model has found exposing the most relevant one from the individual algorithms.

The goal of the dynamic classifier is to bring together of several individual models to improve the accuracy of prediction with respect to an individual static model.

Algorithm 1 Ensemble proposed model

1: **Ensembleproposedmodel**
2: open full dataset file
3: drop incomplete rows
4: application of feature selection
5: **for** each individual ML model
6:   **fit** each individual machine learning model
7:   **validate** each individual machine learning model
8:   **save** accuracy coefficient
9: **fit** dynamic_classifier (x_test =accuracy_coefficient, y_test=attack_categories)
10: prediction = prediction each individual ML model (accuracy-multiclass)
11: **validate** dynamic_classifier (x_validate = prediction)
12: **write** dynamic_classifier (True Positive Rate and Accuracy (multiclass-attack))
---

6. System Analysis and Evaluation

6.1. Analysis of Statics Models

The evaluation metrics presented in this research are derived from the elements of the confusion matrix. The evaluation metric proposed for this research is the true positive rate (TPR) which considers the proportion of real positives correctly predicted, that is, the portion of attacks correctly identified over all possible attacks [12]. In Figure 4, we show the obtained results in terms of TPR of each individual model proposed in the present research for the unbalanced dataset. As we can see some models present a high detection over some types of attacks such as the case of exploits and generic. Instead, in the case of attacks such as analysis, backdoor DoS present a low range of detection, even less than 30% by each class.

In Figure 5, it is presented the obtained results in terms of TPR of each individual model proposed in the present research for the balanced dataset. It is shown how all the models raised the rate of detection over the attacks in comparison with the unbalanced dataset. Despite the increase of rates, in the case of attacks—such as analysis and backdoor exploits—there is still a low rate of detection, even less than 65% for each class. On the other hand, all the models show a decreased the detection rate over class exploits in comparison with the unbalanced dataset. In the case of worms, all the models except SVM (which could not be trained as was exposed in Section 3), raised the rate of detection over the 99%.

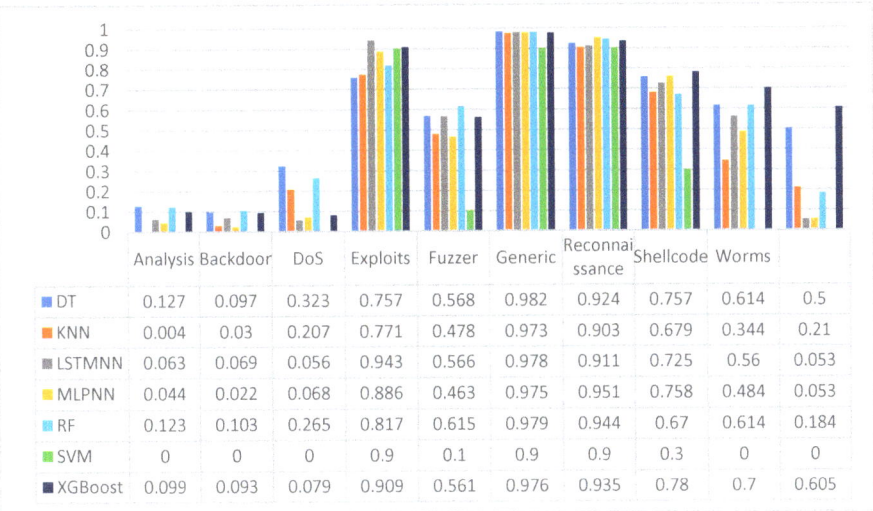

Figure 4. True positive rate metrics for static models applied to the unbalanced dataset.

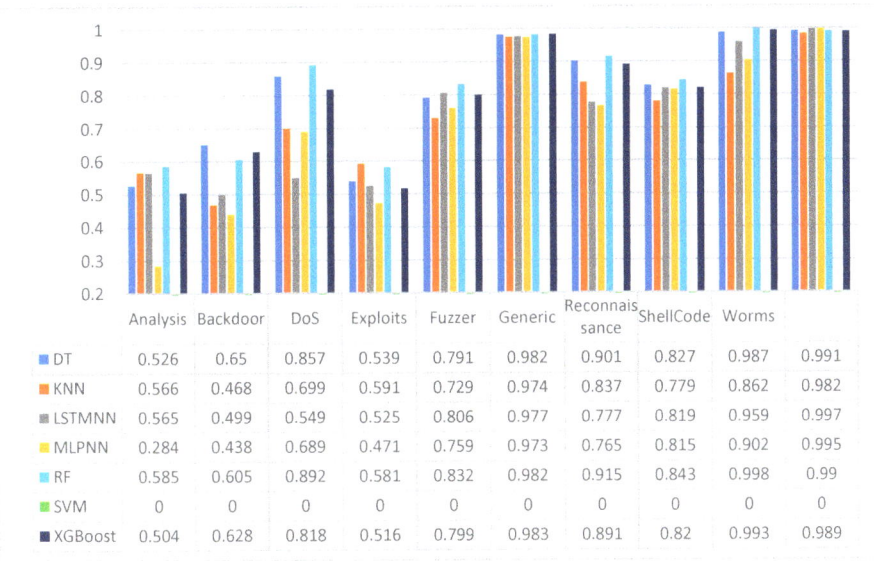

Figure 5. True positive rate metrics for static models applied to the balanced dataset.

In Table 3, we summarize the best results in terms of TPR peer class of attack. The unbalanced dataset presents a low rate of detection in different attacks categories such as analysis, backdoor, and DoS. The improved detection is raised by the balance of the data. All the attack categories improved their detection rate once the data was balanced, except for the exploits and generic categories.

As can be seen in Table 3, different models are more suitable for each different type of attack.

Table 3. Best true positive rate detection per class of attack for balanced and unbalanced datasets applying static models.

Class	Balanced Dataset TPR	Model	Unbalanced Dataset TPR	Model
Analysis	0.585	RF	0.127	DT
Backdoor	0.65	DT	0.103	RF
DoS	0.892	RF	0.323	DT
Exploits	0.591	KNN	0.943	LSTMNN
Fuzzer	0.832	RF	0.615	RF
Generic	0.983	KNN	0.982	DT
Reconnaissance	0.843	RF	0.78	XGBoost
Shellcode	0.998	RF	0.7	XGBoost
Worms	0.997	LSTMNN	0.605	XGBoost

6.2. Dynamic Classifier Results

The dynamic classifier model proposed was based on an ensemble ML model, which consists in obtaining the major benefits per each individual static model, using those introduced in this research work. In Figure 6, we show the confusion matrices for the balanced and unbalanced dataset. The most important differences between both datasets appear on worms, DoS, analysis, and backdoor being practically undetectable using the original dataset. The detection was improved using the SMOTE algorithm, where the lowest rates of detection belongs for analysis and exploits. According to this results, balanced model can improve the accuracy of the ensemble model, since it is able to distinguish better the traffic that is not normal.

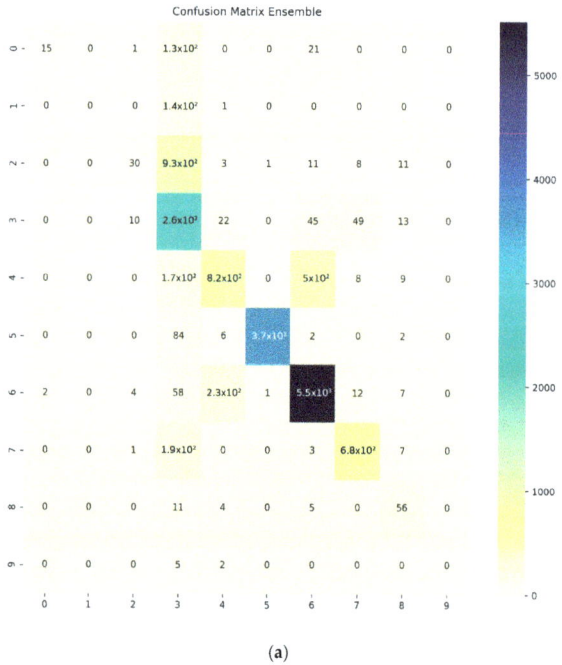

(a)

Figure 6. *Cont.*

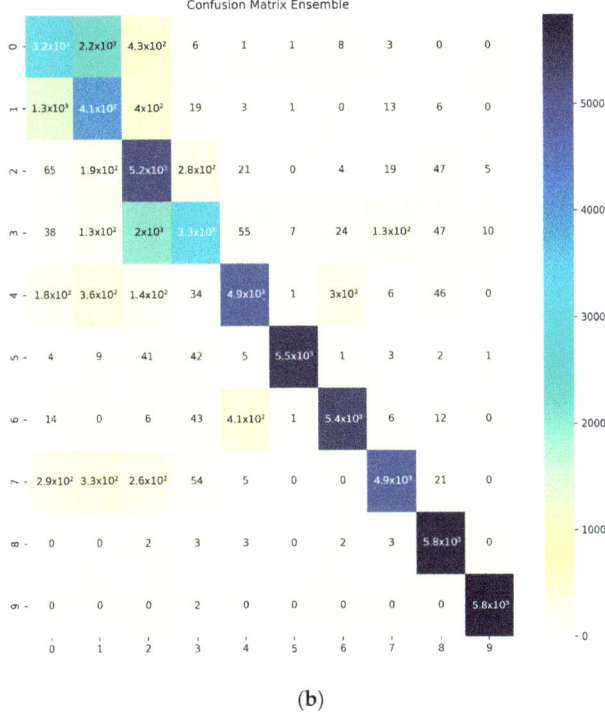

(b)

Figure 6. Confusion matrix: (**a**) balanced dataset; (**b**) unbalanced dataset. The class categories in the figure corresponds to: analysis "0", backdoor "1", DoS "2", exploits "3", fuzzers "4", generic "5", normal traffic "6", reconnaissance "7", shellcode "8", worms "9".

We applied different metrics in order to understand better the results given by the proposed models. Table 4 show the results given by the metric TPR for both datasets balanced an unbalanced. In reference to balanced dataset some attacks present a high detection rate over 90% (worms, shellcode, generic, and DoS). In any case, the dynamic classifier presents an acceptable TPR for some attack categories (reconnaissance, fuzzer, and backdoor). On the other hand, classes such as analysis and exploits present the lowest detection rates.

Table 4. True positive rates by class category obtained by the dynamic classifier.

Class	Balanced Dataset TPR	Unbalanced Dataset TPR
Analysis	0.551	0.089
Backdoor	0.701	0.09
DoS	0.921	0.030
Exploits	0.577	0.95
Fuzzer	0.819	0.601
Generic	0.981	0.995
Reconnaissance	0.835	0.744
Shellcode	0.998	0.737
Worms	0.999	0.65

Table 4 also shows the results given by the classification of the dynamic classifier. The results obtained by some attacks are near to 0% (analysis, backdoor, and DoS). These results are related to the percentage of data by each class, as can be seen Section 3.1.2. On the contrary, the results for the classes exploits and generic and worms were improved in comparison with the results shown in Table 3.

Table 5 shows the results obtained by each individual model proposed and the dynamic classifier. The dynamic classifier improves the results obtained in terms of accuracy up to 5% for the balanced dataset and 3% for the unbalanced dataset being similar for the results obtained the by the F1Score.

Table 5. Comparison between statics models and dynamic classifier in terms of accuracy for balanced and unbalanced dataset.

Model	Balanced Dataset		Unbalanced Dataset	
	Accuracy	F1-Score	Accuracy	F1-Score
KNN	0.739	0.7406	0.779	0.7673
SVM	-	-	0.739	0.685
DT	0.816	0.8066	0.801	0.8062
RF	0.823	0.824	0.828	0.820
XGBoost	0.795	0.796	0.824	0.803
MLPNN	0.709	0.715	0.811	0.784
LSTMNN	0.747	0.754	0.816	0.792
Dynamic Classifier	0.876	0.879	0.851	0.829

As stated in Section 2, there are different approaches for IDS based on ML techniques. Table 6 compares the proposed dynamic classifier for both applied datasets in this research, balanced and unbalanced, with related works. This table shows an improvement in multiclass detection for auto machine learning selection in terms of accuracy in comparison with other studies in the state of the art. The table mention different aspects such as the model, dataset, TPR by attack, and accuracy obtained.

Table 6. Comparison between proposed solution and related works.

Study	Algorithm	Dataset	Data Preprocessing	Feature Selection	Selection Model	Dataset Balancing	TPR	Accuracy
[9]	SVM multiclass	UNSW-NB15	Non-linear scaling method	X	X	X	X	75.77%
[13]	XGBoost multiclass	UNSW-NB15	X	X	X	X	X	86%
[15]	LSTM multiclass	UNSW-NB15	One-hot encoding	X	X	X	X	83%
[23]	Dynamic classifier multiclass	UNSW-NB15	Label encoding	K-best	Voter classifier	X	X	61%
[12]	Dynamic classifier multiclass	NSL-KDD	One-hot encoding	X	Ensemble voting	X	X	85.2
Proposed solution	Dynamic classifier multiclass	UNSW-NB15	Label encoding	Correlation Kendall Coefficient and K-best	Ensemble model	Smote	✓	87.6%

7. Discussion and Conclusions

Most of the diverse studies that proposes IDS based on ML algorithms are based on a static model with an improved performance in terms of accuracy [3,6] with a variation over its hyperparameters. The dataset proposed for this study is the UNSW-NB15 which is considered up to date a benchmark dataset [2]. This dataset has been highly used because its relevance to its recent cyberattacks.

This article exposes through several tests that multiple algorithms can detect significantly better some type of attack over different hyperparameters configurations, as it can be concluded from the data shown in Table 3. Thus, after experimentation with static ML algorithms described in Section 5.2, some categories of attacks presented a better rate of detection as well as RF for attacks—such as analysis, DoS, fuzzers, reconnaissance, and shellcode. However, when comparing these same attacks with the unbalanced dataset, other models allowed obtaining a better detection such as DT, RF, and XGboost.

Following this idea, this research proposes a dynamic auto-selection classifier over different ML models with the objective of obtaining the best capabilities of each individual model to detect cyberattacks. For this purpose, different models purposed in Section 5.3 were studied and tested, obtaining as the best option an ensemble ML based on the XGBoost algorithm [48].

The need of different techniques for data preprocessing and data balancing with the objective to enhance the rate detection was presented in Section 2. We selected the SMOTE to achieve data balancing in our dataset. Also, a feature selection was applied with a correlation index of 0.8 between different features which improved rate detection of the models proposed. Furthermore, two datasets—balanced and unbalanced—were created with the objective to compare the importance of the models proposed.

Several tests and configuration were carried out, with the objective of improving the algorithms' rate detection proposed in Section 3.3. The metric chosen to expose the results is the TPR, which allows evaluate the results in terms of correctly predicted individual classes of attacks. In Figure 4, TPR is taken into account for each individual model for the unbalanced dataset, where the lowest rate of detection was for the categories analysis, backdoor, fuzzers, and worms. In the case of the category attack worms; the algorithms XGBoost and DT could obtain a TPR over 0.5. Other attacks—such as rxploits, generic, and reconnaissance—presented overall an acceptable TPR for each individual algorithm. Comparing the results between Tables 3 and 4, some attacks—such as exploits, fuzzers, generic, and worms—enhanced the detection rate. Instead other categories remaining decreased insignificantly the detection rate. This could be determined by the low numbers of samples for the unbalanced dataset for each class.

On the other hand, for the balanced dataset, static models showed better results in terms of TPR. Figure 5 exposes the enhancing rate detection rate detection up to 50% in comparison with Figure 4. However, the category exploit was the only one decreasing the TPR, up to 40%. The balancing algorithm was successful because it was able to improve the results obtained by the static models. This comparison can be appreciated in Table 3. Furthermore, the application of the dynamic classifier proposed in this research could improve the overall results in terms of TPR. The following classes increased the detection rate: backdoor 5.1%, DoS 2.9%, exploits 8.5%, generic 0.7%, and worms 0.2%. The shellcode class upheld its TPR. Classes such as analysis, fuzzers, and reconnaissance decreased the TPR with a mean of 1.83% which is acceptable, maintaining the highest TPR for the static models proposed.

Finally, Table 5 shows a comparison of the accuracy reached by each individual model and by the dynamic classifier proposed. The dynamic classifier was able to improve the results in 5.3% and 2.3% compared with the best static model for the balanced and unbalanced dataset. The key idea of the model proposed in this research is to auto select, by a ML model, the best rate detection, gathering the individual advantages of each model. We used the method based on an ensemble model based on XGBoost to improve the detection rate. Compared with other related works such as the mentioned in Table 6, it can be seen that our research work achieves a noticeable performance increase in terms of detection rate for IDS based on multiples classes of attacks.

Although our model increases the detection rate, it takes longer time in execution to detect an attack. This happens because the data has to be processed by each individual algorithm and the dynamic classifier model. In a practical scenario, this could delay the time of detection of a possible attack.

For unbalanced classification scenarios, our proposed model also could enhance detection rate as was mentioned above. In practical applications in the area of IDS based on ML algorithms, one of the principal approaches is to improve the quality of the training data, optimizing the features and apply preprocessed methods in order to improve the data quality for the training of the ML algorithms. As for future research lines, some future directions can be considered. The main one is the application of our proposal into a real scenario with real collected data in real-time taking into account the features applied during this research and the proposed models. This can be done with technologies such as Apache Spark Structured Streaming [49] which is designed for real time processing data.

Author Contributions: Conceptualization, V.A.V. and X.L.-N.; Methodology, X.L.-N., C.S.-Z., and V.A.V.; Software, X.L.-N. and C.S.-Z.; Validation, X.L.-N., M.V.-B. and D.R.; Formal analysis, X.L.-N., V.A.V. and M.V.-B.; Investigation, X.L.-N. and C.S.-Z.; Resources, X.L.-N.; Data curation, X.L.-N. and C.S.-Z.; Writing—original draft preparation, X.L.-N.; Writing—review and editing, X.L.-N., V.A.V., M.V.-B. and D.R.; Supervision, V.A.V., M.V.-B. and D.R.; Project administration, V.A.V.; Funding acquisition, V.A.V. All authors have read and agreed to the published version of the manuscript.

Funding: This research received no external funding.

Conflicts of Interest: The authors declare no conflict of interest.

References

1. Masdari, M.; Khezri, H. A survey and taxonomy of the fuzzy signature-based Intrusion Detection Systems. *Appl. Soft Comput.* **2020**, *92*, 106301. [CrossRef]
2. Larriva-Novo, X.A.; Vega-Barbas, M.; Villagra, V.A.; Sanz Rodrigo, M. Evaluation of Cybersecurity Data Set Characteristics for Their Applicability to Neural Networks Algorithms Detecting Cybersecurity Anomalies. *IEEE Access* **2020**, *8*, 9005–9014. [CrossRef]
3. Mishra, P.; Varadharajan, V.; Tupakula, U.; Pilli, E.S. A Detailed Investigation and Analysis of Using Machine Learning Techniques for Intrusion Detection. *IEEE Commun. Surv. Tutor.* **2019**, *21*, 686–728. [CrossRef]
4. Aboueata, N.; Alrasbi, S.; Erbad, A.; Kassler, A.; Bhamare, D. Supervised Machine Learning Techniques for Efficient Network Intrusion Detection. In Proceedings of the 2019 28th International Conference on Computer Communication and Networks (ICCCN), Valencia, Spain, 29 July–1 August 2019; pp. 1–8.
5. Elmasry, W.; Akbulut, A.; Zaim, A.H. Empirical study on multiclass classification-based network intrusion detection. *Comput. Intell.* **2019**, *35*, 919–954. [CrossRef]
6. Hindy, H.; Brosset, D.; Bayne, E.; Seeam, A.; Tachtatzis, C.; Atkinson, R.; Bellekens, X. A Taxonomy and Survey of Intrusion Detection System Design Techniques, Network Threats and Datasets. *arXiv* **2018**, arXiv:1806.03517.
7. Berman, D.S.; Buczak, A.L.; Chavis, J.S.; Corbett, C.L. A Survey of Deep Learning Methods for Cyber Security. *Information* **2019**, *10*, 122. [CrossRef]
8. Teoh, T.T.; Chiew, G.; Franco, E.J.; Ng, P.C.; Benjamin, M.P.; Goh, Y.J. Anomaly detection in cyber security attacks on networks using MLP deep learning. In Proceedings of the 2018 International Conference on Smart Computing and Electronic Enterprise (ICSCEE), Kuala Lumpur, Malaysia, 11–12 July 2018; pp. 1–5.
9. Jing, D.; Chen, H.-B. SVM Based Network Intrusion Detection for the UNSW-NB15 Dataset. In Proceedings of the 2019 IEEE 13th International Conference on ASIC (ASICON), Chongqing, China, 29 October–1 November 2019; pp. 1–4.
10. Gauthama Raman, M.R.; Somu, N.; Jagarapu, S.; Manghnani, T.; Selvam, T.; Krithivasan, K.; Shankar Sriram, V.S. An efficient intrusion detection technique based on support vector machine and improved binary gravitational search algorithm. *Artif. Intell. Rev.* **2020**, *53*, 3255–3286. [CrossRef]
11. Zhiqiang, L.; Mohi-Ud-Din, G.; Bing, L.; Jianchao, L.; Ye, Z.; Zhijun, L. Modeling Network Intrusion Detection System Using Feed-Forward Neural Network Using UNSW-NB15 Dataset. In Proceedings of the 2019 IEEE 7th International Conference on Smart Energy Grid Engineering (SEGE), Oshawa, ON, Canada, 12–14 August 2019; pp. 299–303.
12. Gao, X.; Shan, C.; Hu, C.; Niu, Z.; Liu, Z. An Adaptive Ensemble Machine Learning Model for Intrusion Detection. *IEEE Access* **2019**, *7*, 82512–82521. [CrossRef]
13. Husain, A.; Salem, A.; Jim, C.; Dimitoglou, G. Development of an Efficient Network Intrusion Detection Model Using Extreme Gradient Boosting (XGBoost) on the UNSW-NB15 Dataset. In Proceedings of the 2019 IEEE International Symposium on Signal Processing and Information Technology (ISSPIT), Ajman, UAE, 10–12 December 2019; pp. 1–7.
14. Moustafa, N.; Slay, J. UNSW-NB15: A comprehensive data set for network intrusion detection systems (UNSW-NB15 network data set). In Proceedings of the 2015 military communications and information systems conference (MilCIS), IEEE, Canberra, Australia, 11–12 November 2015; pp. 1–6.
15. Gwon, H.; Lee, C.; Keum, R.; Choi, H. Network Intrusion Detection based on LSTM and Feature Embedding. *arXiv* **2019**, arXiv:1911.11552.

16. Guo, H. A bayesian approach for automatic algorithm selection. In Proceedings of the International Joint Conference on Artificial Intelligence (IJCAI03), Workshop on AI and Autonomic Computing, Acapulco, Mexico, 9–15 August 2003; pp. 1–5.
17. Malkomes, G.; Schaff, C.; Garnett, R. Bayesian optimization for automated model selection. In *Advances in Neural Information Processing Systems 29*; Lee, D.D., Sugiyama, M., Luxburg, U.V., Guyon, I., Garnett, R., Eds.; Curran Associates, Inc.: Red Hook, NY, USA, 2016; pp. 2900–2908.
18. Xu, L.; Hoos, H.H.; Leyton-Brown, K. Hydra: Automatically configuring algorithms for portfolio-based selection. In Proceedings of the Twenty-Fourth AAAI Conference on Artificial Intelligence, Atlanta, GA, USA, 11–15 July 2010; AAAI Press: Palo Alto, CA, USA, 2010; pp. 210–216.
19. Kerschke, P.; Hoos, H.H.; Neumann, F.; Trautmann, H. Automated Algorithm Selection: Survey and Perspectives. *Evol. Comput.* **2018**, *27*, 3–45. [CrossRef]
20. Bischl, B.; Kerschke, P.; Kotthoff, L.; Lindauer, M.; Malitsky, Y.; Fréchette, A.; Hoos, H.; Hutter, F.; Leyton-Brown, K.; Tierney, K.; et al. ASlib: A benchmark library for algorithm selection. *Artif. Intell.* **2016**, *237*, 41–58. [CrossRef]
21. Thornton, C.; Hutter, F.; Hoos, H.H.; Leyton-Brown, K. Auto-WEKA: Combined selection and hyperparameter optimization of classification algorithms. In Proceedings of the 19th ACM SIGKDD International Conference on Knowledge Discovery and Data Mining, Chicago, IL, USA, 11–14 August 2013; Dhillon, I.S., Ed.; Association for Computing Machinery: New York, NY, USA, 2013; pp. 847–855.
22. Kotthoff, L.; Thornton, C.; Hoos, H.H.; Hutter, F.; Leyton-Brown, K. Auto-WEKA 2.0: Automatic model selection and hyperparameter optimization in WEKA. *J. Mach. Learn. Res.* **2017**, *18*, 826–830.
23. Karn, R.R.; Kudva, P.; Elfadel, I.A.M. Dynamic Autoselection and Autotuning of Machine Learning Models for Cloud Network Analytics. *IEEE Trans. Parallel Distrib. Syst.* **2019**, *30*, 1052–1064. [CrossRef]
24. Aceto, G.; Ciuonzo, D.; Montieri, A.; Pescapé, A. Multi-classification approaches for classifying mobile app traffic. *J. Netw. Comput. Appl.* **2018**, *103*, 131–145. [CrossRef]
25. Dainotti, A.; Pescapé, A.; Sansone, C. Early Classification of Network Traffic through Multi-classification. In *Proceedings of the Traffic Monitoring and Analysis*; Domingo-Pascual, J., Shavitt, Y., Uhlig, S., Eds.; Springer: Berlin/Heidelberg, Germany, 2011; pp. 122–135.
26. Kato, K.; Klyuev, V. Development of a network intrusion detection system using Apache Hadoop and Spark. In Proceedings of the 2017 IEEE Conference on Dependable and Secure Computing, Taipei, Taiwan, 7–10 August 2017; pp. 416–423.
27. Terzi, D.S.; Terzi, R.; Sagiroglu, S. Big data analytics for network anomaly detection from netflow data. In Proceedings of the 2017 International Conference on Computer Science and Engineering (UBMK), Jakarta Indonesia, 20–21 July 2017; pp. 592–597.
28. Krovich, D.; Cottrill, A.; Mancini, D.J. A Cloud Based Entitlement Granting Engine. In Proceedings of the National Cyber Summit, Huntsville, AL, USA, 4–6 June 2019; Choo, K.-K.R., Morris, T.H., Peterson, G.L., Eds.; Springer: Cham, Switzerland, 2019; pp. 220–231.
29. Maciá-Fernández, G.; Camacho, J.; Magán-Carrión, R.; García-Teodoro, P.; Theron, R. UGR '16: A new dataset for the evaluation of cyclostationarity-based network IDSs. *Comput. Secur.* **2018**, *73*, 411–424. [CrossRef]
30. Stiawan, D.; Bin Idris, M.Y.; Bamhdi, A.M.; Budiarto, R. CICIDS-2017 Dataset Feature Analysis with Information Gain for Anomaly Detection. *IEEE Access* **2020**, *8*, 132911–132921. [CrossRef]
31. Revathi, S.; Malathi, D.A. A Detailed Analysis on NSL-KDD Dataset Using Various Machine Learning Techniques for Intrusion Detection. *Int. J. Eng. Res. Technol.* **2013**, *2*, 1848–1853.
32. Kumar, V.; Das, A.K.; Sinha, D. Statistical Analysis of the UNSW-NB15 Dataset for Intrusion Detection. In *Computational Intelligence in Pattern Recognition*; Das, A.K., Nayak, J., Naik, B., Dutta, S., Pelusi, D., Eds.; Springer: Singapore, 2020; pp. 279–294.
33. Moustafa, N.; Slay, J. The evaluation of Network Anomaly Detection Systems: Statistical analysis of the UNSW-NB15 data set and the comparison with the KDD99 data set. *Inf. Secur. J. A Glob. Perspect.* **2016**, *25*, 18–31. [CrossRef]
34. Larriva-Novo, X.; Vega-Barbas, M.; Villagrá, V.A.; Rivera, D.; Álvarez-Campana, M.; Berrocal, J. Efficient Distributed Preprocessing Model for Machine Learning-Based Anomaly Detection over Large-Scale Cybersecurity Datasets. *Appl. Sci.* **2020**, *10*, 3430. [CrossRef]

35. Brownlee, J. Supervised and unsupervised machine learning algorithms. *Mach. Learn. Mastery* **2016**, *16*. Available online: https://machinelearningmastery.com/supervised-and-unsupervised-machine-learning-algorithms/ (accessed on 11 August 2020).
36. sklearn.neighbors.KNeighborsClassifier—Scikit-Learn 0.23.2 Documentation. Available online: https://scikit-learn.org/stable/modules/generated/sklearn.neighbors.KNeighborsClassifier.html (accessed on 11 August 2020).
37. sklearn.tree.DecisionTreeClassifier—Scikit-Learn 0.23.2 Documentation. Available online: https://scikit-learn.org/stable/modules/generated/sklearn.tree.DecisionTreeClassifier.html (accessed on 11 August 2020).
38. 3.2.4.3.1. sklearn.ensemble.RandomForestClassifier—Scikit-Learn 0.23.2 Documentation. Available online: https://scikit-learn.org/stable/modules/generated/sklearn.ensemble.RandomForestClassifier.html (accessed on 11 August 2020).
39. sklearn.svm.SVC—Scikit-Learn 0.23.2 Documentation. Available online: https://scikit-learn.org/stable/modules/generated/sklearn.svm.SVC.html (accessed on 11 August 2020).
40. Yıldırım, S. Gradient Boosted Decision Trees-Explained. Available online: https://towardsdatascience.com/gradient-boosted-decision-trees-explained-9259bd8205af (accessed on 11 August 2020).
41. Cszas cszas. Intrusion Detection Models. Available online: https://github.com/cszas/IDSModels (accessed on 10 June 2020).
42. Kathiresan, V.; Sumathi, P. An efficient clustering algorithm based on Z-Score ranking method. In Proceedings of the 2012 International Conference on Computer Communication and Informatics, Coimbatore, India, 10–12 January 2012; pp. 1–4.
43. Wang, J.; Xu, M.; Wang, H.; Zhang, J. Classification of Imbalanced Data by Using the SMOTE Algorithm and Locally Linear Embedding. In Proceedings of the 2006 8th international Conference on Signal Processing, Beijing, China, 16–20 November 2006; Volume 3.
44. Zhang, H.; Huang, L.; Wu, C.Q.; Li, Z. An effective convolutional neural network based on SMOTE and Gaussian mixture model for intrusion detection in imbalanced dataset. *Comput. Netw.* **2020**, *177*, 107315. [CrossRef]
45. Gottwalt, F.; Chang, E.; Dillon, T. CorrCorr: A feature selection method for multivariate correlation network anomaly detection techniques. *Comput. Secur.* **2019**, *83*, 234–245. [CrossRef]
46. Huang, L.; Chiang, D. Better k-best Parsing. In Proceedings of the Ninth International Workshop on Parsing Technology, Vancouver, BC, Canada, 9–10 October 2005; Association for Computational Linguistics: Stroudsburg, PA, USA, 2005; pp. 53–64.
47. sklearn.model_selection.GridSearchCV—Scikit-Learn 0.23.2 Documentation. Available online: https://scikit-learn.org/stable/modules/generated/sklearn.model_selection.GridSearchCV.html (accessed on 12 August 2020).
48. Orlenko, A.; Moore, J.H.; Orzechowski, P.; Olson, R.S.; Cairns, J.; Caraballo, P.J.; Weinshilboum, R.M.; Wang, L.; Breitenstein, M.K. Considerations for automated machine learning in clinical metabolic profiling: Altered homocysteine plasma concentration associated with metformin exposure. In *Biocomputing–World Scientific 2018*; Altman, R.B., Dunker, A.K., Hunter, L., Ritchie, M.D., Murray, T.A., Klein, T.E., Eds.; World Scientific Publishing Co.: Singapore; pp. 460–471. ISBN 978-981-323-552-6.
49. Structured Streaming. In Proceedings of the 2018 International Conference on Management of Data, Houston, TX, USA, 10–15 June 2018; Available online: https://dl.acm.org/doi/abs/10.1145/3183713.3190664 (accessed on 18 October 2020).

Publisher's Note: MDPI stays neutral with regard to jurisdictional claims in published maps and institutional affiliations.

© 2020 by the authors. Licensee MDPI, Basel, Switzerland. This article is an open access article distributed under the terms and conditions of the Creative Commons Attribution (CC BY) license (http://creativecommons.org/licenses/by/4.0/).

Article

Combining K-Means and XGBoost Models for Anomaly Detection Using Log Datasets

João Henriques [1,2,*], Filipe Caldeira [1,3], Tiago Cruz [1] and Paulo Simões [1]

1. Department of Informatics Engineering, University of Coimbra, 3030-290 Coimbra, Portugal; caldeira@estgv.ipv.pt (F.C.); tjcruz@dei.uc.pt (T.C.); psimoes@dei.uc.pt (P.S.)
2. Informatics Department, Polytechnic of Viseu, 3504-510 Viseu, Portugal
3. CISeD—Research Centre in Digital Services, Polytechnic of Viseu, 3504-510 Viseu, Portugal
* Correspondence: jpmh@dei.uc.pt

Received: 24 April 2020; Accepted: 15 July 2020; Published: 17 July 2020

Abstract: Computing and networking systems traditionally record their activity in log files, which have been used for multiple purposes, such as troubleshooting, accounting, post-incident analysis of security breaches, capacity planning and anomaly detection. In earlier systems those log files were processed manually by system administrators, or with the support of basic applications for filtering, compiling and pre-processing the logs for specific purposes. However, as the volume of these log files continues to grow (more logs per system, more systems per domain), it is becoming increasingly difficult to process those logs using traditional tools, especially for less straightforward purposes such as anomaly detection. On the other hand, as systems continue to become more complex, the potential of using large datasets built of logs from heterogeneous sources for detecting anomalies without prior domain knowledge becomes higher. Anomaly detection tools for such scenarios face two challenges. First, devising appropriate data analysis solutions for effectively detecting anomalies from large data sources, possibly without prior domain knowledge. Second, adopting data processing platforms able to cope with the large datasets and complex data analysis algorithms required for such purposes. In this paper we address those challenges by proposing an integrated scalable framework that aims at efficiently detecting anomalous events on large amounts of unlabeled data logs. Detection is supported by clustering and classification methods that take advantage of parallel computing environments. We validate our approach using the the well known NASA Hypertext Transfer Protocol (HTTP) logs datasets. Fourteen features were extracted in order to train a k-means model for separating anomalous and normal events in highly coherent clusters. A second model, making use of the XGBoost system implementing a gradient tree boosting algorithm, uses the previous binary clustered data for producing a set of simple interpretable rules. These rules represent the rationale for generalizing its application over a massive number of unseen events in a distributed computing environment. The classified anomaly events produced by our framework can be used, for instance, as candidates for further forensic and compliance auditing analysis in security management.

Keywords: anomaly detection; clustering; k-means; gradient tree boosting; XGBoost

1. Introduction

Hosts and network systems typically record their detailed activity in log files with specific formats, which are valuable sources for anomaly detection systems. The growing number of hosts per organization and the growing complexity of infrastructures result in an increasingly massive amount of recorded logs available—requiring simpler and cheaper anomaly detection methods. While classic log management applications based on manual or preset rule-based analysis still hold value, they do not scale well with the large volumes of data that are currently available. Moreover, they are limited in terms of exploratory analysis: they fail to detect anomalies not predefined in the rules (i.e., based on prior knowledge) and/or require considerable operator expertise to reach their full potential. This opens the way for the introduction of new approaches, which are less dependent on prior knowledge and human-guided workflows and are able to extract knowledge from large volumes of log data in a scalable and (semi)automated way. Moreover, taking advantage of the available computational resources may also contribute to achieving performance and accuracy for identifying anomalies and retrieving forensic and compliance auditing evidence.

Over the past years, several automated log analysis methods for anomaly detection have been proposed. However, most of those proposals are not suitable to the scale needed for identifying unknown anomalies from the growing high-rate amount of logs being produced and their inherent complexity. In the scope of the ATENA H2020 Project [1,2], we faced this challenge while building a Forensics and Compliance Auditing (FCA) tool able to handle all the logs produced by a typical energy utility infrastructure.

To address such challenges, we researched novel integrated anomaly detection methods employing parallel processing capabilities for improving detection accuracy and efficiency over massive amounts of log records. These methods combine the k-means clustering algorithm [3] and the gradient tree boosting classification algorithm [4] to leverage the filtering capabilities over normal events, in order to concentrate the efforts on the remaining anomaly candidates. Such an approach may greatly contribute to reducing the involved computational complexity.

The characteristics of abnormal system behaviors were obtained by extracting 14 statistical features containing numerical and categorical attributes from the logs. Then, the k-means clustering algorithm was employed to separate anomalous from normal events into two highly coherent clusters. The previous binary clustered data serve as labeled input to produce a gradient tree boosting algorithm implemented by the XGBoost system [5]. Its role is to produce a set of simple rules with the rationale for generalizing the classification of anomalies of a large number of unseen events in a distributed computing environment. K-means, XGBoost and Dask [6] provide the tools for building scalable clustering and classification solutions to find out the candidate events for forensic and compliance auditing analysis.

The rest of this paper is organized as follows. Section 2 discusses background concepts and related work. Section 3 describes the proposed framework. Section 4 presents the validation work and discusses the achieved results, and Section 5 concludes the paper.

2. Background and Related Work

This section starts by providing the reader with the key base concepts related with the scope of our approach. Next, we discuss related work (Section 2.2). Finally, we present the algorithms and tools we adopted in our work, namely k-means (Section 2.3), decision trees (Section 2.4), gradient tree boosting on XGBoost (Section 2.5) and Dask (Section 2.6).

2.1. Base Concepts

By definition, an anomaly is an outlying observation that appears to deviate markedly from other members [7]. Anomalies are typically classified into three types: point anomalies, contextual anomalies and collective anomalies. A point anomaly in data significantly deviates from the average or normal distribution of the rest of the data [8]. A contextual anomaly is identified as anomalous behavior constrained to a specific context, and normal according to other contexts [8]. Collection of data instances may reveal collective anomalies while anomalous behavior may not be depicted when analyzed individually [9]. Time series data include a significant amount of chronologically ordered sequence data sample values retrieved at different instants. Their features include high-dimensionality, dynamicity, high levels of noise, and complexity. Consequently, in the data mining research area, time series data mining was classified as one of the ten most challenging problems [10].

Anomaly detection for application log data faces important challenges due to the inherent unstructured plain text contents, the redundant runtime information and the existence of a significant amount of unbalanced data. Application logs are unstructured and stored as plain text, and their format varies significantly between applications. This lack of structure presents important barriers to data analysis. Moreover, runtime information, such as server IP addresses, may change during execution. Additionally, application log data are designed to record all changes to an application and hence contain data that are significantly unbalanced in comparison to non-anomalous execution. The size and unbalanced nature of log data thus complicate the anomaly detection process.

2.2. Related Work

Various anomaly detection methods have been proposed for applying clustering algorithms to detect unknown abnormal behaviors or potential security attacks.

Some of those proposals have addressed the usage of log analysis as one of the input sources for anomaly detection. Chen and Li [11], for instance, proposed an improved version of the DBSCAN algorithm for detecting anomalies from audit data while updating the detection profile along its execution. Syarif et al. [12] compared five different clustering algorithms and identified those providing the highest detection accuracy. However, they also concluded that those algorithms are not mature enough for practical applications. Hoglund et al. [13], as well as Lichodzijewski et al. [14], constructed host-based anomaly detection systems that applied a self-organizing maps algorithm to evaluate if a user behavior pattern is abnormal.

Clustering techniques, such as the k-means algorithm, are often used by intrusion detection systems for classifying normal or anomalous events. Münz et al. [15] applied the k-means clustering algorithm to feature datasets extracted from raw records, where training data are divided into clusters of time intervals for normal and anomalous traffic. Li and Wang [16] improved a clustering algorithm supported by a traditional means clustering algorithm, in order to achieve efficiency and accuracy when classifying data. Eslamnezhad and Varjani [17] proposed a new detection algorithm to increase the quality of the clustering method based on a MinMax k-means algorithm, overcoming the low sensitivity to initial centers in the k-means algorithm. Ranjan and Sahoo [18] proposed a modified k-medoids clustering algorithm for intrusion detection. The algorithm takes a new approach in selecting the initial medoids, overcoming the means in anomaly intrusion detection and the dependency on initial centroids, number of clusters and irrelevant clusters.

Other authors have used hybrid solutions for log analysis, combining the use of the k-means algorithm with other techniques for improving detection performance. They realized that despite the inherent complex structure and high computational cost, hybrid classifiers can contribute to improving accuracy. Tokanju et al. [19], for instance, took advantage of an integrated signature-based and anomaly-based approach to propose a framework based on frequent patterns. Asif-Iqbal et al. [20] correlated different logs from different sources, supported by clustering techniques, to identify and remove unneeded logs. Hajamydeen et al. [21] classified events in two different stages supported by the same clustering algorithm. Initially, it uses a filtering process to identify the abnormal events, and then it applies it for detecting anomalies. Varuna and Natesan [22] introduced a new hybrid learning method integrating k-means clustering and Naive Bayes classification. Muda et al. [23] proposed k-means clustering and Naive Bayes classifiers in a hybrid learning approach, by using the KDD Cup'99 benchmark dataset for validation. In their approach, instances are separated into potential attacks and normal clusters. Subsequently, they are further classified into more specific categories. Elbasiony et al. [24] used data mining techniques to build a hybrid framework for identifying network misuse and detecting intrusions. They used the random forests algorithm to detect misuses, with k-means as the clustering algorithm for unsupervised anomaly detection. The hybrid approach is achieved by combining the random forests algorithm with the weighted k-means algorithm.

Some research focused on detecting which outliers constitute an anomaly when applying clustering methods [25,26]. Liao and Vemuri [26] computed the membership of data points to a given cluster, supported by the use of Euclidean distance. Breunig et al. [27] stated that some detection proposals weight data points as outliers.

Hybrid approaches have indeed proven quite interesting. However, in general, proposed solutions still take considerable amounts of time to generate models for particular datasets, aggravated by the growth patterns normally associated with log sources in production systems. This situation calls for alternative strategies that are able to improve speed (as well as accuracy and efficiency) by taking advantage of innovative algorithmic approaches together with improved parallelism.

Our work focuses on scalability and interpretability, since the aim is to use it in the forensics and audit compliance contexts already discussed in Section 1. The goal is to be able to sift through data to select candidates for a more detailed analysis or inspection.

Similarly to other works, we also take a hybrid approach for identifying anomalies for log analysis. However, unlike other works, we specifically target speed, agility and interpretability. Our approach allows training and classifying out-of-core datasets in scenarios involving the computation of very large datasets with limited computing resources, parallelizing their processing by distributing them across the available nodes. Therefore, our approach is supported by clustering and classification algorithms that are able to scale and produce interpretable results. Our method works in two stages: first, it starts with the unlabelled dataset, implementing a binary anomalous event classifier through the use of unsupervised learning algorithms; the second stage produces a set of simple rules by considering the previously classified data through the use of supervised learning algorithms. It combines the k-means algorithm for clustering anomalies and gradient tree boosting to produce a simple set of interpretable rules to be parallelized in a distributed environment on classifying a large amount of data.

Next, we present the already existing techniques used by our approach in more detail.

2.3. K-Means

K-means remains one of the most popular clustering methods and one of the most relevant algorithms in data mining [3]. The main advantage of k-means is its simplicity. By starting with a set of randomly chosen initial centers, one procedure assigns each input point to its nearest center and then recomputes the centers given the point assignment [28].

Scaling k-means to massive data is relatively easy, due to its simple iterative nature. Given a set of cluster centers, each point can independently decide which center is closest to it, and given an assignment of points to clusters, computing the optimum center can be performed by simply averaging the points. Indeed, parallel implementations of k-means are readily available [28].

From a theoretical standpoint, k-means is not a good clustering algorithm in terms of efficiency or quality. Thus, the running time can grow exponentially in the worst case [29,30] and even though the final solution is locally optimal, it can be very far away from the global optimum (even under repeated random initializations). Nevertheless, in practice, the speed and simplicity of k-means are attractive. Therefore, recent work has focused on improving its initialization procedure performance in terms of quality and convergence [28].

2.4. Decision Trees

Decision trees is a popular supervised machine learning method that produces regression or classification models in the form of a tree structure containing decisions as nodes, resulting in a set of leaves containing the solution. Decision trees are suitable to be applied to any data without much effort when compared with algorithms such as neural networks. Trees are built top-down from the root node and involve recursive binary splitting. In neural networks, the initial dataset is partitioned into smaller subsets according to their features, while an associated decision tree is incrementally built. Such a splitting process is driven by a greedy algorithm evaluating the best solution at each of those steps and evaluating the maximum loss reduction from the cost function in order to make a split on features. To regulate the complexity of a given model and increase the performance of a given tree, pruning processes are available. Notwithstanding, decision tree learning does not generally provide the best performance in terms of prediction. Some approaches exist in learning decision forests, including bagging [31], random forests [32] and boosted trees [33].

Tree boosting overcomes the above performance problem by the use of an additive model that iteratively builds decision trees to learn decision forests by applying a greedy algorithm (boosting) on top of a decision tree base learner [33–35]. Tree boosting is regarded as one the most effective off-the-shelf nonlinear learning methods for a wide range of application problems [34]. It is also highly effective and widely used for achieving state-of-the-art results on many machine learning challenges hosted by the machine learning competition site Kaggle [36].

Regularized greedy forest is an algorithm that can handle general loss functions with a wider range of applicability, which directly learns decision forests while taking advantage of the tree structure itself, while other methods employ specific loss functions, such as exponential loss function in the case of the Adaboost algorithm [34].

2.5. XGBoost

XGBoost is a scalable system that implements gradient tree boosting and the regularized model so as to prevent overfitting, and simplifies the objective function—for parallelization of the regularized greedy forest algorithm [34]. It is suitable for the development of parallel computing solutions applicable to larger datasets or faster training. Besides processors and memory, it uses disk space to handle data that do not fit into the main memory. To enable out-of-core computation, the data are divided into multiple blocks [5]. The system includes cache access patterns, data compression, and sharding. Its performance relies on a tree learning algorithm, which is able to handle sparse data, and on a weighted quantile sketch procedure. This procedure enables handling instance weights in approximate tree learning and is able to solve real-world scale problems using a minimal amount of resources. Besides the penalty from regularizing the objective function, two techniques prevent overfitting: shrinkage, introduced by Friedman [37], and feature subsampling retrieved from random forests to speed up computations. XGBoost works well in practice and has won several machine learning competitions, such as Kaggle [36], running faster than other popular solutions on a single machine and scaling in distributed or out-of-core settings. It can be easily interpreted, given the tools it provides for finding the important features from the XGBoost model.

2.6. Dask

The Dask parallel computing framework leverages the existing Python ecosystem, including relevant libraries such as "numpy" or "pandas". Dask capabilities are supported by executing graphs to be run by the scheduler component, potentially scaling execution to millions of nodes. Those features are suitable to be applied to out-of-core scenarios (not fitting in memory) on a single machine [6].

Dask is a Python specification representing the computation of directed acyclic graphs of tasks with data dependencies to encode parallel task graph schedules. It extends the easy to adopt NumPy library for leveraging parallel computation over modern hardware. It allows scaling large datasets by using disks that extend the physical memory as out-of-core and parallelize and linearly speedup the code by taking advantage of several cores. The main objective is to parallelize the existing Python software stack without triggering a full rewrite. A Dask cluster includes a central scheduler and several distributed workers. It starts up a XGBoost scheduler and a XGBoost worker within each of the Dask workers sharing the same physical processes and memory spaces.

Dask enables parallel and out-of-core computation by including collections such as arrays, bags and dataframes. It couples blocked algorithms with dynamic and memory-aware task scheduling to achieve a parallel and out-of-core popular NumPy clone [6]. Sharing distributed processes with multiple systems allows usaging of specialized services easily and avoiding large monolithic frameworks.

Dask is often compared with other distributed machine learning libraries, such as H2O [38] or Spark's Machine Learning Library (MLLib) [39]. XGBoost is available in Dask to provide users with a fully featured and efficient solution. The Dask parallel computing approach can handle problems that are more complicated than the map-reduce problem at a lower cost and complexity when compared to solutions such as MLLib, given that most of the problems can be resolved in a single machine. Any function is able to be parallelized by the use of delayed function decorators. Additionally, Dask is substantially lightweight when compared to Spark.

3. Proposed Framework

Motivated by the related work, we propose an integrated method with filtering mechanisms to improve detection accuracy and efficiency in scenarios involving large amounts of logs. This method is supported by the k-means clustering and the gradient tree boosting classification algorithms, as implemented by the XGBoost system. To overcome the limitations of existing anomaly detection methods that spend a significant amount of time building the models for the whole dataset, we built three different tools for improving detection accuracy and efficiency.

This section starts with a formal presentation of the algorithm of the model, followed by a discussion of the three compounding tools used for implementing the proposed approach.

3.1. Description of the Algorithm

The proposed approach is formalized in Algorithm 1, which describes how to combine k-means and XGBoost. The algorithm is implemented as a function that takes as input a set of events E and returns the identification of the anomaly *anomalycluster*, the classified events $ypred^1$, total classified events *totalevents* and the total of those events classified as anomalies *totalanomalies*.

Algorithm 1 Proposed Algorithm

INPUT: E, Event Set

$\quad S \leftarrow \text{CLUSTER}()$
$\quad C \leftarrow \text{CLIENT}(S)$
$\quad G \leftarrow \text{DISTRIBUTEDARRAY}(E)$
$\quad k \leftarrow 2$
$\quad km \leftarrow \text{KMEANS}(C, k)$
$\quad km.\text{TRAIN}(G)$
$\quad Y \leftarrow km.\text{PREDICT}(G)$
$\quad X \leftarrow \text{XGBOOST}(X)$
$\quad X.\text{TRAIN}(Y, Y)$
$\quad ypred \leftarrow X.\text{PREDICT}(G)$
$\quad$ **for all** $i \in ypred$ **do**
$\quad\quad$ **if** $ypred_i > 0.5$ **then**
$\quad\quad\quad ypred_i^1 \leftarrow 1$
$\quad\quad\quad k2 \leftarrow k2 + 1$
$\quad\quad$ **else**
$\quad\quad\quad ypred_i^1 \leftarrow 0$
$\quad\quad\quad k1 \leftarrow k1 + 1$
$\quad\quad$ **end if**
$\quad$ **end for**
$\quad$ **if** $k1 > k2$ **then**
$\quad\quad anomalycluster \leftarrow 1$
$\quad\quad totalanomalies \leftarrow k2$
$\quad$ **else**
$\quad\quad anomalycluster \leftarrow 0$
$\quad\quad totalanomalies \leftarrow k1$
$\quad$ **end if**
$\quad totalevents \leftarrow k1 + k2$

OUTPUT: $ypred^1$, Cluster Predictions
OUTPUT: *anomalycluster*, Identification of the anomaly cluster
OUTPUT: *totalevents*, Total number of events
OUTPUT: *totalanomalies*, Total number o anomalies

It starts by initializing the cluster S and activating the client connection C to the cluster S. Then the distributed array G is prepared from the received events in E. The next step is to initialize the k-means model Km for binary classification in the cluster ($k = 2$) from the distributed array G to separate events into two distinct clusters in Y. Then, the XGBoost model X is initialized with the previously predicted events Y being provided as an input for training in the cluster through the use of the client connection C. The final prediction $ypred$ is achieved from the XGBoost model X. In the next stage, each of those predictions ($i \in ypred$) is classified according to the cluster they belong to in $ypred^1$. Such a classification will be determined by evaluating the total number of events in clusters $k1$ and $k2$, so as to decide which corresponds to the anomaly cluster. To that aim, 0.5 was considered as the threshold to classify events as belonging to clusters 1 or 2 ($ypred_i > 0.5$).

After all events have been classified, the cluster including the fewer number of events ($k1 > k2$) will correspond to the anomaly cluster, and such decision will stored in *anomalycluster*.

3.2. Tools

The framework encompasses three tools that may be independently combined in a cooperative way for normalizing raw data and for producing a model able to achieve evidence for forensic and compliance auditing analysis. The "fca_normalization" tool is used to normalize the raw data, "fca_model" produces the model and "fca_analysis" provides the pieces of evidence for forensic and compliance auditing analysis.

The normalization tool takes as input HTTP raw data logs and normalizes data into a new file. Optionally, the encoded features may also be specified. In case encoding is not provided or in the case of missing feature values, the tool automatically applies an encoding label. The tool can be invoked, for example, by using the following command:

```
python fca_normalization
-in NASA_access_log_Jul95
-in_encoding in_encoding.data
-out logs_NASA.csv
-out_encoding out_encoding.data
```

In this example, "fca_normalization" receives the raw HTTP log data file "NASA_access_log_Jul95" along with the optional encoding file "encoding.data". The output normalized file is saved as "logs_NASA.csv". Finally, the tool optionally defines the encoding table in the "out_encoding.data" file.

The modeling tool takes as input the previously normalized data and builds the XGBoost classification model by making use of the gradient tree boosting algorithm after applying the k-means clustering algorithm. In the example invocation provided next, the input file "logs_NASA.csv" contains the HTTP raw log data and the output model is saved as "fca_xgboost.pkl".

```
python fca_model
-in logs_NASA.csv
-out fca_xgboost.pkl
```

The forensic and compliance auditing analysis tool takes as input the model and the normalized events in order to identify the anomalies. In the invocation example provided next, the input model in read from 'fca_xgboost.pkl', and the normalized data is read from 'logs_NASA.csv'. The final output containing the anomaly events is saved on 'outlier_events.csv'.

```
python fca_analysis
-in_model fca_xgboost.pkl
-in_data logs_NASA.csv
-out outlier_events.csv
```

Table 1 summarizes the inputs and outputs for each tool.

Table 1. Tools' inputs and outputs.

Tool	Input	Output
Normalization	HTTP raw logs data, encoding	Normalized data, encoding
Modelling	Normalized data	Model
Analysis	Model, normalized data	Anomaly events

4. Discussion and Evaluation

This section addresses the validation of the proposed framework. First, we discuss feature extraction. Next, based on the extracted features, we describe the initial application of the k-means clustering algorithm for dividing the dataset into two different clusters. Finally, we discuss how to use the previous clustered data for training a scalable gradient tree boosting implemented by the XGBoost system.

For the sake of readability, throughout this section we extensively use as reference a set of well-known, publicly available datasets [40]. These datasets consist of traces containing two months' worth of all HTTP requests to the NASA Kennedy Space Center WWW server, involving 1,871,988 logged events. This dataset was selected because it is probably the largest log-based dataset publicly available, allowing us to assess our scalability claims.

4.1. Feature Extraction and Data Exploration

To capture the characteristics of the system behaviors, 14 features were extracted, containing both numeric and categorical attributes from the raw log records. The original features in the raw HTTP log records are "IP", "Date", "Request", "Response" and "length". By making use of regular expressions, the most relevant time-related components were extracted from the "date" feature, including "Day", "Month", "Year", "Hour", "Minute" and "Second". From the "Request" field, the "operation", "page" and "method" features were extracted. Then, "Month" names were encoded. Therefore "Year", "Month" and "Day" were composed in the temporary "date" feature in order to retrieve the day of the week ("weekday") and "weekend" features. Next, "Request" and other temporary features were removed from the dataset. Finally, categorical features such as "IP", "page", "operation", "method" and "Response" were encoded, and the dataset was saved in a file.

By exploring the dataset we can achieve the first insights. Figure 1 depicts the covariance of the most representative features, including "length", "Hour", "operation", "method" and "Response". This figure shows an interesting covariance between length and other features.

Figure 2 provides a three-dimensional analysis of the number of events that occurred along the day (from 0 to 24 h) and along each weekday (0 to 6), where days 5 and 6 correspond to Saturday and Sunday, respectively.

4.2. Clustering

Based on the extracted features, we employed the k-means clustering algorithm for grouping log events into two different clusters. The larger cluster gathers the normal events, while the smaller holds the deviations from normal behavior. Therefore, the latter cluster should correspond to the set grouping the anomaly events. In addition, sparse clusters are possibly caused by anomalous activities, which can be labeled as anomaly candidates for further analysis.

Our framework model takes advantage of the initialization k-means|| algorithm (largely inspired by k-means++) to obtain a nearly optimal solution after a logarithmic number of steps. In practice, a constant number of passes suffices [28].

After training this model with 90% of the total number of records and using just the remaining 10% for testing, the model produces a normal cluster containing 185,897 events while the anomaly cluster includes 1301 events, corresponding to 0.06% of the total number of events in the normal cluster.

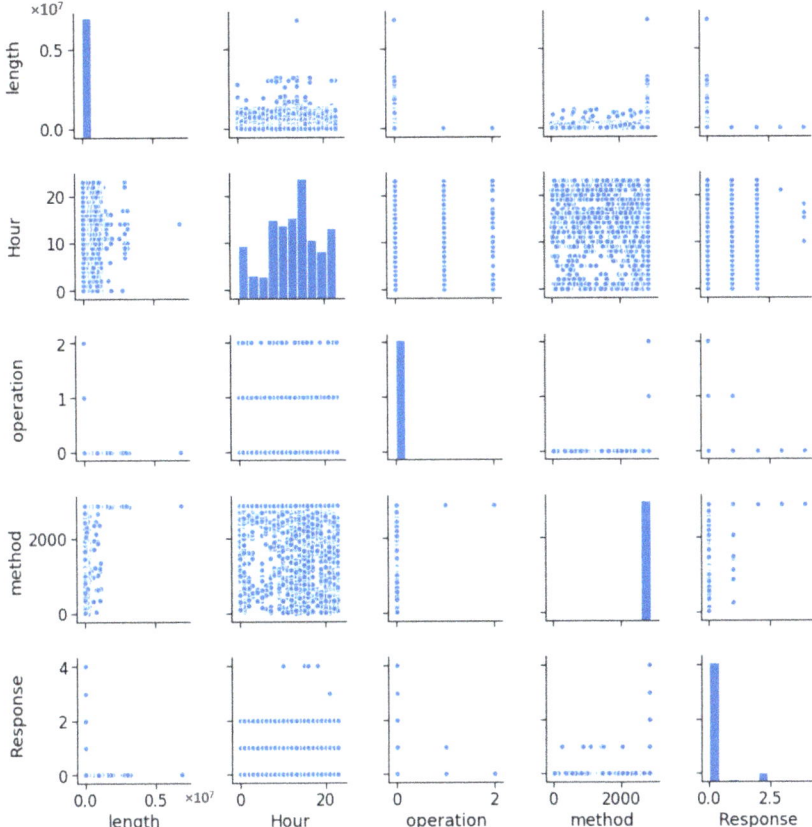

Figure 1. Feature covariance.

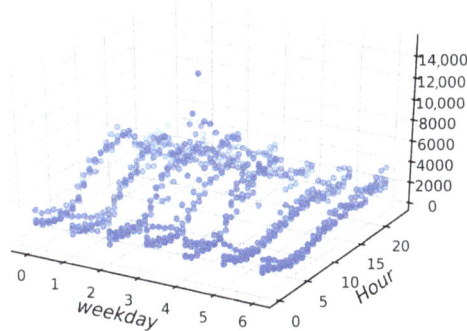

Figure 2. HTTP (Hypertext Transfer Protocol) events over time.

The computed centroids for the two clusters, separating the normal and anomaly events, are the following:

```
[[4.41534608e+04, 0.00000000e+00, 8.12115012e+05, 1.14495884e+01,
0.00000000e+00, 0.00000000e+00, 1.26692042e+01, 2.96762857e+01,
2.94179871e+01, 0.00000000e+00, 1.75010380e+04, 2.84859910e+03,
2.82322741e+00, 2.23245109e-01]
```

```
[4.27877328e+04, 1.63161125e-01, 1.53047043e+04, 1.24323538e+01,
 0.00000000e+00, 0.00000000e+00, 1.26856431e+01, 2.95910303e+01,
 2.94991093e+01, 2.22648078e-03, 1.47567972e+04, 2.84883391e+03,
 2.68136168e+00, 1.93607622e-01]]
```

4.3. Classification

Classification results from the application of the gradient tree boosting algorithm implemented by the XGBoost system, which is the second and final stage of our model. The resulting tree can be linearized into decision rules, where the outcome is the content of the leaf node, and the conditions along the path form a conjunction in the if clause.

The results of this stage were validated by comparing if the number of events classified as anomalies is equal to the number of events belonging to the anomaly cluster. This condition was verified for 1301 events. The predict function for XGBoost outputs probabilities by default and not actual class labels. To calculate accuracy we converted them to 0 and 1 labels, where a 0.5 probability corresponds to the threshold. XGBoost is able to correctly classify all the test data according to the k-means clustering algorithm. Figure 3 depicts the importance of the XGBoost features, according to the F-score metric.

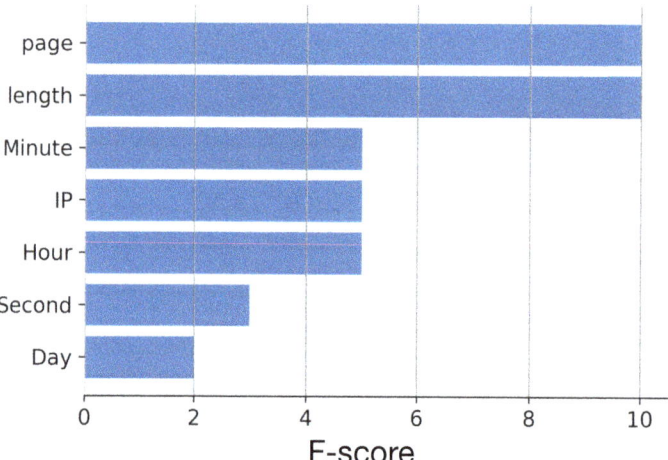

Figure 3. Features importance.

This classification model produces a set of rules providing the rationale for generalizing to unseen events, as shown in Figure 4. The leaf values depicted in the figure are converted into probabilities by applying the logistic function.

Figure 5 depicts the covariance of "length" and "page", which are the two most important features computed by the final model. The events tagged as anomalous are highlighted in red color.

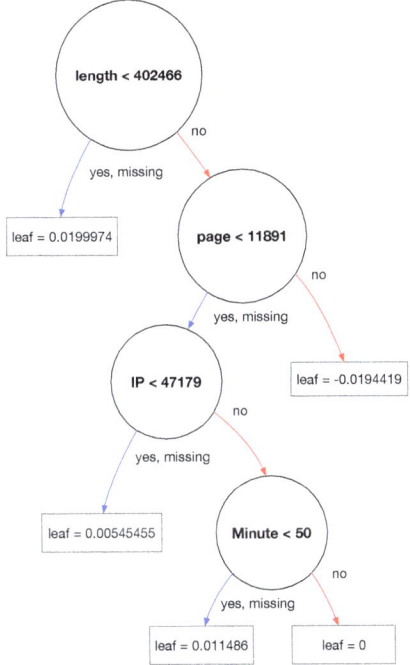

Figure 4. Decision tree.

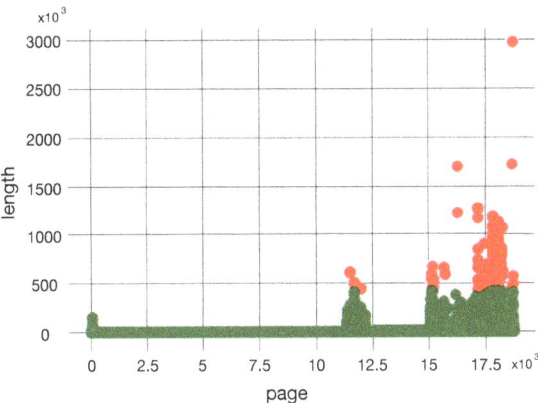

Figure 5. Page and length covariance.

4.4. Parallelization

The proposed framework makes use of the k-means algorithm and the XGBoost system, which are designed to scale in a distributed environment supported by available parallel computing capabilities. Such an approach comes in line with a Big Data scenario.

Our approach is supported by the use of parallel computing capabilities available in the Python "Dask" library. More specifically, the "dataframe" component is able to manage out-of-core datasets along the execution pipeline, since the features are extracted until the clustering and classification models are implemented. Figure 6 provides an example of the kind of graphs Dask is able to produce when reading and splitting a dataset. The Dask libraries "dask_ml" and "dask_xgboost" provide the implementation of popular machine learning algorithms, such as k-means and XGBoost, which support the framework models.

Our experiment involved a simple cluster formed by just two workers in a single node with two cores while the total available memory was 13.66 GB.

To study the framework model's ability to scale in order to cope with large datasets in a reasonable time, two experiments were performed using the parsed NASA HTTP logs dataset. Due to constrained laboratory resources, those experiments were limited to the use of two cores in a single node. As a setup configuration, the Dask chunk size was set to 50,000 events. The model's ability to scale was assessed by comparing its performance under different configurations. To determine the model performance, running time (in milliseconds) was considered throughout the training and predict steps for both k-means and XGBoost stages in accordance with the model topology.

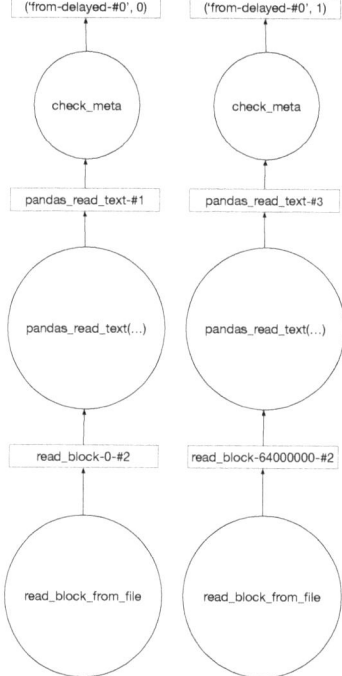

Figure 6. Paralellized Dask graphs.

The first experiment aimed at determining the parallel approach performance, compared with the sequential approach—considering non-Dask sklearn as the sequential approach and Dask as the parallel approach. As a setup configuration, the Dask framework included a single worker and two threads. The running time was measured along the four steps previously defined for the two stages. Those sequential steps include the train (1) and predict (2) steps for the k-means stage, followed by the train (3) and predict (4) steps for XGBoost stage. The running time for each framework, along those running steps, is provided in Figure 7. The achieved results show that the Dask framework outperforms the non-Dask sklearn framework, especially in the case of the training steps.

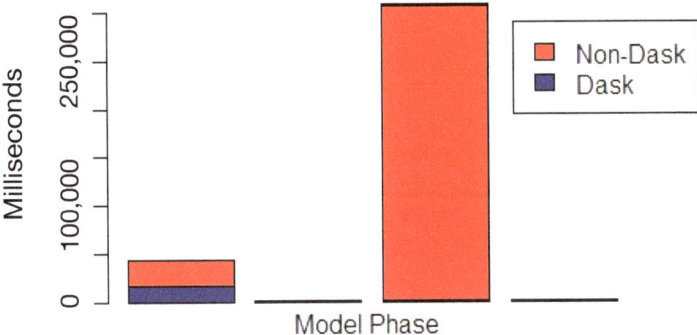

Figure 7. Sequential (non-Dask) vs. parallel (Dask) comparison.

The second experiment evaluated the parallelization capability of the Dask framework under different configurations, such as the number of workers and threads per worker, by measuring the aggregated running time along the topological steps. Figure 8 compares the performance for one and two running workers, while increasing the number of threads per worker from one to ten. Measurements showed that one worker outperforms two workers. Increasing the number of workers did not improve performance, while increasing the number of threads contributes to improved performance until a given threshold is reached. Finally, it was also possible to depict higher performance running over an even number of threads in comparison to the odd ones—due to the less optimal parallelization gains that occur when splitting an odd number of threads into two cores.

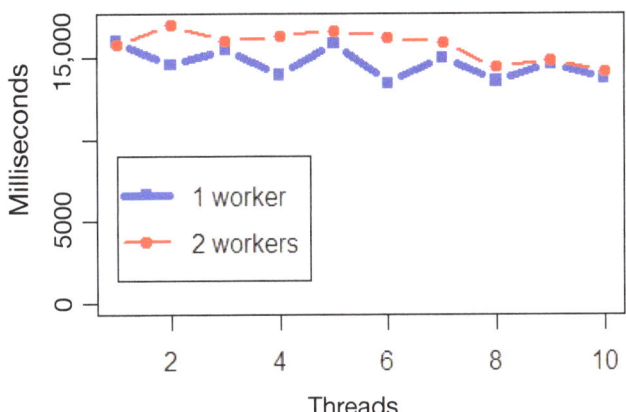

Figure 8. Dask parallel comparison.

4.5. Discussion

The presented framework method relies on two stages. The clustering model is the output of the first stage and serves as the input for the classification stage. Therefore, this approach allows starting from initial unlabelled data for obtaining the interpretable meaningful rules with the rationale for classifying unseen events. Those rules are simple to understand, interpret and visualize, requiring relatively little data preparation effort. Additionally, the described algorithms can easily handle heterogeneous data containing different features produced by different sources. Although the initial nature of our problem is not a classification problem, this approach may be adapted to different scenarios where labeled data are not available. This way, it becomes possible to convert an unsupervised into a supervised learning scenario and take advantage of the use of classification algorithms.

The decision to select the k-means algorithm and XGBoost system, both supported by the Dask library for parallel processing, was driven by requirements in terms of scaling and interpretability when working with limited resources. This decision enabled the application of this framework to larger datasets in order to highlight the anomalous events. Given the inherent nature of the problem being addressed through the use of the unsupervised learning approach, it is not trivial to evaluate the framework model's accuracy in the scope of this paper. An alternative option would be to compare the achieved results with those provided in the existing literature. However, to the best of our knowledge, there are no anomaly detection research works addressing the NASA HTTP logs.

The obtained results highlight the obviously normal events in highly coherent clusters, with a minor subset of events being classified as anomalies for further forensic and compliance auditing analysis. The model interpretability is indirectly validated by the produced decision rule set already provided in Figure 4, which implicitly shows how the model identifies classes. Figure 7 also shows the performance of the parallel approach compared with the sequential approach, and Figure 8 highlights the parallelization capabilities of the Dask library in processing out-of-core datasets.

Designing the framework with independent tools makes it possible to reuse them over different scenarios. For example, the same modeling tool can be combined with a different normalization tool for processing a different data source. Additionally, these framework tools can be applied to the context of the aforementioned ATENA project in order to identify anomaly events from massive logs. This approach can be independently applied to different datasets in a first stage, allowing to correlate them as heterogeneous sources in a second stage.

The achieved results demonstrate the capability of the proposed method in terms of finding a set of interpretable rules that can parallelized and applied in scale.

5. Conclusions and Future Work

In this paper we proposed a framework that takes a parallel computing approach for identifying anomaly events in massive log files. In its first stage, our method uses the k-means algorithm to separate anomalies from normal events. In the second stage, a gradient tree boosting classification model, implemented using the XGBoost system, produces the interpretable meaningful rationale rule set for generalizing its application to a massive number of unseen events. This approach is suitable for application in the context of out-of-core datasets in cases where log sources are so massive that it becomes impossible to use more traditional approaches.

The proposed method was presented, and the achieved results demonstrated its applicability to producing simple and interpretable rules for highlighting anomalies in application log data to scale and in a distributed environment. Such an approach makes it suitable to be applied in the fields of forensics and audit compliance.

Regarding future work, we plan to explore the application of collective anomaly detection over time series summarized data logs and the application of Bayesian networks as the classification model component, and evaluate the method's capability of producing scalable and interpretable models. We also plan to explore the map-reduce model as a way to achieve higher parallelism performance on data preparation.

Author Contributions: Conceptualization, J.H., F.C., T.C. and P.S.; Funding acquisition, P.S.; Software and Experiments, J.H.; Supervision, F.C., T.C. and P.S.; Writing—original draft, J.H., F.C., T.C. and P.S.; Writing—review editing, F.C., T.C. and P.S. All authors have read and agreed to the published version of the manuscript.

Funding: This work was partially funded by the National Funds through the FCT—Foundation for Science and Technology, I.P., and the European Social Fund, through the Regional Operational Program Centro 2020, within the scope of the projects UIDB/05583/2020 and CISUC UID/CEC/00326/2020. Furthermore, we would like to thank the Research Center in Digital Services (CISeD) and the Polytechnic of Viseu for their support.

Conflicts of Interest: The authors declare no conflicts of interest.

References

1. Adamsky, F.; Aubigny, M.; Battisti, F.; Carli, M.; Cimorelli, F.; Cruz, T.; Giorgio, A.D.; Foglietta, C.; Galli, A.; Giuseppi, A.; et al. Integrated protection of industrial control systems from cyber-attacks: The ATENA approach. *Int. J. Crit. Infrastruct. Prot.* **2018**, *21*, 72–82. [CrossRef]
2. Rosa, L.; Proença, J.; Henriques, J.; Graveto, V.; Cruz, T.; Simões, P.; Caldeira, F.; Monteiro, E. An Evolved Security Architecture for Distributed Industrial Automation and Control Systems. In Proceedings of the 16th European Conference on Cyber Warfare and Security (ECCWS 2017), Dublin, Ireland, 29–30 June 2017; pp. 380–390.
3. Wu, X.; Kumar, V.; Quinlan, J.R.; Ghosh, J.; Yang, Q.; Motoda, H.; McLachlan, G.J.; Ng, A.; Liu, B.; Philip, S.Y.; et al. Top 10 algorithms in data mining. *Knowl. Inf. Syst.* **2008**, *14*, 1–37. [CrossRef]
4. Friedman, J.H. Greedy function approximation: A gradient boosting machine. *Ann. Stat.* **2001**, *29*, 1189–1232. [CrossRef]
5. Chen, T.; Guestrin, C. Xgboost: A scalable tree boosting system. In Proceedings of the 22nd ACM Sigkdd International Conference on Knowledge Discovery and Data Mining, San Francisco, CA, USA, 13–17 August 2016; pp. 785–794.
6. Rocklin, M. Dask: Parallel computation with blocked algorithms and task scheduling. In Proceedings of the 14th Python in Science Conference, Austin, TX, USA, 6–12 July 2015; pp. 130–136.
7. Grubbs, F.E. Procedures for detecting outlying observations in samples. *Technometrics* **1969**, *11*, 1–21. [CrossRef]
8. Gogoi, P.; Bhattacharyya, D.; Borah, B.; Kalita, J.K. A survey of outlier detection methods in network anomaly identification. *Comput. J.* **2011**, *54*, 570–588. [CrossRef]
9. Zheng, Y.; Zhang, H.; Yu, Y. Detecting collective anomalies from multiple spatio-temporal datasets across different domains. In Proceedings of the 23rd SIGSPATIAL international conference on advances in geographic information systems, Seattle, WA, USA, 3–6 November 2015; pp. 1–10.
10. Yang, Q.; Wu, X. 10 challenging problems in data mining research. *Int. J. Inf. Technol. Decis. Making* **2006**, *5*, 597–604. [CrossRef]
11. Chen, Z.; Li, Y.F. Anomaly detection based on enhanced DBScan algorithm. *Procedia Eng.* **2011**, *15*, 178–182. [CrossRef]
12. Syarif, I.; Prugel-Bennett, A.; Wills, G. Unsupervised clustering approach for network anomaly detection. In Proceedings of the International Conference on Networked Digital Technologies, Dubai, United Arab Emirates, 24–26 April 2012; pp. 135–145.
13. Hoglund, A.J.; Hatonen, K.; Sorvari, A.S. A computer host-based user anomaly detection system using the self-organizing map. In Proceedings of the IEEE-INNS-ENNS International Joint Conference on Neural Networks (IJCNN 2000), Como, Italy, 27 July 2000; pp. 411–416.
14. Lichodzijewski, P.; Zincir-Heywood, A.N.; Heywood, M.I. Host-based intrusion detection using self-organizing maps. In Proceedings of the 2002 International Joint Conference on Neural Networks (IJCNN'02), Honolulu, HI, USA, 12–17 May 2002; pp. 1714–1719.
15. Münz, G.; Li, S.; Carle, G. Traffic Anomaly Detection Using K-Means Clustering. GI/ITG Workshop MMBnet, 2007; pp. 13–14. Available online: https://pdfs.semanticscholar.org/634e/2f1a20755e7ab18e8e8094f48e140a32dacd.pdf (accessed on 15 June 2017).
16. Tian, L.; Jianwen, W. Research on network intrusion detection system based on improved k-means clustering algorithm. In Proceedings of the 2009 International Forum on Computer Science-Technology and Applications, Chongqing, China, 25–27 December 2009; pp. 76–79.
17. Eslamnezhad, M.; Varjani, A.Y. Intrusion detection based on MinMax K-means clustering. In Proceedings of the 7'th International Symposium on Telecommunications (IST'2014), Tehran, Iran, 9–11 September 2014; pp. 804–808.
18. Ranjan, R.; Sahoo, G. A new clustering approach for anomaly intrusion detection. *Int. J. Data Min. Knowl. Manage. Process* **2014**, *4*, 29–38. [CrossRef]
19. Makanju, A.; Zincir-Heywood, A.N.; Milios, E.E. Investigating event log analysis with minimum apriori information. In Proceedings of the 2013 IFIP/IEEE International Symposium on Integrated Network Management (IM 2013), Ghent, Belgium, 27–31 May 2013; pp. 962–968.

20. Asif-Iqbal, H.; Udzir, N.I.; Mahmod, R.; Ghani, A.A.A. Filtering events using clustering in heterogeneous security logs. *Inf. Technol. J.* **2011**, *10*, 798–806. [CrossRef]
21. Hajamydeen, A.I.; Udzir, N.I.; Mahmod, R.; GHANI, A.A.A. An unsupervised heterogeneous log-based framework for anomaly detection. *Turkish J. Electr. Eng. Comput. Sci.* **2016**, *24*, 1117–1134. [CrossRef]
22. Varuna, S.; Natesan, P. An integration of k-means clustering and naïve bayes classifier for Intrusion Detection. In Proceedings of the 2015 3rd International Conference on Signal Processing, Communication and Networking, ICSCN 2015, Tamil Nadu, India, 26–28 March 2015; pp. 1–5.
23. Muda, Z.; Yassin, W.; Sulaiman, M.; Udzir, N. K-means clustering and naive bayes classification for intrusion detection. *J. IT Asia* **2016**, *4*, 13–25. [CrossRef]
24. Elbasiony, R.M.; Sallam, E.A.; Eltobely, T.E.; Fahmy, M.M. A hybrid network intrusion detection framework based on random forests and weighted k-means. *Ain Shams Eng. J.* **2013**, *4*, 753–762. [CrossRef]
25. Sequeira, K.; Zaki, M. ADMIT: Anomaly-based data mining for intrusions. In Proceedings of the Eighth ACM SIGKDD International Conference on Knowledge Discovery and Data Mining, Edmonton, AB, Canada, 23–26 July 2002; pp. 386–395.
26. Liao, Y.; Vemuri, V.R. Use of k-nearest neighbor classifier for intrusion detection. *Comput. Secur.* **2002**, *21*, 439–448. [CrossRef]
27. Breunig, M.M.; Kriegel, H.P.; Ng, R.T.; Sander, J. LOF: Identifying density-based local outliers. In Proceedings of the 2000 ACM SIGMOD International Conference on Management of Data, Dallas, TX, USA, 16–18 May 2000; pp. 93–104.
28. Bahmani, B.; Moseley, B.; Vattani, A.; Kumar, R.; Vassilvitskii, S. Scalable k-means++. *Proc. VLDB Endowment* **2012**, *5*, 622–633. [CrossRef]
29. Vattani, A. K-means requires exponentially many iterations even in the plane. *Discrete Comput. Geom.* **2011**, *45*, 596–616. [CrossRef]
30. Arthur, D.; Vassilvitskii, S. How slow is the k-means method? In Proceedings of the twenty-second annual symposium on Computational geometry, Sedona, AZ, USA, 5–7 June 2006; pp. 144–153.
31. Breiman, L. Bagging predictors. *Mach. Learn.* **1996**, *24*, 123–140. [CrossRef]
32. Breiman, L. Random forests. *Mach. Learn.* **2001**, *45*, 5–32. [CrossRef]
33. Friedman, J.; Hastie, T.; Tibshirani, R. Additive logistic regression: A statistical view of boosting (with discussion and a rejoinder by the authors). *Ann. Statist.* **2000**, *28*, 337–407. [CrossRef]
34. Johnson, R.; Zhang, T. Learning nonlinear functions using regularized greedy forest. *IEEE Trans. Pattern Anal. Mach. Intell.* **2014**, *36*, 942–954. [CrossRef]
35. He, X.; Pan, J.; Jin, O.; Xu, T.; Liu, B.; Xu, T.; Shi, Y.; Atallah, A.; Herbrich, R.; Bowers, S.; et al. Practical lessons from predicting clicks on ads at facebook. In Proceedings of the Eighth International Workshop on Data Mining for Online Advertising, New York, NY, USA, 24 August 2014; pp. 1–9.
36. Kaggle. Available online: www.Kaggle.com (accessed on 16 July 2018).
37. Friedman, J.H. Stochastic gradient boosting. *Comput. Stat. Data Anal.* **2002**, *38*, 367–378. [CrossRef]
38. H2O.ai. H2O Framework for Machine Learning. Available online: http://docs.h2o.ai/h2o/latest-stable/h2o-docs/index.html (accessed on 15 February 2020).
39. Meng, X.; Bradley, J.; Yavuz, B.; Sparks, E.; Venkataraman, S.; Liu, D.; Freeman, J.; Tsai, D.; Amde, M.; Owen, S.; et al. MLlib: Machine Learning in Apache Spark. *J. Mach. Learn. Res.* **2016**, *17*, 1235–1241.
40. NASA. NASA HTTP. Available online: http://ita.ee.lbl.gov/html/contrib/NASA-HTTP.html (accessed on 1 July 2018).

© 2020 by the authors. Licensee MDPI, Basel, Switzerland. This article is an open access article distributed under the terms and conditions of the Creative Commons Attribution (CC BY) license (http://creativecommons.org/licenses/by/4.0/).

Article

Utilising Deep Learning Techniques for Effective Zero-Day Attack Detection

Hanan Hindy [1,*], Robert Atkinson [2], Christos Tachtatzis [2], Jean-Noël Colin [3], Ethan Bayne [1] and Xavier Bellekens [2]

1. Division of Cybersecurity, Abertay University, Dundee DD1 1HG, UK; e.bayne@abertay.ac.uk
2. Electronic and Electrical Engineering Department, University of Strathclyde, Glasgow G1 1XQ, UK; robert.atkinson@strath.ac.uk (R.A.); christos.tachtatzis@strath.ac.uk (C.T.); xavier.bellekens@strath.ac.uk (X.B.)
3. InfoSec Research Team, University of Namur, 5000 Namur, Belgium; jean-noel.colin@unamur.be
* Correspondence: hananhindy@ieee.org

Received: 14 September 2020; Accepted: 3 October 2020; Published: 14 October 2020

Abstract: Machine Learning (ML) and Deep Learning (DL) have been used for building Intrusion Detection Systems (IDS). The increase in both the number and sheer variety of new cyber-attacks poses a tremendous challenge for IDS solutions that rely on a database of historical attack signatures. Therefore, the industrial pull for robust IDSs that are capable of flagging zero-day attacks is growing. Current outlier-based zero-day detection research suffers from high false-negative rates, thus limiting their practical use and performance. This paper proposes an autoencoder implementation for detecting zero-day attacks. The aim is to build an IDS model with high recall while keeping the miss rate (false-negatives) to an acceptable minimum. Two well-known IDS datasets are used for evaluation—CICIDS2017 and NSL-KDD. In order to demonstrate the efficacy of our model, we compare its results against a One-Class Support Vector Machine (SVM). The manuscript highlights the performance of a One-Class SVM when zero-day attacks are distinctive from normal behaviour. The proposed model benefits greatly from autoencoders encoding-decoding capabilities. The results show that autoencoders are well-suited at detecting complex zero-day attacks. The results demonstrate a zero-day detection accuracy of 89–99% for the NSL-KDD dataset and 75–98% for the CICIDS2017 dataset. Finally, the paper outlines the observed trade-off between recall and fallout.

Keywords: autoencoder; artificial neural network; one-class support vector machine; intrusion detection; zero-day attacks; CICIDS2017; NSL-KDD

1. Introduction

Central to tackling the exponential rise in cyber-attacks [1,2], is Intrusion Detection Systems (IDS) systems that are capable of detecting zero-day cyber-attacks. Machine Learning (ML) techniques have been extensively utilised for designing and building robust IDS [3,4]. However, while current IDS can achieve high detection accuracy for known attacks, they often fail to detect new, zero-day attacks. This is due to the limitations of current IDS, which rely on pre-defined patterns and signatures. Moreover, current IDS suffer from high false-positive rates, thus limiting the performance and their practical use in real-life deployments. As a result, large numbers of zero-day attacks remain undetected, which escalate their consequences (denial of service, stolen customer details, etc.).

According to Chapman [5], a zero-day attack is defined as "a traffic pattern of interest that, in general, has no matching patterns in malware or attack detection elements in the network" [5]. The implications of zero-day attacks in real-world are discussed by Bilge and Dumitras [6]. Their research focuses on studying their impact and prevalence. The authors highlight that zero-day attacks are significantly more prevalent than suspected, demonstrating that, out of their 18 analysed attacks, 11 (61%) were previously unknown [6]. Their findings showed that a zero-day attack can exist for a substantial period of time (average of 10 months [6]) before they are detected and can compromise systems during that period. Additionally, Nguyen and Reddi [7] refer to a statistical study that shows that 62% of the attacks are identified after compromising systems. Moreover, the number of zero-day attacks in 2019 exceeds the previous three years [8]. All of these considerations highlight the clear and urgent need for more effective attack detection models.

One of the main research directions to detect zero-day attacks relies on detecting outliers (i.e., instances/occurrences that vary from benign traffic). However, the main drawback of the available outlier-based detection techniques is their relatively low accuracy rates as a result of both high false-positive rates and high false-negative rates. As discussed, the high false-negative rates leave the system vulnerable to attack, while the high false-positive rates needlessly consume the time of cyber security operation centres; indeed, only 28% of investigated intrusions are real [9]. Ficke et al. [10] emphasise the limitations that false-negative could bring to IDS development, for example, it reduces IDS effectiveness.

Sharma et al. [11] proposed a framework to detect zero-day attacks in Internet of Things (IoT) networks. They rely on a distributed diagnosis system for detection. Sun et al. [12] proposed a Bayesian probabilistic model to detect zero-day attack paths. The authors visualised attacks in a graph-like structure and introduced a prototype to identify attacks. Zhou and Pezaros [13] evaluated six different supervised ML techniques; using the CIC-AWS-2018 dataset. The authors use decision tree, random forest, k-nearest neighbour, multilayer perceptron, quadratic discriminant analysis, and gaussian naïve bayes classifiers. The authors do not fully detail how these supervised ML techniques are trained on benign traffic solely to be utilised for unknown attacks detection or how zero-day (previously unseen) attacks are simulated and detected. Moreover, transfer learning is used to detect zero-day attacks. Zhao et al. [14] used transfer learning to map the connection between known and zero-day attacks [14]. Sameera and Shashi [15] used deep transudative transfer learning to detect zero-day attacks.

Furthermore, ML is used to address Zero-day malware detection. For example, Abri et al. evaluated the effectiveness of using different ML techniques (Support Vector Machine (SVM), Naïve Bayes, Multi-Layer Perceptron, Decision trees, k-Nearest Neighbour, and Random Forests) to detect zero-day malware [16], while Kim et al. [17] proposed the use of Deep-Convolutional Generative Adversarial Network (DCGAN).

In this paper, we propose utilising the capabilities of Deep Learning (DL) to serve as outlier detection for zero-day attacks with high recall. The main goal is to build a lightweight intrusion detection model that can detect new (unknown) intrusions and zero-day attacks, with a high recall (true-positive rate) and low fallout (false-positive rate). Accordingly, having a high detection capability of zero-day attacks will help to reduce the complications and issues that are associated with new attacks.

The contributions of this work are threefold;

- Proposing and implementing an original and effective autoencoders model for zero-day detection IDS.
- Building an outlier detection One-Class SVM model.
- Comparing the performance of the One-Class SVM model as a baseline outlier-based detector to the proposed Autoencoder model.

The rest of the paper is organised as follows; the background is presented in Section 2, Section 3 discusses the related work showing the results and approaches of recent IDS research. Section 4 describes the datasets that are used and how zero-day attacks are simulated. In Section 5, the proposed models are explained. Section 6 presents the experimental results and findings. Finally, the paper is concluded in Section 7.

2. Background

In this section, the models utilised in this investigation are discussed. Section 2.1 describes the deep-learning based autoencoder model and Section 2.2 describes an unsupervised variant of a support vector machine model.

2.1. Autoencoders

The model that is proposed in this manuscript principally benefits from the autoencoder characteristics and attributes. The objective is that the autoencoder acts as a light-weight outlier detector, which could then be used for zero-day attacks detection, as further discussed in Section 5.2.

Rumelhart et al. [18] first introduced autoencoders in order to overcome the back propagation in unsupervised context using the input as the target. Autoencoders are categorised as self-supervised, since the input and the output are particularly the same [19]. As defined by Goodfellow et al. [20], an Autoencoder is "a neural network that is trained to attempt to copy its input to its output" [20]. Figure 1 illustrates the basic architecture of an autoencoder. The architecture of an autoencoder and the number of hidden layers differ based on the domain and the usage scenario.

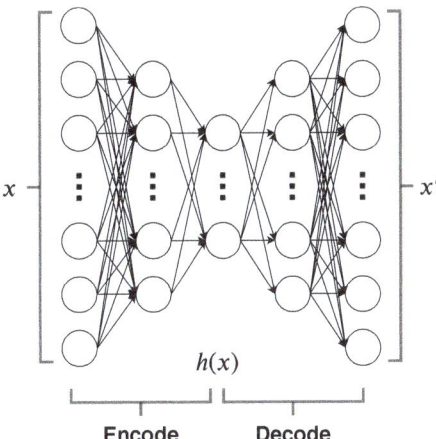

Figure 1. Autoencoder Architecture.

Formally, given an input $\mathcal{X}$, an autoencoder is trained in order to minimise the reconstruction error, as demonstrated in Equation (1) [19].

$$\begin{aligned} \phi &: \mathcal{X} \to \mathcal{F} \\ \psi &: \mathcal{F} \to \mathcal{X} \\ \phi, \psi &= \underset{\phi, \psi}{\operatorname{argmin}} ||\mathcal{X} - (\phi \circ \psi)\mathcal{X}||^2 \end{aligned} \quad (1)$$

such that ϕ and ψ represent the encoding and decoding functions, respectively.

Commonly, the reconstruction error of an input x is represented as the difference between x and x', such that:

$$x' = g(f(x))$$

where $f(x)$ is the encoding function, constructing the encoded vector of x; $g(x)$ is the decoding function, restoring x to its initial value

The reconstruction error is defined by a function that represents the difference between the input x and the reconstructed input x'. Mean square error and mean absolute error are common functions that are used in order to calculate the reconstruction error, as shown in Equations (2) and (3), respectively.

$$MSE = \sum_{i=1}^{N} (x' - x)^2 \tag{2}$$

$$MAE = \sum_{i=1}^{N} |x' - x| \tag{3}$$

Supposing that the encoding function $f(x)$ is single layer network with a linear function, the Autoencoder is viewed as equivalent to Principal Components Analysis (PCA) [21].

Autoencoders were originally used for dimensionality reduction and feature learning [22,23]. However, many other applications have been recently proposed. These applications include: word semantics [24], image compression [25], image anomaly detection [26], denoising [27], and others.

2.2. One-Class SVM

The SVM is one of the most well-established supervised ML techniques. Given the training samples, an SVM is trained to construct a hyperplane in a high-dimensional space that best separates the classes [28]. This hyperplane is a line in two dimensions case, a plane in the case of three dimensions (3D) or higher dimensions (n-dimensional). When data are not linearly separable, a kernel is used to map the input features/data to a non-linear higher dimensional space, in which a hyperplane would best separate the classes. The SVM kernels include; linear, polynomial, Gaussian, and Radial Basis Function (RBF).

Formally, given two classes, the minimisation problem of SVM is represented, as shown in Equation (4) [29].

$$\min_{w \in \mathbb{R}^d} \|w\|^2 + C \sum_{i}^{N} \max(0, 1 - y_i f(x_i)) \tag{4}$$

$$f(x_i) = (w^T x_i + b)$$

where C is a regularisation parameter to represent the trade-off between ensuring that x_i is on the expected side of the plane and increasing the margin. Based on Equation (3), the data points fall in one of three places based on $y_i f(x_i)$. If $y_i f(x_i)$ is greater than 1, then the point is outside the margin and it does not contribute to the loss. If $y_i f(x_i)$ equals 1, then the point is on the margin. Finally, if $y_i f(x_i)$ is less than 1, then the point contributes to the loss, as it is on the wrong side [30].

In contrast to its supervised counterpart, the One-Class SVM is an unsupervised variant. It is defined as a model that is capable of detecting "Novelty" [31]. The goal of One-Class SVM is to fit a hyperplane that acts as a boundary which best includes all the training data and excludes any other data point. The result of training a One-Class SVM is seen as a spherically shaped boundary [32]. Because One-Class SVM is considered to be one of the most established outlier-based ML techniques, it provides an ideal comparison for assessing the performance of a deep neural network based autoencoder.

Formally, given a class with instances $\{x_1, ..., x_N\}$, and a mapping function $\varphi()$ that maps the features to a space H, the goal of One-Class SVM is to fit a hyperplane Π in H that has the largest distance to the origin, and all $\varphi(x_i)$ lie at the opposite side of hyper-plane to the origin [33].

3. Related Work

IDS is defined as "a system or software that monitors a network or systems for malicious activity". Generally, IDSs can either be Network Intrusion Detection System (NIDS) or Host Intrusion Detection System (HIDS). NIDS monitors the network and communication while HIDS monitors the internal operation and log files [34]. Based on their detection techniques, IDSs are classified into Signature-based IDS, which relies on known signatures of prior and known attacks, and Anomaly-based IDS, which relies on patterns [35]. When compared to signature-based IDS, anomaly-based IDS perform better with complex attacks and unknown attacks.

In the past decade, researchers developed multiple techniques in order to enhance the robustness of anomaly-based IDS. Subsequent to a long period of using statistical methods to detect cyber anomalies and attacks, the need for ML emerged. Because of the sophistication of cyber-attacks, statistical methods were rendered inadequate to handle their complexity. Therefore, with the advancement of ML and DL in other domains (i.e., image and video processing, natural language processing, etc.), the researchers adopted these techniques for cyber use. Nguyen and Reddi [7] discuss the importance and benefit ML can provide to cybersecurity by granting a 'robust resistance' against attacks.

Based on the analysis of recent IDS research [36], ML has dominated the IDS research in the past decade. The analysis shows that Artificial Neural Networks (ANN), SVM, and k-means are the prevailing algorithms. Buczak and Guven [37] analyse the trends and complexity of different ML and DL techniques used for IDS. Moreover, recent research is directed towards the use of DL to analyse network traffic, due to the DL capabilities of handling complex patterns. Due to the complexity and challenges that are associated with encrypted traffic, building robust and reliable DL-based IDSs is crucial. Aceto et al. [38] describe this advancement in traffic encryption, as it 'defeats traditional techniques' that relies on packet-based and port-based data. At the beginning of 2019, 87% of traffic was encrypted [39], which emphasises on the growth and, thus the need for corresponding IDs. Research has utilised flow-based features as the building block for training and analysing IDSs in order to handle encrypted and non-encrypted traffic. The benefit of flow-based features, when compared to packet-based ones, relies on the fact that they can be used with both encrypted and unencrypted traffic and also they characterise high-level patterns of network communications. New DL approaches have recently been used to build robust and reliable IDS. One of these techniques is autoencoders.

In the cyber security domain, autoencoders are used for feature engineering and learning. Kunang et al. [40] used autoencoders for feature extraction, features are then passed into an SVM for classification. KDD Cup'99 and NSL-KDD datasets are both used for evaluation. The evaluation of the model, using autoencoder for feature extraction and SVM for multi-class classification, has an overall accuracy of 86.96% and precision of 88.65%. The different classes accuracies show a poor performance, as follows: 97.91%, 88.07%, 12.78%, 8.12%, and 97.47% for DoS, probe, R2L, U2R, and normal, respectively; a precision of 99.45%, 78.12%, 97.57%, 50%, and 81.59% for DoS, probe, R2L, U2R, and normal, respectively.

Kherlenchimeg and Nakaya [41] use a sparse autoencoder to extract features. The latent representation (the bottleneck layer of the autoencoder) is fed into a Recurrent Neural Network (RNN) for classification. The accuracy of the IDS model while using the NSL-KDD dataset is 80%. Similarity, Shaikh and Shashikala [42] focus on the detection of DoS attacks. They utilise a Stacked Autoencoder with an LSTM network for the classification. Using the NSL-KDD dataset, the overall detection accuracy is 94.3% and a false positive rate of 5.7%.

Abolhasanzadeh [43] used autoencoders for dimensionality reduction and the extraction of bottleneck feature. The experiments were evaluated while using the NSL-KDD dataset. In a similar fashion, Niyaz et al. [44] used autoencoders for unsupervised feature learning. They used the NSL-KDD dataset. Additionally, AL-Hawawreh et al. [45] used deep autoencoders and trained them on benign traffic in order to deduce the most important feature representation to be used in their deep feed-forward ANN.

Shone et al. [46] use a Stacked Non-Symmetric Deep Autoencoder to refine and learn the complex relationships between features that are then used in the classification using random forest technique. The authors used both the KDD Cup'99 and NSL-KDD datasets.

Farahnakian and Heikkonen [47] used a deep autoencoder to classify attacks. The deep autoencoder is fed into a single supervised layer for classification. Using the KDD Cup'99 dataset, the highest accuracies are 96.53% and 94.71% for binary and multi-class classification respectively.

In agreement with our manuscript, Meidan et al. [48] utilised the encoding-decoding capabilities of autoencoders to learn normal behaviour in IoT networks setup. Subsequently, IoT botnet and malicious behaviour are detected when the autoencoder reconstruction fails. Bovenzi et al. [49] emphasised the need for adaptive ML models to cope with the heterogeneity and unpredictability of IoT networks. The authors propose a two-stage IDS model, where they leverage the autoencoder capabilities in the first stage of their IDS.

4. Datasets

Two mainstream IDS datasets are chosen in order to evaluate the proposed models. The first is the CICIDS2017 dataset [50], which was developed by the Canadian Institute for Cybersecurity (CIC). The CICIDS2017 dataset covers a wide range of recent insider and outsider attacks. It comprises a diverse coverage of protocols and attacks variations and, finally, it is provided in a raw format which enables researchers the flexibility of processing the dataset. Therefore, the CICIDS2017 dataset is well-suited for evaluating the proposed models.

The CICIDS2017 dataset is a recording of a five-day benign, insider and outsider attacks traffic. The recorded PCAPs are made available. Table 1 summarises the traffic recorded per day. The raw files of the CICIDS2017 dataset are pre-processed, as described in Section 5.1. The full CICIDS2017 description and analysis is available in [51,52].

Table 1. CICIDS2017 attacks.

Day	Traffic
Monday	Benign
Tuesday	SSH & FTP Brute Force
Wednesday	DoS/DDoS & Heartbleed
Thursday	Web Attack (Brute Force, XSS, Sql Injection) & Infiltration
Friday	Botnet, Portscan & DDoS

The second dataset is the NSL-KDD [53]. NSL-KDD was released by the CIC in order to overcome the problems of the KDD Cup'99 dataset [54]. The KDD Cup'99 dataset was the dataset of choice for evaluating more than 50% of the past decade IDS [36], followed by the NSL-KDD dataset, which was used for evaluating over 17% of IDS. However, the KDD Cup'99 has multiple drawbacks, as discussed thoroughly in [55]. These drawbacks include class imbalance and redundant records. Additionally, Siddique et al. [56] discussed the warnings provided to the UCI lab advising not to use KDD Cup'99 dataset in further IDS research. Consequently, NSL-KDD fits for the evaluation purpose of this manuscript, as well as the comparison with relevant research.

The NSL-KDD dataset covers normal/benign traffic and four cyber-attack classes, namely, Denial of Service (DoS), probing, Remote to Local (R2L), and User to Root (U2R). The NSL-KDD dataset is available in two files 'KDDTrain+.csv' and test file 'KDDTest+.csv'. Similar to the KDD Cup'99, the NSL-KDD dataset is provided in comma separated value (csv) feature files. Each instance is represented with its feature values alongside the class label. The feature files undergo categorical features encoding to be appropriate for ML usage. The KDD Cup'99 and NSL-KDD datasets are analysed in [54]; furthermore, NSL-KDD is studied in [57].

5. Methodology, Approach and Proposed Models

In this section, the pre-processing of the datasets is discussed, followed by the explanation of the proposed, showing both the training and evaluation processes. Subsequently, Section 6 details the evaluation and results.

5.1. CICIDS2017 Pre-Processing

The process that is involved in preparing the CICIDS2017 dataset for use is described as follows. Firstly, '.pcap' files of the CICIDS2017 dataset are split based on the attack type and the timestamps provided by the dataset. This process results in a separate '.pcap' file for each attack class. Secondly, the '.pcap' files are processed to generate bidirectional flows features. As highlighted by Rezaei and Liu [58], with the advancement and complexity of networks and relying on encrypted traffic, features need to be suitable for both encrypted and unencrypted traffic analysis. The authors also indicate that flow-based features are better-suited for modern IDS development. Based on the analysis of recent IDSs by Aceto et al. [38], flow and bidirectional flow features are the most commonly used. Thirdly, features with high correlation are dropped in order to minimise model instability. Algorithm 1 describes the process of dropping highly correlated features. A threshold of '0.9' is used. Features with correlation less than the threshold are used for training. Finally, features are scaled using a Standard Scalar. It is important to mention that only benign instances are used in selecting the features and scaling in order to ensure zero influence of attack instances.

Algorithm 1 Drop correlated features

Input: Benign Data 2D Array, N, Correlation Threshold
Output: Benign Data 2D Array, Dropped Columns

1: $correlation\_matrix \leftarrow data.corr().abs()$
2: $upper\_matrix \leftarrow correlation\_matrix[i,j]$ \qquad $\{i,j \in N : i <= j\}$
3: $dropped \leftarrow i\{i \in N : correlation\_matrix[i,*] > threshold\}$
4: $data \leftarrow data.drop\_columns(dropped)$
5: **return** $data, dropped$

As aforementioned, the goal is to train models using benign traffic and evaluate their performance to detect attacks. Therefore, normal/benign traffic solely is used for training. The normal instances are divided into 75% for training and 25% for testing/validation [59] by using sklearn train_test_split function with the shuffling option set to True (https://scikit-learn.org/stable/modules/generated/sklearn.model_selection.train_test_split.html). Furthermore, each of the attack classes then mimics a zero-day attack, thus assessing the ability of the model to detect its abnormality. Because the NSL-KDD dataset is split into training and testing, attacks in both files are used for evaluation.

5.2. Autoencoder-Based Model

The building block for the proposed Autoencoder is an Artificial Neural Network (ANN). For hyper-parameter optimisation, random search [60] is used in order to select the architecture of the network, number of epochs, and learning rate. Random search is known to converge faster than grid search to a semi-optimal set of parameters. It is also proved to be better than grid search when a small number of parameters are needed [61]. Finally, it limits the possibility of obtaining over-fitted parameters.

Once the hyper-parameters are investigated, the model is trained, as detailed in Algorithm 2. First, the benign instances are split into 75%:25% for training and validation, respectively. Subsequently, the model is initialised using the optimal ANN architecture (number of layers and number of hidden neurons per layer). Finally, the model is trained for n number of epochs. The loss and accuracy curves are observed in order to verify that the autoencoder convergence.

Once the model converges, as rendered in Figure 2, the model is evaluated using Algorithm 3. An attack instance is flagged as a zero-day attack if the Mean Squared Error (MSE) (reconstruction error) of the decoded (x') and the original instance (x) is larger than a given threshold. For the purpose of evaluation, multiple thresholds are assessed: 0.05, 0.1, 0.15. These thresholds are chosen based on the value that is chosen by the random search hyper-parameter optimisation. The threshold plays an important role in deciding the value at which an instance is considered a zero-day attack, i.e., what MSE between x' and x is within the acceptable range.

Algorithm 2 Autoencoder Training

    Input: benign_data, ANN_architecture, regularisation_value, num_epochs
    Output: Trained Autoencoder

1: $training = 75\% \; i \in benign\_data$
2: $testing = benign\_data \cap training$
3: $autoencoder \leftarrow build\_autoencoder(ANN\_Architecture, regularisation\_value)$
4: $batch\_size \leftarrow 1024$
5: $autoencoder.train(batch\_size, num\_epochs, training, testing)$
6: **return** $autoencoder$

Algorithm 3 Evaluation

    Input: Trained Autoencoder, attack, thresholds
    Output: Detection accuracies

1: $detection\_accuracies \leftarrow \{\}$
2: $predictions \leftarrow model.predict(attack)$
3: **for** $th \in thresholds$ **do**
4:     $accuracy \leftarrow (mse(predictions, attack) > th)/len(attack)$
5:     $detection\_accuracies.add(threshold, accuracy)$
6: **end for**
7: **return** $detection\_accuracies$

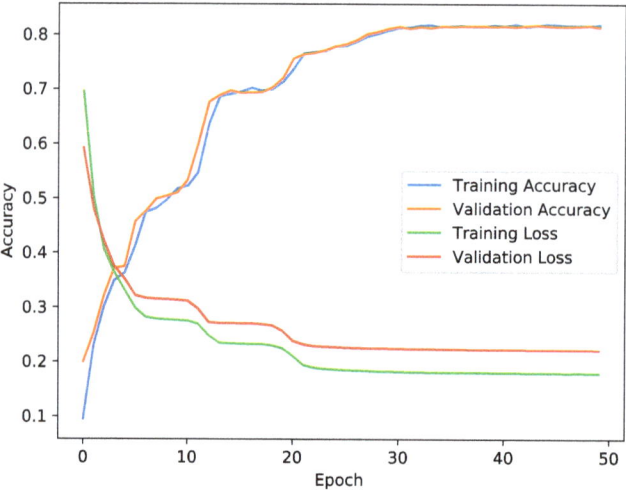

Figure 2. Autoencoder Convergence Curves.

5.3. One-Class SVM Based Model

One-Class SVM is trained using the benign instances. In order to train the One-Class SVM, a 'ν' value was specified. As defined by Chen et al., "$\nu \in [0, 1]$, which is the lower and upper bound on the number of examples that are support vectors and that lie on the wrong side of the hyperplane, respectively." [62]. The ν default value is 0.5, which includes 50% of the training sample in the hyperplane. However, for the purpose of this experiment, multiple ν values were chosen (0.2, 0.15, 0.1). These values were used in otder to evaluate and assess the autoencoder performance.

Algorithm 4 shows the process of training the One-Class SVM mode. Similar to the model that is discussed in Section 5.2, 75% of the benign samples are used to fit the One-Class SVM model. Unlike the Autoencoder model, where the evaluation relies on a threshold, a One-Class SVM trained model outputs a binary value {0,1}. The output represents whether an instance belongs to the class to which the SVM is fit. Hence, each attack is evaluated based on how many instances are predicted with a '0' SVM output.

Algorithm 4 One-Class SVM Model

Input: benign_data, nu_value
Output: Trained SVM

1: $training = 75\% \, i \in benign\_data$
2: $testing = benign\_data \cap training$
3: $oneclasssvm \leftarrow OneClassSVM(nu\_value, 'rbf')$
4: $oneclasssvm.fit(training)$
5: **return** $oneclasssvm$

---

6. Experimental Results

6.1. CICIDS2017 Autoencoder Results

As mentioned, 75% of the benign instances is used to train the Autoencoder. The autoencoder optimised architecture for the CICIDS2017 dataset is comprised from an ANN network with 18 neurons in both the input and the output layers and 3 hidden layers with 15, 9, 15 neurons respectively. The optimal batch size is 1024. Other optimised parameters include mean square error loss, L2 regularisation of 0.0001 and for 50 epochs.

Figure 3 summarises the autoencoder accuracy of all CICIDS2017 classes. It is crucial to note that accuracy is defined differently for benign. Unlike attacks, for benign class, the accuracy represents the rate of instances not classified as zero-day (i.e., benign) which reflects the specificity and for the attack classes it represents the recall.

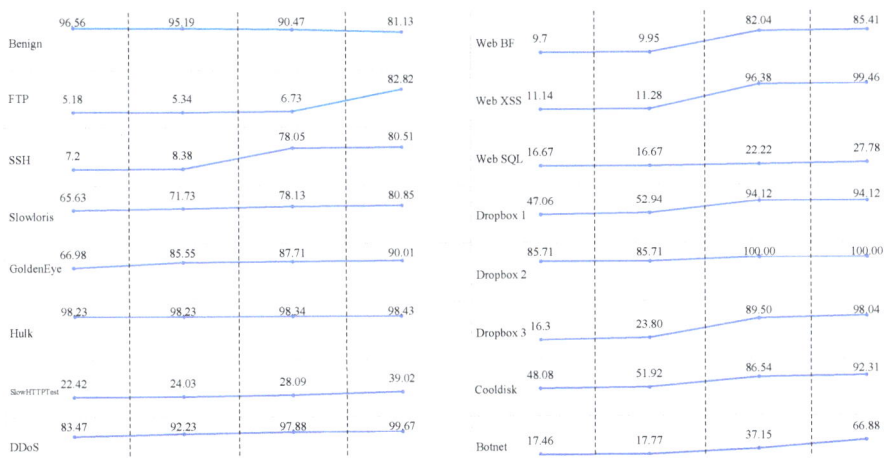

Figure 3. CICIDS2017 Autoencoder Detection Results Summary Per Class.

By observing Figure 3, benign accuracy is 95.19%, 90.47% and 81.13% for a threshold of 0.15, 0.1 and 0.05 respectively. Moreover, for the different attack detection accuracy, it is observed that there are three categories. Firstly, attacks that are very different from benign (for example, Hulk and DDoS), the detection accuracy is high regardless the threshold [92–99%]. Secondly, classes that are slightly different from benign (for example, SSH Brute-force and Port scanning), an accuracy rise is observed for lower thresholds. This emphasise the threshold's role. Thirdly, classes that are not distinguishable from benign traffic, they are detected but with a lower accuracy (for example, Botnet, SQL Injection and DoS-SlowHTTPTest).

By observing Figure 3, different categories can be seen, (a) classes with a stable detection accuracy (i.e., line), and (b) classes with a prompt rise in detection accuracy in the right-most slice (0.05 threshold). Finally, the benign accuracy (top left) falls within an acceptable range with different thresholds.

6.2. CICIDS2017 One-Class SVM Results

Table 2 summarises the One-Class SVM results. By observing the One-Class SVM results, two assertions are identified, (a) the detection accuracy is not affected significantly by changing ν value, and (b) the classes with high detection accuracy in the Autoencoder results (Figure 3 are also detected by the One-Class SVM; however, the One-Class SVM fails to detect the two other categories (rise in detection accuracy with small thresholds and low detection accuracy). This is due to the limitations of the One-Class SVM algorithm which attempts to fit a spherical hyperplane to separate benign class from other classes, however, classes that fall into this hyperplane will always be classified as benign/normal.

This can further be visualised in Figure 4. One-Class SVM is well suited for flagging recognisable zero-day attacks. However, autoencoders are better suited for complex zero-day attacks as the performance rank is significantly higher. Furthermore, Figure 4 shows a class by class comparison of the performance of autoencoder versus One-Class SVM. Figure 4a plots the results using One-Class SVM $\nu = 0.2$ and autoencoder threshold of 0.05, while Figure 4b plots the results using One-Class SVM $\nu = 0.09$ and autoencoder threshold of 0.1.

Table 2. CICIDS2017 One-Class Support Vector Machine (SVM) Results.

Class	Accuracy		
ν	0.2	0.15	0.1
Benign (Validation)	89.81%	84.84%	79.71%
FTP Bruteforce	10.19%	15.16%	20.29%
SSH Bruteforce	79.51%	80.26%	80.95%
DoS (Slowloris)	7.66%	8.38%	10.37%
DoS (GoldenEye)	71.87%	72.39%	72.85%
DoS (Hulk)	90.69%	91.35%	91.55%
DoS (Slowhttps)	98.59%	98.66%	98.71%
DDoS	39.35%	39.94%	40.96%
Heartbleed	99.49%	99.54%	99.58%
Web BF	21.1%	23.41%	35.84%
Web XSS	9.58%	9.76%	10.13%
Web SQL	5.77%	6.31%	6.85%
Infiltration - Dropbox 1	38.89%	38.89%	38.89%
Infiltration - Dropbox 2	29.41%	35.29%	35.29%
Infiltration - Dropbox 3	57.14%	57.14%	57.14%
Infiltration - Cooldisk	92.15%	93.8%	94.91%
Botnet	44.23%	46.15%	50%
PortScan	59.27%	60.04%	63.43%

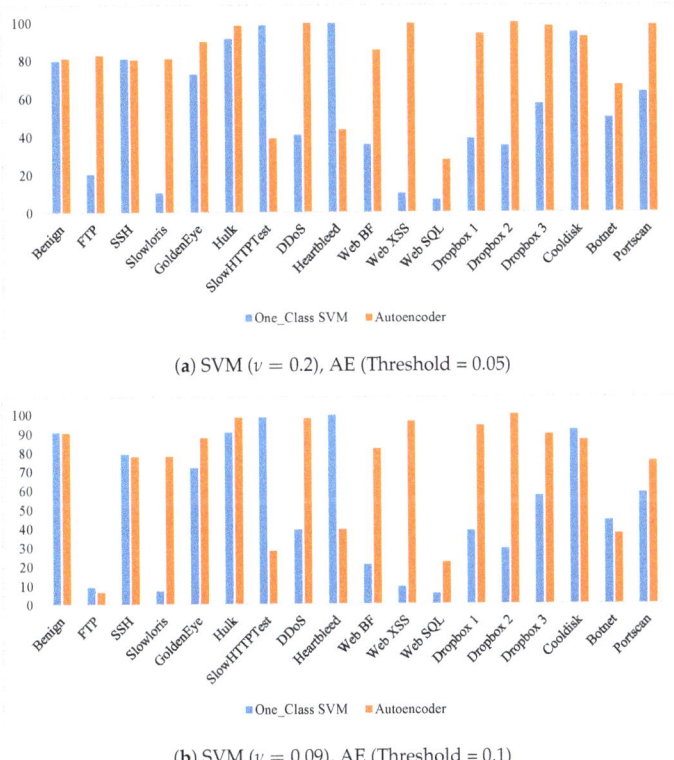

(a) SVM ($\nu = 0.2$), AE (Threshold = 0.05)

(b) SVM ($\nu = 0.09$), AE (Threshold = 0.1)

Figure 4. CICIDS2017 Autoencoder, One-Class SVM Comparison.

6.3. NSL-KDD Results

The autoencoder optimised architecture for the NSL-KDD dataset is comprised from an ANN network with 122 neurons in both the input and output layers and three hidden layers with 100, 60, 100 neurons, respectively. The optimal batch size is 1024. Other optimised parameters include mean absolute error loss, L2 regularisation of 0.001, and for 50 epochs.

Table 3 shows the autoencoder results for the NSL-KDD dataset. As aforementioned, attacks in both the KDDTrain+ and KDDTest+ files are used in order to evaluate the model. Similar to the results that are discussed in Section 6.1, the trade-off between the threshold choice and the true negative rate is observed.

Furthermore, when compared to the most recent available autoencoder implementation for detecting zero-day attacks in the literature [63], the autoencoder that is proposed in this manuscript largely outperforms the performances of [63]. The work proposed by Gharib et al. [63] used a hybrid two stage autoencoder to detect normal and abnormal traffic. Training on KDDTrain+ file and testing on KDDTest+, the overall accuracy of their proposed model is 90.17%, whereas the proposed autoencoder in this manuscript the overall accuracy is 91.84%, 92.96%, and 91.84% using a threshold of 0.3, 0.25, and 0.2, respectively. It is important to highlight that Gharib et al. [63] do not mention details regarding how they define anomalies or zero-day attacks or the classes they use in the testing process. Moreover, as summarised in Table 4, it is shown that the proposed approach in this manuscript outperforms the Denoising Autoencoder that is proposed in [64], specifically with the KDDTest+ instances with the authors accuracy is capped by 88%, while this manuscript reaches 93%.

Table 3. NSL-KDD Autoencoder Results.

Class	Accuracy		
Threshold	0.3	0.25	0.2
KDDTrain+.csv			
Normal (Validation)	78.81%	77.63%	78.81%
DoS	98.15%	98.16%	98.15%
Probe	99.89%	99.94%	99.89%
R2L	83.12%	96.48%	83.12%
U2R	84.62%	100%	84.62%
KDDTest+.csv			
Normal	84.82%	84.42%	84.82%
DoS	94.67%	94.67%	94.67%
Probe	100%	100%	100%
R2L	95.95%	96.5%	95.95%
U2R	83.78%	89.19%	83.78%

Table 4. NSL-KDD Performance Comparison with Recent Literature. (Highest accuracy in **Bold**).

Year	Reference	Approach	Train:Test % of KDDTrain+	KDDTrain+ Accuracy	KDDTest+ Accuracy
This paper		AE th = 0.3	75 : 25	88.92%	91.84%
		AE th = **0.25**		**94.44%**	**92.96%**
		AE th = 0.2		88.92%	91.84%
2019	[63]	2 AEs	-	-	90.17%
2017	[64]	AE	80 : 20	93.62%	88.28%
		Denoising AE		94.35%	88.65%

Table 5 summarises the NSL-KDD One-Class SVM results. The results show a similar detection trend. This is due to the limited number and variance of attacks that are covered by the NSL-KDD dataset.

Table 5. NSL-KDD One-Class SVM Results.

Class	Accuracy		
ν	0.2	0.15	0.1
KDDTrain+.csv			
Normal (Validation)	89.9%	85.14%	80.54%
DoS	98.13%	98.14%	98.14%
Probe	97.74%	98.77%	99.52%
R2L	49.35%	52.26%	81.71%
U2R	78.85%	80.77%	82.69%
KDDTest+.csv			
Normal	88.12%	86.02%	84.72%
DoS	94.67%	94.67%	94.69%
Probe	99.55%	99.91%	100%
R2L	80.17%	82.22%	90.31%
U2R	78.38%	78.38%	83.78%

7. Conclusions and Future Work

The work that is presented in this manuscript proposes a new outlier-based zero-day cyber-attacks detection. The main goal was to develop an intelligent IDS model that is capable of detecting zero-day cyber-attacks with a high detection accuracy while overcoming the limitations of currently available IDS. This manuscript purposes and evaluates an autoencoder model to detect zero-day attacks. The idea is inspired by the encoding-decoding capability of autoencoders.

The results show high detection accuracy for the autoencoder model for both the CICIDS2017 and the NSL-KDD. The CICIDS2017 zero-day detection accuracy reaches 90.01%, 98.43%, 98.47%, and 99.67% for DoS (GoldenEye), DoS (Hulk), Port scanning, and DDoS attacks. Moreover, the NSL-KDD detection accuracy reached 92.96%, which outperforms the only available zero-day autoencoder-based detection manuscript [63].

Furthermore, the autoencoder model is compared to an unsupervised outlier-based ML technique; One-Class SVM. One-Class SVM is a prominent unsupervised ML technique that detects outliers. The one-class SVM mode presents its effectiveness in detecting zero-day attacks for NSL-KDD datasets and the distinctive ones from the CICIDS2017 dataset. When compared to One-Class SVM, autoencoder demonstrates its surpassing detection accuracy. Furthermore, both of the models demonstrate low miss rate (false-positives). Future work involves evaluating the proposed models with datasets that cover special purpose network IDS (e.g., IoT and Critical Infrastructure networks), which will comprise insights into adapting the proposed models, as well as proposing and adapting other ML techniques to use for zero-day attack detection. The source code for building and evaluating the proposed models will be made available through an open-source GitHub repository.

Author Contributions: Conceptualization, H.H., R.A. and X.B.; Formal analysis, H.H., R.A. and J.-N.C.; Investigation, H.H.; Methodology, H.H., R.A., C.T. and X.B.; Project administration, X.B.; Software, H.H.; Supervision, E.B. and X.B.; Validation, R.A., C.T., E.B. and X.B.; Writing—original draft, H.H.; Writing—review & editing, R.A., J.-N.C., E.B. and X.B. All authors have read and agreed to the published version of the manuscript.

Funding: This research received no external funding.

Conflicts of Interest: The authors declare no conflict of interest.

References

1. Kaloudi, N.; Li, J. The AI-Based Cyber Threat Landscape: A Survey. *ACM Comput. Surv.* **2020**, *53*. [CrossRef]
2. Hindy, H.; Hodo, E.; Bayne, E.; Seeam, A.; Atkinson, R.; Bellekens, X. A Taxonomy of Malicious Traffic for Intrusion Detection Systems. In Proceedings of the 2018 International Conference On Cyber Situational Awareness, Data Analytics and Assessment (Cyber SA), Glasgow, UK, 11–12 June 2018; pp. 1–4.
3. Khraisat, A.; Gondal, I.; Vamplew, P.; Kamruzzaman, J. Survey of Intrusion Detection Systems: Techniques, Datasets and Challenges. *Cybersecurity* **2019**, *2*, 20. [CrossRef]
4. Hindy, H.; Brosset, D.; Bayne, E.; Seeam, A.; Tachtatzis, C.; Atkinson, R.; Bellekens, X. A Taxonomy and Survey of Intrusion Detection System Design Techniques, Network Threats and Datasets. *arXiv* **2018**, arXiv:1806.03517.
5. Chapman, C. Chapter 1—Introduction to Practical Security and Performance Testing. In *Network Performance and Security*; Chapman, C., Ed.; Syngress: Boston, MA, USA, 2016; pp. 1–14. [CrossRef]
6. Bilge, L.; Dumitraş, T. Before We Knew It: An Empirical Study of Zero-Day Attacks in the Real World. In Proceedings of the 2012 ACM Conference on Computer and Communications Security (CCS '12), Raleigh, NC, USA, 16–18 October 2012; pp. 833–844. [CrossRef]
7. Nguyen, T.T.; Reddi, V.J. Deep Reinforcement Learning for Cyber Security. *arXiv* **2019**, arXiv:1906.05799.
8. Metrick, K.; Najafi, P.; Semrau, J. *Zero-Day Exploitation Increasingly Demonstrates Access to Money, Rather than Skill—Intelligence for Vulnerability Management*; Part One; FireEye Inc.: Milpitas, CA, USA, 2020.
9. Cisco. Cisco 2017 Annual Cyber Security Report. 2017. Available online: https://www.grouppbs.com/wp-content/uploads/2017/02/Cisco_2017_ACR_PDF.pdf (accessed on 20 July 2020).
10. Ficke, E.; Schweitzer, K.M.; Bateman, R.M.; Xu, S. Analyzing Root Causes of Intrusion Detection False-Negatives: Methodology and Case Study. In Proceedings of the 2019 IEEE Military Communications Conference (MILCOM), Norfolk, VA, USA, 12–14 November 2019; pp. 1–6.
11. Sharma, V.; Kim, J.; Kwon, S.; You, I.; Lee, K.; Yim, K. A Framework for Mitigating Zero-Day Attacks in IoT. *arXiv* **2018**, arXiv:1804.05549.

12. Sun, X.; Dai, J.; Liu, P.; Singhal, A.; Yen, J. Using Bayesian Networks for Probabilistic Identification of Zero-Day Attack Paths. *IEEE Trans. Inf. Forensics Secur.* **2018**, *13*, 2506–2521. [CrossRef]
13. Zhou, Q.; Pezaros, D. Evaluation of Machine Learning Classifiers for Zero-Day Intrusion Detection–An Analysis on CIC-AWS-2018 dataset. *arXiv* **2019**, arXiv:1905.03685.
14. Zhao, J.; Shetty, S.; Pan, J.W.; Kamhoua, C.; Kwiat, K. Transfer Learning for Detecting Unknown Network Attacks. *EURASIP J. Inf. Secur.* **2019**, *2019*, 1. [CrossRef]
15. Sameera, N.; Shashi, M. Deep Transductive Transfer Learning Framework for Zero-Day Attack Detection. *ICT Express* **2020**. [CrossRef]
16. Abri, F.; Siami-Namini, S.; Khanghah, M.A.; Soltani, F.M.; Namin, A.S. The Performance of Machine and Deep Learning Classifiers in Detecting Zero-Day Vulnerabilities. *arXiv* **2019**, arXiv:1911.09586.
17. Kim, J.Y.; Bu, S.J.; Cho, S.B. Zero-day Malware Detection using Transferred Generative Adversarial Networks based on Deep Autoencoders. *Inf. Sci.* **2018**, *460–461*, 83–102. [CrossRef]
18. Rumelhart, D.E.; Hinton, G.E.; Williams, R.J. *Learning Internal Representations by Error Propagation*; Technical Report; California Univ San Diego La Jolla Inst for Cognitive Science: San Diego, CA, USA, 1985.
19. Stewart, M. Comprehensive Introduction to Autoencoders. 2019. Available online: https://towardsdatascience.com/generating-images-with-autoencoders-77fd3a8dd368 (accessed on 21 July 2020).
20. Goodfellow, I.; Bengio, Y.; Courville, A. *Deep Learning*; MIT Press: Cambridge, MA, USA, 2016.
21. Barber, D. *Implicit Representation Networks*; Technical Report; Department of Computer Science, University College London: London, UK, 2014.
22. Hinton, G.E.; Salakhutdinov, R.R. Reducing the Dimensionality of Data with Neural Networks. *Science* **2006**, *313*, 504–507. [CrossRef] [PubMed]
23. Zabalza, J.; Ren, J.; Zheng, J.; Zhao, H.; Qing, C.; Yang, Z.; Du, P.; Marshall, S. Novel Segmented Stacked Autoencoder for Effective Dimensionality Reduction and Feature Extraction in Hyperspectral Imaging. *Neurocomputing* **2016**, *185*, 1–10. [CrossRef]
24. Liou, C.Y.; Cheng, W.C.; Liou, J.W.; Liou, D.R. Autoencoder for Words. *Neurocomputing* **2014**, *139*, 84–96. [CrossRef]
25. Theis, L.; Shi, W.; Cunningham, A.; Huszár, F. Lossy Image Compression with Compressive Autoencoders. *arXiv* **2017**, arXiv:1703.00395.
26. Zhou, C.; Paffenroth, R.C. Anomaly Detection with Robust Deep Autoencoders. In Proceedings of the 23rd ACM SIGKDD International Conference on Knowledge Discovery and Data Mining, Halifax, NS, Canada, 13–17 August 2017; pp. 665–674.
27. Creswell, A.; Bharath, A.A. Denoising Adversarial Autoencoders. *IEEE Trans. Neural Netw. Learn. Syst.* **2018**, *30*, 968–984. [CrossRef] [PubMed]
28. Cortes, C.; Vapnik, V. Support-Vector Networks. *Mach. Learn.* **1995**, *20*, 273–297. [CrossRef]
29. Ng, A. Part V: Support Vector Machines | CS229 Lecture Notes. 2000. Available online: http://cs229.stanford.edu/notes/cs229-notes3.pdf (accessed on 21 July 2020).
30. Zisserman, A. The SVM classifier | C19 Machine Learning. 2015. Available online: https://www.robots.ox.ac.uk/~az/lectures/ml/lect2.pdf (accessed on 21 July 2020).
31. Schölkopf, B.; Williamson, R.C.; Smola, A.J.; Shawe-Taylor, J.; Platt, J.C. Support Vector Method for Novelty Detection. *Adv. Neural Inf. Process. Syst.* **2000**, *12*, 582–588.
32. Tax, D.M.; Duin, R.P. Support Vector Data Description. *Mach. Learn.* **2004**, *54*, 45–66. [CrossRef]
33. Wang, S.; Liu, Q.; Zhu, E.; Porikli, F.; Yin, J. Hyperparameter Selection of One-Class Support Vector Machine by Self-Adaptive Data Shifting. *Pattern Recognit.* **2018**, *74*, 198–211. [CrossRef]
34. Hodo, E.; Bellekens, X.; Hamilton, A.; Tachtatzis, C.; Atkinson, R. Shallow and Deep Networks Intrusion Detection System: A Taxonomy and Survey. *arXiv* **2017**, arXiv:1701.02145.
35. Hamed, T.; Ernst, J.B.; Kremer, S.C. A Survey and Taxonomy of Classifiers of Intrusion Detection Systems. In *Computer and Network Security Essentials*; Daimi, K., Ed.; Springer International Publishing: Cham, Switzerland, 2018; pp. 21–39. [CrossRef]
36. Hindy, H.; Brosset, D.; Bayne, E.; Seeam, A.K.; Tachtatzis, C.; Atkinson, R.; Bellekens, X. A Taxonomy of Network Threats and the Effect of Current Datasets on Intrusion Detection Systems. *IEEE Access* **2020**, *8*, 104650–104675. [CrossRef]
37. Buczak, A.L.; Guven, E. A Survey of Data Mining and Machine Learning Methods for Cyber Security Intrusion Detection. *IEEE Commun. Surv. Tutor.* **2016**, *18*, 1153–1176. [CrossRef]

38. Aceto, G.; Ciuonzo, D.; Montieri, A.; Pescapé, A. Toward effective mobile encrypted traffic classification through deep learning. *Neurocomputing* **2020**, *409*, 306–315. [CrossRef]
39. Rashid, F.Y. Encryption, Privacy in the Internet Trends Report | Decipher. 2019. Available online: https://duo.com/decipher/encryption-privacy-in-the-internet-trends-report (accessed on 14 September 2020).
40. Kunang, Y.N.; Nurmaini, S.; Stiawan, D.; Zarkasi, A.; Jasmir, F. Automatic Features Extraction Using Autoencoder in Intrusion Detection System. In Proceedings of the 2018 International Conference on Electrical Engineering and Computer Science (ICECOS), Pangkal Pinang, Indonesia, 2–4 October 2018; pp. 219–224.
41. Kherlenchimeg, Z.; Nakaya, N. Network Intrusion Classifier Using Autoencoder with Recurrent Neural Network. In Proceedings of the Fourth International Conference on Electronics and Software Science (ICESS2018), Takamatsu, Japan, 5–7 November 2018; pp. 94–100.
42. Shaikh, R.A.; Shashikala, S. An Autoencoder and LSTM based Intrusion Detection Approach Against Denial of Service Attacks. In Proceedings of the 2019 1st International Conference on Advances in Information Technology (ICAIT), Chickmagalur, India, 25–27 July 2019; pp. 406–410.
43. Abolhasanzadeh, B. Nonlinear Dimensionality Reduction for Intrusion Detection using Auto-Encoder Bottleneck Features. In Proceedings of the 2015 7th Conference on Information and Knowledge Technology (IKT), Urmia, Iran, 26–28 May 2015; pp. 1–5.
44. Javaid, A.; Niyaz, Q.; Sun, W.; Alam, M. A Deep Learning Approach for Network Intrusion Detection System. In Proceedings of the 9th EAI International Conference on Bio-Inspired Information and Communications Technologies (Formerly BIONETICS), New York, NY, USA, 3–5 December 2016; pp. 21–26.
45. AL-Hawawreh, M.; Moustafa, N.; Sitnikova, E. Identification of Malicious Activities In Industrial Internet of Things Based On Deep Learning Models. *J. Inf. Secur. Appl.* **2018**, *41*, 1–11. [CrossRef]
46. Shone, N.; Ngoc, T.N.; Phai, V.D.; Shi, Q. A Deep Learning Approach To Network Intrusion Detection. *IEEE Trans. Emerg. Top. Comput. Intell.* **2018**, *2*, 41–50. [CrossRef]
47. Farahnakian, F.; Heikkonen, J. A Deep Auto-encoder Based Approach for Intrusion Detection System. In Proceedings of the 2018 20th International Conference on Advanced Communication Technology (ICACT), Chuncheon, Korea, 11–14 February 2018; p. 1.
48. Meidan, Y.; Bohadana, M.; Mathov, Y.; Mirsky, Y.; Shabtai, A.; Breitenbacher, D.; Elovici, Y. N-BaIoT—Network-Based Detection of IoT Botnet Attacks Using Deep Autoencoders. *IEEE Pervasive Comput.* **2018**, *17*, 12–22. [CrossRef]
49. Bovenzi, G.; Aceto, G.; Ciuonzo, D.; Persico, V.; Pescapé, A. A Hierarchical Hybrid Intrusion Detection Approach in IoT Scenarios. Available online: https://www.researchgate.net/profile/Domenico_Ciuonzo/publication/344076571_A_Hierarchical_Hybrid_Intrusion_Detection_Approach_in_IoT_Scenarios/links/5f512e5092851c250b8e934c/A-Hierarchical-Hybrid-Intrusion-Detection-Approach-in-IoT-Scenarios.pdf (accessed on 14 September 2020).
50. Canadian Institute for Cybersecurity. Intrusion Detection Evaluation Dataset (CICIDS2017). 2017. Available online: http://www.unb.ca/cic/datasets/ids-2017.html (accessed on 15 June 2018).
51. Panigrahi, R.; Borah, S. A Detailed Analysis of CICIDS2017 Dataset for Designing Intrusion Detection Systems. *Int. J. Eng. Technol.* **2018**, *7*, 479–482.
52. Sharafaldin, I.; Habibi Lashkari, A.; Ghorbani, A.A. A Detailed Analysis of the CICIDS2017 Data Set. In *Information Systems Security and Privacy*; Mori, P., Furnell, S., Camp, O., Eds.; Springer International Publishing: Cham, Switzerland, 2019; pp. 172–188.
53. Canadian Institute for Cybersecurity. NSL-KDD Dataset. Available online: http://www.unb.ca/cic/datasets/nsl.html (accessed on 15 June 2018).
54. Tavallaee, M.; Bagheri, E.; Lu, W.; Ghorbani, A.A. A Detailed Analysis of the KDD CUP 99 Data Set. In Proceedings of the 2009 IEEE Symposium on Computational Intelligence for Security and Defense Applications, Ottawa, ON, Canada, 8–10 July 2009; pp. 1–6.
55. Tobi, A.M.A.; Duncan, I. KDD 1999 Generation Faults: A Review And Analysis. *J. Cyber Secur. Technol.* **2018**, 1–37. [CrossRef]
56. Siddique, K.; Akhtar, Z.; Khan, F.A.; Kim, Y. Kdd Cup 99 Data Sets: A Perspective on the Role of Data Sets in Network Intrusion Detection Research. *Computer* **2019**, *52*, 41–51.
57. Bala, R.; Nagpal, R. A Review on KDD CUP99 and NSL-KDD Dataset. *Int. J. Adv. Res. Comput. Sci.* **2019**, *10*, 64. [CrossRef]

58. Rezaei, S.; Liu, X. Deep Learning for Encrypted Traffic Classification: An Overview. *IEEE Commun. Mag.* **2019**, *57*, 76–81. [CrossRef]
59. Guggisberg, S. How to Split a Dataframe into Train and Test Set with Python. 2020. Available online: https://towardsdatascience.com/how-to-split-a-dataframe-into-train-and-test-set-with-python-eaa1630ca7b3 (accessed on 17 August 2020).
60. Bergstra, J.; Bengio, Y. Random Search for Hyper-parameter Optimization. *J. Mach. Learn. Res.* **2012**, *13*, 281–305.
61. Liashchynskyi, P.; Liashchynskyi, P. Grid Search, Random Search, Genetic Algorithm: A Big Comparison for NAS. *arXiv* **2019**, arXiv:1912.06059.
62. Chen, P.H.; Lin, C.J.; Schölkopf, B. A Tutorial on ν-Support Vector Machines. *Appl. Stoch. Model. Bus. Ind.* **2005**, *21*, 111–136. [CrossRef]
63. Gharib, M.; Mohammadi, B.; Dastgerdi, S.H.; Sabokrou, M. AutoIDS: Auto-encoder Based Method for Intrusion Detection System. *arXiv* **2019**, arXiv:1911.03306.
64. Aygun, R.C.; Yavuz, A.G. Network Anomaly Detection with Stochastically Improved Autoencoder Based Models. In Proceedings of the 2017 IEEE 4th International Conference on Cyber Security and Cloud Computing (CSCloud), New York, NY, USA, 26–28 June 2017; pp. 193–198.

Publisher's Note: MDPI stays neutral with regard to jurisdictional claims in published maps and institutional affiliations.

© 2020 by the authors. Licensee MDPI, Basel, Switzerland. This article is an open access article distributed under the terms and conditions of the Creative Commons Attribution (CC BY) license (http://creativecommons.org/licenses/by/4.0/).

Article

Intelligent On-Off Web Defacement Attacks and Random Monitoring-Based Detection Algorithms

Youngho Cho

Department of Computer Engineering, Graduate School of National Defense Management, Korea National Defense University, Nonsan 33021, Korea; youngho@kndu.ac.kr

Received: 26 October 2019; Accepted: 11 November 2019; Published: 13 November 2019

Abstract: Recent cyberattacks armed with various ICT (information and communication technology) techniques are becoming advanced, sophisticated and intelligent. In security research field and practice, it is a common and reasonable assumption that attackers are intelligent enough to discover security vulnerabilities of security defense mechanisms and thus avoid the defense systems' detection and prevention activities. Web defacement attacks refer to a series of attacks that illegally modify web pages for malicious purposes, and are one of the serious ongoing cyber threats that occur globally. Detection methods against such attacks can be classified into either server-based approaches or client-based approaches, and there are pros and cons for each approach. From our extensive survey on existing client-based defense methods, we found a critical security vulnerability which can be exploited by intelligent attackers. In this paper, we report the security vulnerability in existing client-based detection methods with a fixed monitoring cycle and present novel intelligent on-off web defacement attacks exploiting such vulnerability. Next, we propose to use a random monitoring strategy as a promising countermeasure against such attacks, and design two random monitoring defense algorithms: (1) Uniform Random Monitoring Algorithm (URMA), and (2) Attack Damage-Based Random Monitoring Algorithm (ADRMA). In addition, we present extensive experiment results to validate our idea and show the detection performance of our random monitoring algorithms. According to our experiment results, our random monitoring detection algorithms can quickly detect various intelligent web defacement on-off attacks (AM1, AM2, and AM3), and thus do not allow huge attack damage in terms of the number of defaced slots when compared with an existing fixed periodic monitoring algorithm (FPMA).

Keywords: web defacement attack; on-off strategy; random monitoring algorithm; web security

1. Introduction

Web defacement attacks refer to a series of attacks that illegally modifies web pages in unauthorized manners for malicious purposes. According to recent statistics provided by ZONE-H [1], 500,000 websites over the world were defaced only in 2018, and around 100,000 defaced-websites were reported during the first quarter of 2019. Detail reports on major web defacement incidents can be found in [2].

Typical types of web defacement attacks vary from changing main images of websites to launching drive-by-download attacks that stealthily inject a malicious link into a web page through which malwares are automatically downloaded to web users' devices which accessed the defaced web pages [3,4]. Recently, the latter type is more often reported because the attacker can construct a large-scale botnet that consists of compromised personal computers, laptops, smartphones, appliances, and Internet of Things (IoT) devices. With the botnet, attackers can easily achieve their intended goals, such as launching distributed denial-of-service (DDoS) attacks to certain websites.

In general, web defacement attacks are performed as follows (see Figure 1). First, the attacker (A) maliciously modifies one or more web pages (or source codes) stored in the web server (WS) by exploiting security vulnerabilities of the WS. For example, A injects a malicious link (downloader) to malwares stored in malicious server (M) which cooperates with A. Such malicious link is injected in a way that system administrators or normal users cannot easily identify its existence within the defaced web page. Next, when a web user (U) accesses the WS, U is automatically connected to the external malicious server M through the injected malicious link and then malwares are downloaded to U from M; these processes proceed such that U does not know that they are happening, and the number of U (victims) can be hundreds, thousands, or even more depending on the popularity of web services provided by the WS. Once the U is infected with downloaded malwares, U becomes a bot which is under the control of A (or a bot master). After that, A starts launching its actual intended secondary attacks such as extracting critical information from U or DDoS attacks by using the botnet that consists of many Us (bots) [4].

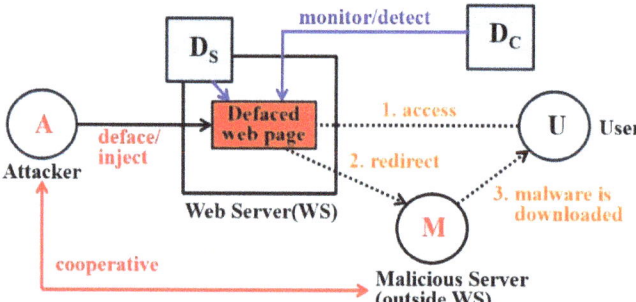

Figure 1. An illustration of web defacement attacks and existing detection systems; WS: web server; A: attacker; Ds: server-based detection system; Dc: client-based detection system; M: malicious server.

As shown in Figure 1, existing detection approaches against web defacement attacks can be classified into either server-based detection approach or client-based detection approach [5–21].

In the server-based detection approach [5,6], the detection system (D_S) is installed in the WS, and it regularly monitors web pages in the WS and checks if they are modified in unauthorized ways. Once the D_S detects modified web pages by attackers, the D_S raises an alarm and reports it to a server administrator or CERT (Computer Emergency Response Team) for further investigation and timely response. To check unauthorized modification of web pages in the WS, various file integrity monitoring methods [22–25] can be used. However, when the attacker successfully defaced web pages of the WS, the WS cannot be trusted because the attacker may have some or full control over the WS in that it is common for attackers (hackers) to try to obtain the root privilege of the WS, and then install backdoors in the WS after hacking applications of the WS. For example, the attacker can obtain operating system root privilege by launching various privilege escalation attacks [26,27]. Once the attacker gains root privilege of the WS, the attacker can disable security software such as the file integrity checker or local monitoring tool. Consequently, there is no guarantee that server-based detection methods work properly when web defacement attacks are successfully launched.

On the other hand, in the client-based detection approach [7–21], the client-based detection system (D_C) is located outside the server WS and monitors web pages in the WS remotely. D_C behaves as a common web user U; D_C periodically accesses the WS and collects web pages from the WS. After downloading web pages from the WS, D_C checks if they are defaced by using various detection techniques. Since D_C is located outside the WS, its detection process can be more trustful compared with the server-based approach. In addition, the client-based approach has some advantage over the server-based approach such that it can detect web defacement attacks performed in man-in-the-middle position between the web server WS and the client U.

Meanwhile, most existing researches on client-based detection mainly focused on either proposing new detection methods or improving attack detection accuracy [7–21]. Interestingly, to the best of our knowledge based on our extensive survey, there are no studies that explain how frequently their monitoring and detection processes should be performed. Most existing client-based detection methods simply monitor web pages with a fixed periodic monitoring cycle (interval) or they do not even mention about the monitoring cycle. However, such fixed monitoring cycles can be seriously vulnerable to intelligent attackers because detection systems with a fixed monitoring cycle can be completely avoided by intelligent attackers. In this paper, we first justify why fixed periodic monitoring should not be used by introducing *intelligent on-off web defacement attacks* that can completely avoid client-based web defacement detection systems with a fixed monitoring cycle. Since we introduced intelligent on-off web defacement attacks newly in this paper, we cannot provide real industry incident cases and reports. The primary goal of this study is to let security researchers and engineers understand this potential cyber threat to the Internet, and thus let them motivate more research and develop effective security mechanisms to defend against such attacks in advance.

Our contributions in this paper are as the following:

- We introduce a new intelligent on-off web defacement attack model that can completely avoid existing client-based detection methods using fixed periodic monitoring.
- We propose to use a random monitoring defense strategy against intelligent on-off web defacement attacks as a promising countermeasure, and conduct a simple probabilistic analysis that shows how random monitoring defense strategy can be effective in detecting such attacks.
- We devise two random monitoring algorithms, the Uniform Random Monitoring Algorithm (URMA) and the Attack Damage-based Random Monitoring Algorithm (ADRMA), against intelligent on-off web defacement attacks and provide extensive experiment results that show their detection performance by comparing with a fixed periodic monitoring algorithm (FPMA).

The rest of this paper is organized as follows. In Section 2, we review existing client-based detection methods against web defacement attacks. In Section 3, we introduce a new intelligent on-off web defacement attack. In Section 4, we justify that the random monitoring strategy can effectively defend against the intelligent on-off attack strategy, and propose two random monitoring detection algorithms (URMA and ADRMA). In Section 5, we conduct extensive experiments to show the performance of our proposed algorithms by comparing an existing fixed periodic monitoring algorithm (FPMA). Finally, we conclude with future research directions in Section 6.

2. Related Works

In this section, we briefly introduce previous studies on client-based monitoring and detection methods against web defacement attacks. With the recent advancements and popularity of machine learning (ML) techniques, many studies using various ML techniques have been conducted in this area as follows.

Borgolte et al. [10] proposed Meerkat which is a web defacement detection system using various techniques used in the computer vision field. In the training stage, Meerkat extracts high-level features from screenshots of monitored web pages by using image processing techniques and ML techniques together, and then generates a set of features of monitored web pages; Meerkat works based on a deep neural network in this stage. In the monitoring stage, Meerkat uses generated features of monitored web pages to examine whether current monitored web pages are defaced.

Medvet et al. [12] used a genetic programming technique to learn monitored web pages without any prior knowledge (learning phase), and to monitor the corresponding web pages at pre-determined intervals (monitoring phase). In addition, Bartoli et al. [16,17] proposed Goldrake which is a framework that uses sensors and alarms to automatically check remote web resources' integrity.

Kim et al. [9] proposed an n-gram based detection method that uses N-Gram-based Index Distance (NGID) to validate dynamic web pages. In [19], they proposed a defense mechanism for detecting web pages in a remote site and two threshold adjustment methods to lower false alarm rate.

Hoang and Nguyen [18] proposed a hybrid defacement detection model that is designed based on the combination of the ML-based detection and the signature-based detection.

Davanzo et al. [4] assessed the performance of several anomaly detection approaches designed based on ML techniques in terms of false positive ratio and false negative ratio. They conducted extensive experiments by using around 300 dynamic web pages for three months.

In addition to ML-based detection methods, various client-based detection methods have been proposed as follows.

Kim et al. [7] proposed a website falsification detection method in which web crawlers regularly collect web pages from a website, extract codes and images from the collected web pages, and determine whether web pages are defaced by analyzing the extracted codes and images in terms of similarity.

Masango et al. [13] proposed a WDIMT (Web Defacement and Intrusion Monitoring Tool) that operates like a web vulnerability scanner. When web defacement is detected, WDIMT can automatically recover the defaced web page by using its original web file stored before it is defaced. Similarly, Kals et al. [14] proposed Secubat, which is designed based on penetration testing techniques.

Park and Cho [11] proposed CREMS (Client-based Real-time wEb defacement Monitoring and detection System) that periodically examines web pages to see if they are defaced. Specifically, CREMS compares each web page's source codes every second and measures similarity after comparison. If the measured similarity value is below a certain threshold, CREMS raises an alarm and reports it to system administrators for further investigation. In addition, by using its source code matching algorithm, CREMS can locate the exact place where malicious codes are injected within a defaced web page.

According to our extensive survey, most previous works can be classified into either proposing new web defacement detection methods or improving detection accuracy of existing detection approaches. Interestingly, we observed that no studies clearly described how their monitoring cycles are set or should be set. For example, some detection methods [8,11,16,20,21] monitor web pages with a periodic or fixed monitoring cycle without clear explanations, and some works [7,9,10,12,15,18,19] did not even explain in detail about their monitoring method and cycle.

Meanwhile, recent advances of information and communication technology (ICT) techniques including AI (artificial intelligence) and ML (machine learning) techniques make cyberattacks more intelligent and sophisticated. Consequently, we should not ignore the possibility that attackers can avoid or nullify existing security systems by exploiting vulnerabilities that can be discovered by analyzing their operational patterns and behaviors [4].

For this reason, in this paper we introduce intelligent on-off web defacement attacks in order to show how existing client-based web defacement detection systems using fixed monitoring cycle can be vulnerable and then discuss our defense strategy and methods against such attacks.

3. Intelligent On-Off Web Defacement Attack

In this section, we explain why existing client-based detection approaches with fixed monitoring cycle can be easily, completely nullified by on-off attack strategy, and then we introduce a new intelligent web defacement attack model based on the on-off attack strategy.

3.1. The Security Weakness of Client-Based Detection Methods with a Fixed Monitoring Cycle

First, we describe a general description of client-based defense approaches with fixed monitoring cycle. Figure 2 shows an example of a web defacement detection system with a fixed monitoring cycle c = 10 seconds, which means that the detection system monitors and examines a web page every 10 seconds. We assume that the monitoring cycle c necessarily exists since no detection systems can monitor continuously due to their limited computing resources, the complexity of monitoring algorithms, etc. In this example, we assume that the unit is second for simplicity, but depending on detection systems, the unit of monitoring cycle can be second, milli-second, or even smaller. In addition, if we consider each monitoring cycle as a monitoring round (MR), then one MR consists of 10 time slots. As described in Figure 2, if the first monitoring activity is done at the first monitoring slot (ms_1), every $(10t + 1)$-th time slot will be examined by the detection system where $t = 1, 2, \ldots, \infty$. A simple fixed periodic monitoring algorithm (**FPMA**) is described in Algorithm 1.

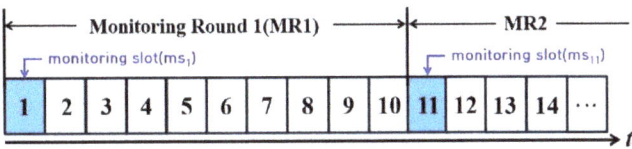

Figure 2. An example of a client-based web defacement detection system with a fixed monitoring cycle c (c = 10 seconds).

Algorithm 1: Fixed Periodic Monitoring Algorithm (FPMA).

Input:
  Number of slots: n
  Current time: $t_{current}$
  Start time of current monitoring round (MR): $t_{MRstart}$
  Fixed monitoring slot: ms_{fixed}
Output:
  Detection result: *detection_result*

1: begin
2:     while $(t_{MRstart} \leq t_{current} \leq (t_{MRstart} + n - 1))$:
3:         if $t_{current} == (t_{MRstart} + ms_{fixed} - 1)$:
4:             // monitor( ) checks if web pages are defaced
5:             *detection_result* ← monitor()
6:         else:
7:             // monitor( ) is not performed
8:             continue
9: end

Next, we describe how an intelligent web defacement attacker with an on-off attack strategy can avoid and nullify the above monitoring mechanism. We have the following assumptions (AS1-AS4):

- AS1: Attacker can discover some security vulnerabilities of its target web server such as **WS**
- AS2: Defender (client-based web defacement monitoring system located outside **WS**) monitors web pages stored in **WS** periodically (every 10 seconds)
- AS3: Attacker can modify web pages in **WS** by AS1
- AS4: Attacker can figure out monitoring cycle c and previous monitoring slots at the time t

Based on the above assumptions (AS1-AS4), the attacker can also figure out the next monitoring slots (blue-colored slots) at t. Figure 3 shows every possible monitoring slots (red-colored slots) at which the attacker can safely launch web defacement attacks to the victim server. Thus, except for the monitoring slots, the attacker can deface web pages at the red-colored time slots (non-monitoring slots). When we define *Attack Success Rate* $(ASR)(\%) = \frac{Num.\ of\ defaced\ time\ slots}{Num.\ of\ all\ time\ slots} \times 100$, ASR is 90% in this example; in other words, the attacker is able to deface its target web pages for 90% of time even without being detected by the monitoring system. Note that a defaced (time) slot means that the web deface attack is successfully launched at that time slot, and we use the term for the rest of this paper. Moreover, instead of defacing web pages for all time slots, the attacker can selectively choose some part of time slots for defacement in an on-off manner. In this case, ASR will decrease according to the amount of chosen attacking slots, but it will become more difficult to detect such attacks. In this paper, we name this type of attack as *intelligent on-off web defacement attack* and describe the attack model with a generalized algorithm (Algorithm 2) in Section 3.2.

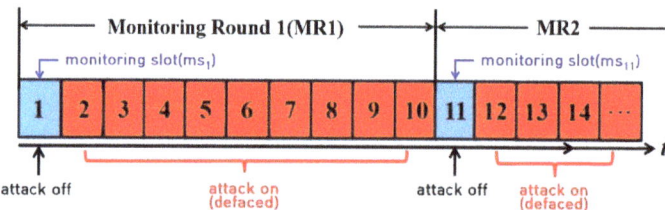

Figure 3. A description of an intelligent on-off web defacement attack; this intelligent attacker defaces web pages only for the red-colored time slots safely by avoiding the blue-colored monitoring slots.

3.2. Attack model: Intelligent on-off web defacement attacks

When the intelligent on-off web defacement attacker successfully defaced a certain web page *WP*, let $WP_{original}$ be the original web page of *WP* and $WP_{defaced}$ be the defaced web page of *WP*. To avoid being captured by a client-based web defacement detection with a fixed monitoring cycle, the intelligent on-off web defacement attacker acts as follows (see Algorithm 2).

- Attacker stores $WP_{original}$ before defacing it;
- To avoid a monitoring slot, the attacker calculates (or estimates) the next monitoring slot ms_{next} by a detection system based on current time $t_{current}$, monitoring cycle c, and previous monitoring slot $ms_{previous}$;
- When $t_{current}$ is not ms_{next}, attacker defaces *WP*;
- When $t_{current}$ is ms_{next}, attacker does not deface *WP*; if the web page is already defaced, attacker replaces $WP_{defaced}$ with $WP_{original}$ to avoid being captured by defender.

Algorithm 2: Intelligent On-Off Web Defacement Attack

Input:
  Current time: $t_{current}$
  Previous monitoring time (slot): $ms_{previous}$
  Monitoring cycle (fixed): c
  Original web page: $WP_{original}$
  Defaced web page: $WP_{defaced}$
Output:
  State of web page: WP_{state}

1: begin
2:    while (true):
3:      if ($t_{current} - ms_{previous}$) != c:
4:        $WP_{state} \leftarrow WP_{defaced}$ # attack mode is on
5:      else:
6:        $WP_{state} \leftarrow WP_{original}$ # attack mode is off
7: end

---

4. Random Monitoring-Based Defense Strategy and Two Detection Algorithms

In this section, we claim that a random monitoring strategy can effectively defend against intelligent on-off web defacement attacks by conducting a simple probabilistic analysis, and design two web defacement attack detection algorithms based on the random monitoring strategy.

4.1. Defense Strategy: Random Monitoring

In many computer and network security problems, it is often assumed that attackers are in superior positions than defenders. For example, defenders have limited resources but need to care for many defense spots (weak points) of their assets while attackers are able to successfully launch attacks if they can exploit at least one vulnerability of defenders' assets. For this reason, many computer and network security problems are formulated as unfair games between the attacker and the defender [28–30].

One of the effective defense strategies is to assign defenders' limited small defense resources to large defense spots in random ways, so that attackers cannot figure out which spots will be monitored [29,30]. For example, in [30], a defender uses a random patrol strategy to capture attackers in many defense spots because the defender cannot patrol all patrol spots at the same time, and a fixed periodic patrol method can be easily avoided by attackers. As another effective defense method, moving target defense (MTD) has been actively studied to defend against attackers targeting our assets, such as network devices and data, by moving the assets (or changing the locations of the assets) randomly and frequently and to thus make it very difficult for attackers to accurately target assets when they want [31–34]. In this paper, we will use the former random defense strategy to detect the intelligent on-off web defacement attacks because our research focus is to detect attackers rather than avoiding attackers; we note that studying the latter MTD in this research problem is out of the scope of this paper, but MTD techniques can be very effective for protecting our assets from attackers.

We now justify why the random monitoring strategy can effectively defend against the intelligent on-off web defacement attacks by using a simple probabilistic analysis. Variables and notations used in the analysis are as follows:

- n: the size of monitoring round (MR) or the size of the monitoring cycle; thus, n is the number of slots that consist of one MR. Each slot in MR can be identified by an index such as $s_1, s_2, \ldots, s_i, \ldots, s_n$. As we explained in Section 3.1, n can vary depending on the performance of defense systems. Given n, detection system can monitor only once at s_j where $j \in [1, n]$.
- S_{DS}: A finite set of all possible slots from which the defender chooses one slot during one MR; Thus, given n, $S_{DS} = \{s_1, s_2, \ldots, s_n\}$ and the cardinality of S_{DS} ($|S_{DS}|$) = n.
- S_{AS}: A finite set of all possible combinations of slots from which the attacker chooses one or more slots for launching defacement attacks during one MR. Thus, given n, $S_{AS} = \{s_1, s_2, \ldots, s_n, (s_1, s_2), (s_2, s_3), \ldots, (s_{n-1}, s_n), \ldots, (s_1, s_2, \ldots, s_n)\}$ and $|S_{AS}| = 2^n - 1$; S_{AS} = the power set of S_{DS} - null set $\emptyset$.
- Random variable X: slots that the attacker chooses
- Random variable Y: one slot that the defender chooses
- Let $P[X = s_i^+]$ be the probability that X contains s_i.

By the definition, two random variables X and Y are independent each other. Since the total number of elements of S_{AS} is $2^n - 1$, the probability p that the attacker will be detected during one MR can be obtained by:

$$p = \sum_{i \in S_{AS}} P[Y = i] P[X = i^+] = \frac{(2^{n-1} - 1)}{(2^n - 1)}. \quad (1)$$

By using Equation (1), the probability $p(r)$ that the attacker will be detected during r consecutive MRs can be obtained by:

$$p(r) = 1 - (1-p)^r = 1 - \left(1 - \frac{(2^{n-1} - 1)}{(2^n - 1)}\right)^r = 1 - \left(\frac{2^{n-1}}{2^n - 1}\right)^r \quad (2)$$

When $n = 10$, $p \simeq 0.4995$ according to Equation (1). According to Equation (2), $p(r)$ becomes higher than 0.87 when $r \geq 3$. As shown in Figure 4, as r grows, $p(r)$ grows quickly and eventually converges to 1. Meanwhile, even if the intelligent on-off web defacement attacker knows that the defender is monitoring only one slot during one MR, it is very unlikely for the attacker to avoid being detected for a long time (many MRs) when the attacker keeps defacing web pages in on-off manner. Consequently, the random monitoring strategy can effectively defend against intelligent on-off web defacement attacks.

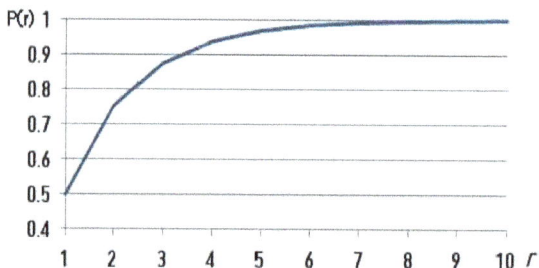

Figure 4. *P(r)*, the probability that the attacker will be captured by random monitoring method when the number of MR = *r* and the size of MR (*n*) = 10.

4.2. Design of Two Detection Abased on Random Monitoring Strategy

Based on the random monitoring strategy, we design two detection algorithms against intelligent on-off web defacement attacks: 1) Uniform Random Monitoring Algorithm (**URMA**) and 2) Attack Damage-based Random Monitoring Algorithm (**ADRMA**). In this study, our goal is not to design the best random monitoring algorithm in terms of detection performance, but to show you various ways of designing random monitoring algorithms. For the detection performance of our algorithms, we will explain in Section 5.

4.2.1. Uniform Random Monitoring Algorithm (URMA)

The uniform random monitoring algorithm (URMA) chooses one slot per one MR in a uniform manner and checks if web pages are attacked in the chosen slot.

As described in Algorithm 3 below, when each MR starts, URMA first selects one slot from *n* slots according to the uniform distribution; the probability that each slot is selected is $1/n$. Next, if current time *t* is equal to the chosen slot, monitoring operation is performed to see whether web pages are defaced. For remaining slots, monitoring operation is not performed by assumptions we mentioned in Section 3.

Algorithm 3: Uniform Random Monitoring Algorithm (URMA)
Input: Number of slots: *n* Current time: $t_{current}$ Start time of current MR: $t_{MRstart}$ **Output:** Detection result: *detection_result*
1: begin
2: if $t_{current} == t_{MRstart} - 1$:
3: *ms* ← choose one slot for detection slot
4: according to uniform (1, *n*)
5: while ($t_{MRstart} \leq t_{current} \leq (t_{MRstart} + n - 1)$):
6: if $t_{current} == (t_{MRstart} + ms - 1)$:
7: // monitor( ) checks if web pages are defaced
8: *detection_result* ← monitor( )
9: else:
10: // monitor( ) is not performed
11: continue
12: end

4.2.2. Attack Damage-Based Random Monitoring Algorithm (ADRMA)

When defenders' resources are limited, they need to wisely use their resources to defend against attackers. One of such ways is that defenders use their defense resources such that the amount of attack damages introduced by attackers can be minimized. Based on this rationale, unlike URMA using uniform randomness against attackers, we design a different random monitoring algorithm ADRMA that considers a factor of attack damage introduced by attacks.

For simplicity, let us consider a case where the size of MR is three (that is, $n = 3$). Then,

- $S_{DS} = \{1, 2, 3\}$
- S_{AS} = all subsets of S_{DS} – null set $\emptyset$ = {1, 2, 3, (1, 2), (1, 3), (2, 3), (1, 2, 3)}.

In this example, the total number of attack combinations that can be selected by the web defacement on-off attacker is 7 (= $2^3 - 1$) except $\emptyset$; we do not consider $\emptyset$ because it means that the attacker does not launch attacks. When the defender chooses defense slot d and the attacker chooses attack slots $S_{AS}(i)$ from S_{AS} where i is an index of attack combination, let $D_{Attack=S_{AS(i)}}(d)$ be the amount of attack damage introduced by the attacker. Then, given d and $S_{AS}(i)$, $D_{Attack=S_{AS(i)}}(d)$ is obtained by summing up the amount of attack damage introduced by each attack slot of $S_{AS}(i)$ as:

$$D_{Attack=S_{AS(i)}}(d) = \sum_{i \in S_{AS}} D_{Attack=i}(d) \qquad (3)$$

For example, assuming that the amount of attack damage for a single slot is 1, if the attacker launches web defacement attacks at slot 1 and slot 3 and the defender monitors slot 3, $D_{Attack=S_{AS}(5)}(3) = D_{Attack=1}(3) + D_{Attack=3}(3) = 1 + 0 = 1$ (see the blue-colored column in Table 1). That is, the amount of attack damage ($D_{Attack=1}(3)$) is 1 since the attack slot 1 is not monitored and thus attack at slot 1 is valid, and the amount of attack damage ($D_{Attack=3}(3)$) is 0 since the attack slot 3 is monitored and thus attack at slot 3 is invalid. On the other hand, when slot 1 is chosen for defense slot, $D_{Attack=S_{AS}(5)}(1) = 0$ because the attacker will be captured at slot 1 which is monitored by the defender (defense slot = 1).

Table 1 shows the attack damages calculated for each combination of d and $S_{AS}(i)$ by this manner. We can see that $D_{Attack}(1) = 4$, $D_{Attack}(2) = 6$ and $D_{Attack}(3) = 8$, and as d grows, $D_{Attack}(d)$ also grows. Meanwhile, a rational defender should not always choose the first slot for monitoring since the attacker may not choose the first slot always.

Table 1. Attack damages given a defense slot and attack slots (when the size of MR = 3).

d	$S_{AS}(i)$							$D_{Attack}(d)$	
	$S_{AS}(1)$	$S_{AS}(2)$	$S_{AS}(3)$	$S_{AS}(4)$	$S_{AS}(5)$	$S_{AS}(6)$	$S_{AS}(7)$	Σ	ratio
	1	2	3	(1,2)	(1,3)	(2,3)	(1,2,3)		
1	0	1	1	0	0	2	0	4	2
2	1	0	1	1	2	0	1	6	3
3	1	1	0	2	1	1	2	8	4

When n is given, the attacker can consider choosing one from at most $2^n - 1$ combinations of attack slots. Given n and d, $D_{Attack}(d)$ can be easily, efficiently calculated by the below Equation (4), which operates in $O(1)$ in terms of algorithmic time complexity. Derivation of Equation (4) is shown in Appendix A.

$$D_{Attack}(d) = \begin{cases} (n-1)2^{n-2} & \text{for } d = 1, \\ (n+d-2)2^{n-2} & \text{for } 2 \leq d < n, \\ (n-1)2^{n-1} & \text{for } d = n. \end{cases} \qquad (4)$$

● **Design of Attack Damage-based Random Monitoring Algorithm (ADRMA)**

Now we design a random monitoring algorithm by using $D_{Attack}(d)$. As shown in Table 1, $D_{Attack}(1) = 4$, $D_{Attack}(2) = 6$ and $D_{Attack}(3) = 8$. The ratio of $D_{Attack}(1)$, $D_{Attack}(2)$, $D_{Attack}(3)$ is 2

: 3 : 4. The higher $D_A(d)$, the larger damages the defender will be likely to get. **ADRMA** chooses a defense slot according to the inverse ratio of $D_{Attack}(d)$. In this approach, a slot with lower $D_{Attack}(d)$ will be more likely chosen as a defense slot than a slot with higher $D_{Attack}(d)$.

As shown in Algorithm 4, ADRMA works in two steps. In Step 1, given d and n, ADRMA calculates attack damage $D_{Attack}(d)$ for each slot according to the Equation (4). In Step 2, each time MR starts, ADRMA chooses one from n slots randomly according to the inverse ratio of $D_{Attack}(d)$. After that, ADRMA checks if web pages are defaced at the chosen defense slot and does not check for the remaining slots.

Algorithm 4: Attack Damage-Based Random Monitoring Algorithm (ADRMA)

Input:
 Number of slots: n
 Current time: $t_{current}$
 Start time of current MR: $t_{MRstart}$
Output:
 Attack damage D_A
 Defense slot ds
 Detection result: *detection_result*

1:   begin
2:     // Step 1: Calculate D_{Attack} to choose a defense slot
3:     for each d in $[1, n]$:
4:       if $d == 1$:
5:         $D_{Attack}(d) = (n-1)2^{n-2}$
6:       if $2 \leq d < n$:
7:         $D_{Attack}(d) = (n+d-2)2^{n-2}$
8:       if $d == n$:
9:         $D_{Attack}(d) = (n-1)2^{n-1}$
10:    // Step 2: Choose a defense slot and Monitor
11:   $ds \leftarrow$ choose one slot randomly by using $D_{Attack}(d)$
12:       such that a slot with lower $D_{Attack}(d)$ will be
13:       more likely chosen as a defense slot than a slot
14:       with higher $D_{Attack}(d)$.
15:   while ($t_{MRstart} \leq t_{current} \leq t_{MRstart} + n - 1$):
16:     if $t_{current} == (t_{MRstart} + ds - 1)$:
17:       // monitor( ) checks if web pages are defaced
18:       *detection_result* $\leftarrow$ monitor( )
19:     else:
20:       // monitor( ) is not performed
21:       continue
22:   end

---

5. Experiment Results

5.1. Experimental Objectives and Methods

5.1.1. Purpose of Experiments

The main purpose of conducting experiments here is to show how effectively our two random monitoring algorithms (URMA and ADRMA) work against various intelligent on-off web defacement attacks.

For this purpose, with Python 3 programming language, we implemented three intelligent on-off web defacement attack models (AM1, AM2, and AM3) based on Algorithm 2 as we described below in detail. In addition, we implemented our two random monitoring algorithms URMA and ADRMA according to Algorithm 3 and Algorithm 4, respectively. Moreover, to compare with our random monitoring algorithms, we implemented a simple fixed periodic monitoring algorithm (FPMA) according to Algorithm 1, and for simplicity FPMA always monitors the first slot of each monitoring round (MR).

5.1.2. Three Intelligent On-Off Attack Models

- **AM1** (most aggressive): In this attack model, we assume that the attacker knows how FPMA operates but does not have any knowledge about our random-monitoring algorithms. In AM3, the attacker is very aggressive such that it tries to deface all slots except the first slot of each MR monitored by FPMA.
- **AM2** (moderately aggressive): In this attack model, we assume that the attacker knows not only FPMA but also the existence of our random-monitoring algorithms. Unlike AM1, the attacker in AM2 does not attack all safe slots. Instead, the attack tries to deface one or more slots randomly until he/she is detected. Specifically, the attacker will decide whether it deface each slot according to attack rate R_A. For example, if attack rate $R_A = 80\%$, the attacker will launch defacement attack at each slot with the probability = 0.8. Thus, the higher R_A is, the more aggressively the attacker defaces. In our experiment, we used $R_A = 80\%$, 60% and 40%.
- **AM3** (least aggressive): Like AM2, we assume that the attacker knows not only FPMA but also the existence of our random-monitoring algorithms. Unlike AM2, the attacker in AM3 randomly chooses only one slot for each MR until he/she is detected by our random monitoring algorithms as the following. Assuming that the size of MR = n and slot 1 is the monitoring slot by FPMA, slot i will be more likely chosen by the attacker than slot j where $i \geq j$ and $2 \leq i, j \leq n$. This attack model is designed by considering that the attacker may think that slot 2 just after slot 1 is the most safe slot for launching defacement attacks because slot 2 is the most distant slot to the next monitoring slot (slot $1 + n$) while slot n is the most dangerous slot at which the attacker may be detected by FPMA.

5.1.3. Experimental Methods and Metrics

In our experiments, one experiment proceeds as follows. First, each monitoring round MR (whose size |MR| = n) starts, the attacker randomly chooses one attack combination that consists of one or more slots for launching the defacement attack according to three attack models (AM1, AM2, and AM3), and also the defender chooses one defense slot according to three monitoring algorithms (FPMA, URMA, and ADRMA). After that, each experiment checks whether the launched attack is monitored at the chosen defense slot; that is, we conclude the launched attack is monitored if any slot of the chosen attack combination by the attacker matches the chosen defense slot by the defender. Finally, each experiment terminates either when the attack is monitored by all three monitoring algorithms (FPMA, URMA and ADRMA) or when the number of monitoring rounds reaches 100 rounds; as we will explain later in Section 5.2, the latter condition is necessary since FPMA could not detect any of implemented intelligent attack models while our random monitoring algorithms URMA and ADRMA could detect all attack models successfully within a couple of monitoring rounds. We conducted all our experiments on our laptop (with Intel 7th Gen Core i5 and RAM 4GB) by running simulation programs which we implemented with Python 3.

For experiment result analysis, we use the following experiment metrics (N_{MR}, N_{ES}, N_{DS}, N_{AD}, and $AADR$) to compare three monitoring algorithms in the presence of three attack models:

(1) The number of elapsed monitoring rounds until the attacker is detected (N_{MR}) and the number of elapsed slots until the attacker is detected (N_{ES}): These two metrics explain how quickly a monitoring algorithm can detect the attacker in terms of attack detection speed; recall that one monitoring round consists of n slots.

(2) **The number of defaced slots until the attacker is detected (N_{DS}):** This metric explains how long the attacker can successfully launch the web defacement attack until he/she is detected by a monitoring algorithm. That is, N_{DS} indicates the amount of damage caused by the attacker and N_{DS} is measured by counting the total number of slots that the attacker has defaced successfully until the attacker is detected by a monitoring algorithm.

(3) **The number of successful attack detections for each monitoring round m ($N_{AD}(m)$) and accumulated attack detection rate by monitoring round m ($AADR(m)$):** These metrics measure how successfully and quickly a monitoring algorithm can detect the attack as monitoring round m increases. By using $N_{AD}(m)$, we can obtain $AADR(m)$ by

$$AADR(m) = \frac{\sum_{i=1}^{m} N_{AD}(i)}{Total\ num.\ of\ launched\ attacks} \quad (5)$$

Then, by definition, if a monitoring algorithm could successfully detect the attacker by monitoring round k, $\sum_{i=1}^{k} AADR(i) = 1$. We will use this metric to see how attack detection rate changes as the monitoring round m grows.

We conducted 10,000 experiments and measured the average value of the above metrics. For the size of monitoring round MR (|MR| or n), we used 5 and 10. We consider one slot as the base time unit in our experiments (e.g., one slot = one second).

5.2. Experiment Results and Analysis

Table 2 shows results obtained by conducting extensive experiments according to the experimental methods described in Section 5.1. We explain the results and our analysis on them as follows.

Table 2. Experiment results.

Size of Monitoring Round |MR|		|MR| = 5			|MR| = 10		
Attack Models	Metrics	FPMA	Proposed Algorithms		FPMA	Proposed Algorithms	
			URMA	ADRMA		URMA	ADRMA
AM1	Elapsed MR (N_{MR})	Not detected	1	1	Not detected	1	1
	Elapsed Slots (N_{ES})	Not detected	3.5	3.31	Not detected	6.05	5.48
	Defaced Slots (N_{DS})	400	1.5	1.31	900	4.05	3.48
AM2 ($R_A = 80$)	Elapsed MR (N_{MR})	Not detected	1.25	1.24	Not detected	1.24	1.24
	Elapsed Slots (N_{ES})	Not detected	4.73	4.5	Not detected	8.38	8.38
	Defaced Slots (N_{DS})	323.24	1.8	1.63	727.57	4.76	4.37
AM2 ($R_A = 60$)	Elapsed MR (N_{MR})	Not detected	1.64	1.63	Not detected	1.63	1.64
	Elapsed Slots (N_{ES})	Not detected	6.66	6.46	Not detected	12.23	12.01
	Defaced Slots (N_{DS})	244.75	2.08	1.98	545.57	5.43	5.3
AM2 ($R_A = 40$)	Elapsed MR (N_{MR})	Not detected	2.27	2.31	Not detected	2.47	2.46
	Elapsed Slots (N_{ES})	Not detected	9.84	9.85	Not detected	20.72	20.11
	Defaced Slots (N_{DS})	174.29	2.38	2.37	364.36	6.38	6.12
AM2 ($R_A = 20$)	Elapsed MR (N_{MR})	Not detected	3.31	3.34	Not detected	4.61	4.64
	Elapsed Slots (N_{ES})	Not detected	15.03	15	Not detected	42.07	41.86
	Defaced Slots (N_{DS})	121.34	2.81	2.83	195.16	7.2	7.19
AM3	Elapsed MR (N_{MR})	Not detected	3.91	3.69	Not detected	8.9	8.6
	Elapsed Slots (N_{ES})	Not detected	17.55	16.27	Not detected	84.34	80.91
	Defaced Slots (N_{DS})	100	2.91	2.69	100	7.9	7.6

First, FPMA could not detect all intelligent on-off web defacement attacks (AM1, AM2, and AM3) in our experiments as shown in Table 2. This is because all intelligent attack models in our experiments are designed according to Algorithm 1 such that the attacker knows exactly which slots FPMA will monitor and thus is able to avoid FPMA's monitoring. As shown in Table 2 and Figure 5, the average N_{DS} (defaced slots) varies according to attack models. As we explained in experimental methods, although FPMA could not detect attacks, we measured the average N_{DS} when 100 monitoring rounds elapsed because without this condition, experiments will not stop and N_{DS} will continue to grow endlessly.

The result shows that AM1 has the highest N_{DS} because it is the most aggressive attack model in our experiments while AM3 has the lowest N_{DS} because it is the least aggressive attack model. For AM2, as attack rate R_A decreases from 80% to 20%, N_{DS} also decreases almost linearly because R_A for each slot decreases. In addition, except AM3 where only one slot is attacked regardless of the size of MR, N_{DS} when |MR| = 10 is much larger than N_{DS} when |MR| = 5 because the number of successful defaced slots per one MR is 4 when |MR| = 5 and 9 when |MR| = 10, respectively. Consequently, we can see that as the size of MR grows, the attack damage will also grow.

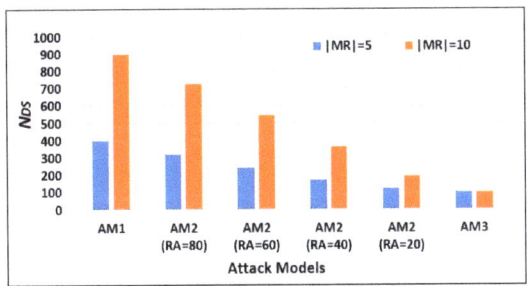

Figure 5. The number of defaced slots until the attacker is detected (N_{DS}) by FPMA in the presence of various attack models (AM1, AM2(R_A = 80, 60, 40, and 20%), and AM3); FPMA cannot detect all type of attacks.

Second, unlike FPMA, all our proposed algorithms (URMA and ADRMA) can successfully detect all type of attacks (AM1, AM2, and AM3) in our experiments. For attack detection speed, as shown in Figure 6, as the attack rate for each lot grows (AM3 → AM2(R_A = 20%) → AM2(R_A = 40%) → AM2(R_A = 60%) → AM2(R_A = 80%) → AM1), N_{MR} also decreases. This means the attack detection speed of both our monitoring algorithms also increases. This result is clear because as the number of attack slots grows according to the attack rate, the possibility that the attacker will be detected also increases. Meanwhile, regardless of |MR| (= 5 or 10), N_{MR} is the same when AM1 is used because the AM1-based attacker defaces all slots except slot 1 and all our algorithms capture the attacker at slot 2. On the other hand, as the attack rate for each slot decreases (AM2(R_A = 80%) → AM2(R_A = 60%) → AM2(R_A = 40%) → AM2(R_A = 20%) → AM3), the difference of N_{MR} when |MR| = 5 and |MR| = 10 becomes lager as described in Figure 6. When AM3 is used, the attacker launches web defacement attacks randomly at only one slot per one MR, the detection speed is relatively slow, but the attack damage is not very high; we will explain the reason below in detail. Figure 7 shows the number of elapsed slots until the attacker is detected (N_{ES}) when |MR| = 10, which is similar with the results of N_{MR}.

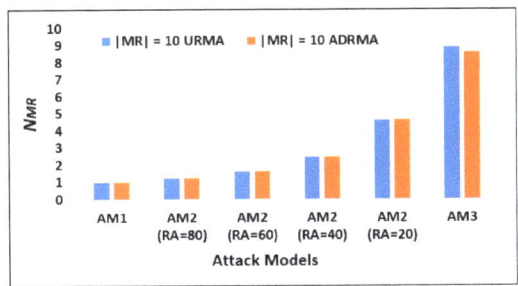

Figure 6. The number of elapsed monitoring rounds until the attacker is detected (N_{MR}) when |MR| = 10.

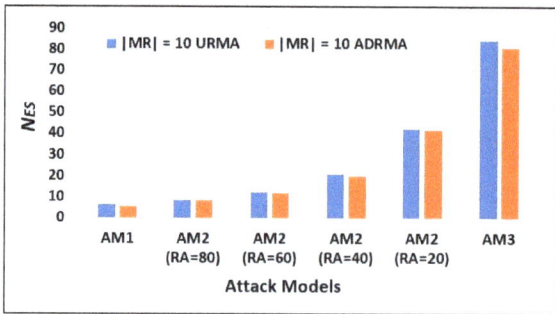

Figure 7. The number of elapsed slots until the attacker is detected (N_{ES}) when $|MR| = 10$.

Third, in addition to N_{MR} and N_{ES}, $N_{AD}(m)$ and $AADR(m)$ show the attack detection speed of monitoring algorithms according to monitoring round m, and as shown in Figures 8 and 9, our monitoring algorithms could detect most of attacks in early stage of monitoring rounds, especially in the presence of aggressive attacks models (AM1 or AM2(R_A = 80%)). In particular, Figure 9 shows how the accumulated attack detection rate $AADR(m)$ of our two monitoring algorithms changes when AM2 is used. We can see that as m grows, $AADR(m)$ converges to 1 quickly although the growth rate of $AADR(m)$ can vary according to attack models. Intuitively, as the attack rate increases, the growth rate of $AADR(m)$ also increases.

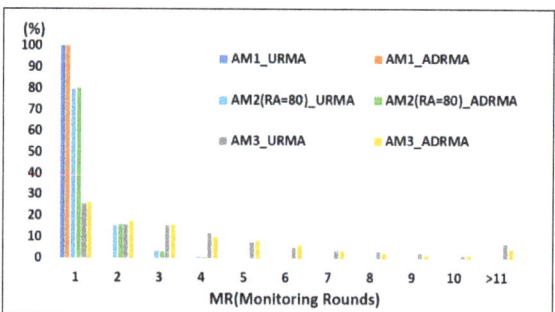

Figure 8. The number of successful attack detections for each monitoring round m ($N_{AD}[m]$) by URMA and ADRMA when $|MR| = 10$.

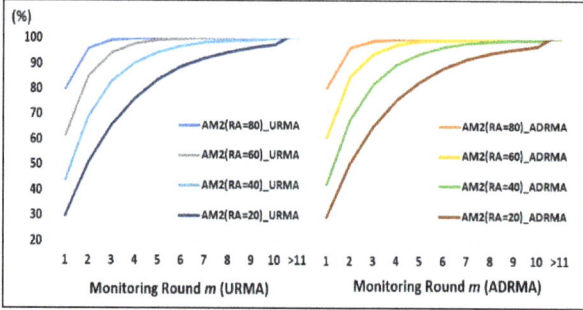

Figure 9. Accumulated attack detection rate by monitoring round m ($AADR(m)$) in the presence of AM2-based attacks.

Fourth, as shown in Table 2, N_{DS} (the number of defaced slots until the attacker is detected) shows that all attack models could not make huge damage (many defaced slots) in the presence of our random monitoring algorithms, especially compared with the case where FPMA cannot detect attacks at all and thus the attack damage (N_{DS}) continue to grow endlessly. As shown in the below Figure 10, regardless of |MR|, all attack models could not deface more than eight slots in our experiments. Among all attack models, AM3 could make the largest attack damage, but even AM3-based attacker could deface only 7.9 slots at most and then captured by our monitoring algorithm (URMA). This is because random monitoring defense strategy works effectively against intelligent on-off web defacement attack models by quickly capturing such attacks.

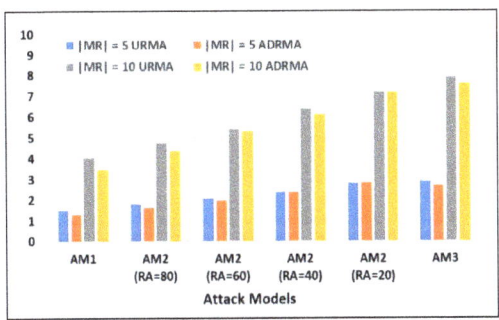

Figure 10. The number of defaced slots until the attacker is detected (N_{DS}) by URMA and ADRMA.

Fifth, ADRMA allows slightly smaller attack damage (the number of defaced slots) than URMA in most attack models used in our experiments (see Table 2 and Figure 10). This is because ADRMA was originally designed to randomly choose a defense slot such that the amount of attack damage can be reduced as we discussed in Section 4. For example, as you can see in Table 2, for |MR| = 10, when compared with URMA, ADRMA could reduce N_{DS} by 8.19% and 14.07% when AM2($R_A = 80$) and AM1 was used, respectively. This means that the detection performance of random monitoring algorithms can vary according to their design characteristics. Nevertheless, we can see that both our random monitoring algorithms could effectively defend against all attack models in our experiments by allowing very small deface slots.

Last, AM3-based attack, which is the least aggressive attack used in our experiments, could keep launching attacks successfully about 3.69~8.7 times longer than AM1-based attack before being detected by our random monitoring algorithms (see N_{MR} in Table 2 and Figure 6). This is because AM3 is designed to choose only one slot for each monitoring round and thus the possibility that it can be captured is much smaller than other aggressive attack models that choose one or more slots according to their design characteristics. Consequently, AM3-based attacks were able to make larger attack damage by at most 118% than AM1-based attacks in terms of the number of defaced slots N_{DS}. For attackers, the stealthy attack strategy like AM3 can be better to make larger attack damage to defenders than the aggressive attack strategy like AM1 or AM2($R_A = 80$). Nevertheless, they will be caught quickly if they continue to launch attacks even in the presence of random monitoring algorithms.

6. Conclusions and Future Works

In this paper, we first reported that existing client-based web defacement detection methods with a fixed monitoring cycle can be vulnerable to intelligent on-off web defacement attacks. Next, we proposed to use random monitoring defense strategy as a promising countermeasure against the intelligent on-off web defacement attacks by providing a probabilistic analysis on how such strategy can be effective in detecting on-off attacks. In addition, we devised two random monitoring algorithms based on the random monitoring strategy and provided extensive experiment results to validate our approach and to show the detection performance of our random monitoring algorithms. According to our experiment results, our proposed random monitoring algorithms can detect various intelligent web defacement on-off attacks very quickly while their detection performance slightly vary depending on their design characteristics.

Our future research directions are as follows.

First, we will develop a client-based web defacement detection system with our random monitoring algorithms after further advancing and optimizing their detection performance. To this end, we will deploy it in a real network environment and conduct real-time case studies to see how serious intelligent on-off web defacement attacks can be in the real network environment. Based on the analysis results and findings of case studies, we will further improve our random monitoring detection methods to make our system more feasible and efficient. In addition, we will study a hybrid web defacement defense mechanism that combines client-based methods and server-based methods to better defend against various web defacement attacks.

Next, we will further investigate potential security vulnerabilities of current web defacement detection systems that can be exploited for intelligent attackers to avoid them including our random monitoring algorithms. For example, intelligent attackers may try to find the security weaknesses of random function (or randomness extractor) used in random defense systems. If we inappropriately use a random function with a fixed seed value or use a weak random function with known security vulnerabilities carelessly, it can be possible for the attacker to figure out next random monitoring slots with high probability in advance and then simply avoid random monitoring detection systems. Therefore, we should not only use a strong random function but also protect the random function from intelligent attackers.

Last, we would like to extend our research by investigating more broad range of detection and surveillance systems used in various networks. We want to see if our random detection strategy and algorithms can help them better defend against adversaries who want to actively avoid such defense systems. Especially, we are interested in examining various detection and surveillance systems used in VANET (Vehicular Ad-hoc NETwork) and IoT (Internet of Things) environments [35–37].

Funding: This research received no external funding.

Acknowledgments: An earlier version of this paper was presented and selected as one of outstanding papers at the Conference on Information Security and Cryptography-Summer (CISC-S) in June 2017, South Korea. The author would like to thank reviewers for their valuable comments and constructive suggestions.

Conflicts of Interest: The authors declare no conflict of interest.

Appendix A

Derivation of $D_{ATTACK}(d)$ in Equation (4)

Let the size of MR be n and $AD(i)$ be the summation of attack damage that is made by every possible attack combination at a certain slot i (see Figure A1). Recall the definition of $D_{Attack}(d)$ in (3). If there is no defense slot, for all i, $AD(i) = 2^{n-1}$ because there are no attack damage for the half of slots that the attacker will not choose (when attack combination = 1, 2, 3, 4). For example, Figure A1a shows an example when $n = 3$ and there is no defense slot, and in this case $AD(1) = AD(2) = AD(3) = 4$ $\left(= 2^{3-1}\right)$.

(1) for $d = 1$

Figure A1b shows an example when $n = 3$ and $d = 1$. When the slot 1 is chosen as the defense slot, $AD(1) = 0$ because either the attacker is captured at the slot 1 (when attack combination = 5, 6, 7, 8) or there is no attack damage (when attack combination = 1, 2, 3, 4). In addition, for each of remaining slots, the amount of attack damage is reduced to 2^{n-2} because there are no attack damages for the half of attack combinations since the attacker will be captured at the slot 1 (when attack combination = 5, 6, 7, 8). Consequently, since the number of the remaining slots except slot 1 is $n - 1$, $D_{Attack}(d = 1) = \sum_{i=1}^{n-1} AD(i) = (n-1)2^{n-2}$.

attack combination	slot 1	slot 2	slot 3
1	0	0	0
2	0	0	1
3	0	1	0
4	0	1	1
5	1	0	0
6	1	0	1
7	1	1	0
8	1	1	1
attack damage	4	4	4

(a) d is not chosen

attack combination	slot 1	slot 2	slot 3
1	0	0	0
2	0	0	1
3	0	1	0
4	0	1	1
5	1	0	0
6	1	0	1
7	1	1	0
8	1	1	1
attack damage	0	2	2

(b) d = 1

attack combination	slot 1	slot 2	slot 3
1	0	0	0
2	0	0	1
3	0	1	0
4	0	1	1
5	1	0	0
6	1	0	1
7	1	1	0
8	1	1	1
attack damage	4	4	0

(c) d = n

attack combination	slot 1	slot 2	slot 3
1	0	0	0
2	0	0	1
3	0	1	0
4	0	1	1
5	1	0	0
6	1	0	1
7	1	1	0
8	1	1	1
attack damage	4	0	2

(d) 1 < d < n

Figure A1. An example that shows how D_{Attack} is calculated given n (=3) and d (=0, 1, 2, 3). For example, in (**b**), the yellow-colored slots are valid attack slots when d = 1, and thus D_{Attack} can be calculated as the number of yellow-colored slots by assuming the base unit of attack damage for defacing one slot is 1.

(2) for $d = n$

If the last slot n is chosen as a defense slot, the attack can make attack damage for all previous slots from slot 1 to slot $n - 1$ except the last slot n. Therefore, the total amount of attack damage $D_{Attack}(d = n)$ is $(n - 1)2^{n-1}$ which can be calculated by multiplying 2^{n-1} (the amount of damage from each slot) by $n - 1$ (the number of previous slots). Figure A1d shows an example when $n = 3$ and $d = 3$ (the last slot in this case).

(3) for $2 \leq d < n$

As shown in Figure A1c, if the defender chooses one slot j between slot 2 and slot $n - 1$ for defense, we need to consider attack damage before and after the slot j as follows. First, the amount of attack damage for each slot before slot j is 2^{n-1} and the number of slots before slot j is $j - 1$. Next, the amount of attack damage for each slot after slot j is 2^{n-2} which is the half of the former case because attack combination 3, 4, 7 and 8 are excluded additionally and the number of slots after the slot j is $n - j$. Therefore, $D_{Attack}(2 \leq d < n) = (d - 1)2^{n-1} + (n - d)2^{n-2} = (n + d - 2)2^{n-2}$.

References

1. Zone-H.org. Available online: http://www.zone-h.org/stats/ymd/ (accessed on 15 April 2019).
2. Banff Cyber Technologies. Defacement, B.I.o.W. Available online: https://www.banffcyber.com/knowledge-base/articles/business-implications-web-defacement/ (accessed on 20 January 2019).
3. Bartoli, A.; Davanzo, G.; Medvet, E. The Reaction Time to Web Site Defacements. *Internet Comput.* **2009**, *13*, 52–58. [CrossRef]
4. Davanzo, G.; Medvet, E.; Bartoli, A. Anomaly Detection Technique for a Web Defacement Monitoring Service. *Expert Syst. Appl.* **2011**, *38*, 12521–12530. [CrossRef]
5. Kim, G.H.; Spafford, E.H. Design and Implementation of Tripwire: A File System Integrity Checker. In Proceedings of the 2nd ACM Conference on Computer and Communications Security, Fairfax, VA, USA, 19 November 1993; pp. 18–29.
6. Ganger, A.P.; Pennington, A.G.; Strunk, J.D.; Griffin, J.L.; Soules, C.A.N.; Goodson, G.R.; Ganger, G.R. Storage-based Intrusion Detection: Watching Storage Activity for Suspicious Behavior. In Proceedings of the 12th USENIX Security Symposium, Washington, DC, USA, 4–8 Auguest 2003; pp. 1–15.
7. Kim, K.; Choi, S.-S.; Park, H.-S.; Ko, S.-J.; Song, J.-S. Website Falsification Detection System Based on Image and Code Analysis for Enhanced Security Monitoring and Response. *J. Korea Inst. Inf. Secur. Cryptol.* **2014**, *24*, 871–883. [CrossRef]
8. Medvet, E.; Bartoli, A. On the Effects of Learning Set Corruption in Anomaly-Based Detection of Web Defacements. In *Detection of Intrusions and Malware, and Vulnerability Assessment (DIMVA), Lucerne, Switzerland, 12 July 2007*; Springer: Berlin/Heidelberg, Germany, 2007; pp. 65–78.
9. Kim, W.; Joo, M.; Lee, E.; Lee, D.; Park, E.; Kim, S. N-gram-based dynamic web page defacement validation. In Proceedings of the Information Security Applications, 5th International Workshop, WISA 2004, Jeju Island, Korea, 23–25 August 2004.
10. Borgolte, K.; Kruegel, C.; Vigna, G. Meerkat: Detecting Website Defacements through Image-based Object Recognition. In Proceedings of the 24th USENIX Conference on Security Symposium, Washington, DC, USA, 12–14 August 2015; pp. 595–610.
11. Park, H.; Cho, Y. CREMS: Client-based Real-time wEb defacement Monitoring and detection System. In Proceedings of the Conference on Information Security and Cryptography-Summer (CSIC-S), Asan, Korea, 3–5 June 2017; pp. 657–658.
12. Medvet, E.; Fillon, C.; Bartoli, A. Detection of Web Defacements by means of Genetic Programming. In Proceedings of the IEEE International Symposium on Information Assurance and Security, Manchester, UK, 29–31 August 2007; pp. 227–234.
13. Masango, M.; Francois, M.; Palesa, A.; Bokang, M. Web Defacement and Intrusion Monitoring Tool: WDIMT. In Proceedings of the International Conference on Cyberworlds, Chester, UK, 20–22 September 2017; pp. 72–79.
14. Kals, S.; Kirda, E.; Kruegel, C.; Jovanovic, N. Secubat: A Web Vulnerability Scanner. In Proceedings of the International Conference on World Wide Web, Edinburgh, Scotland, 23–26 May 2006; pp. 247–256.
15. Kanti, T.; Richariya, V. Implementing a web browser with web defacement detection techniques. *World Comput. Sci. Inf. Technol. J.* **2011**, *1*, 307–310.
16. Bartoli, A.; Davanzo, G.; Medvet, E. A Framework for Large-Scale Detection of Web Site Defacements. *ACM Trans. Internet Technol.* **2010**, *10*, 10–37. [CrossRef]
17. Bartoli, A.; Medvet, E. Automatic Integrity Checks for Remote Web Resources. *IEEE Internet Comput.* **2006**, *10*, 56–62. [CrossRef]
18. Hoang, X.D.; Nguyen, N.T. Detecting Website Defacements Based on Machine Learning Techniques and Attack Signatures. *Computers* **2019**, *8*, 35. [CrossRef]
19. Kim, W.; Lee, J.; Park, E.; Kim, S. Advanced Mechanism for Reducing False Alarm Rate in Web Page Defacement Detection. In Proceedings of the International Workshop on Information Security Applications (WISA), Jeju Island, Korea, 28–30 August 2006.
20. Bergadano, F.; Carretto, F.; Cogno, F.; Ragno, D. Defacement Detection with Passive Adversaries. *Algorithms* **2019**, *12*, 150. [CrossRef]
21. WebOrion Defacement Monitor. Available online: https://www.banffcyber.com/weborion-defacement-monitor/ (accessed on 23 August 2019).

22. Julianto, S.M.; Munir, R. Intrusion detection against unauthorized file modification by integrity checking and recovery with HW/SW platforms using programmable system-on-chip (SoC). In Proceedings of the International Conference on Information and Communications Technology (ICOIACT), Yogyakarta, Indonesia, 6–8 March 2018; pp. 174–179.
23. Shi, B.; Li, B.; Cui, L.; Ouyang, L. Vanguard: A Cache-Level Sensitive File Integrity Monitoring System in Virtual Machine Environment. *IEEE Access* **2018**, *6*, 38567–38577. [CrossRef]
24. Smith, C.L. *AIDE-Advanced Intrusion Detection Environment*; Pacific Northwest Nat. Lab.: Richland, WA, USA, 2013.
25. Li, S.; Xiao, L.; Qin, G.; Ruan, L.; Su, S. COW-IMM A Novel Integrity Measurement Method Based on Copy-on-Write for File in Virtual Machine. *IEEE Access* **2018**, *6*, 51776–51790. [CrossRef]
26. Qiang, W.; Yang, J.; Jin, H.; Shi, X. PrivGuard: Protecting Sensitive Kernel Data From Privilege Escalation Attacks. *IEEE Access* **2018**, *6*, 46584–46594. [CrossRef]
27. O'Leary, M. Privilege Escalation in Linux. In *Cyber Operations*; Apress: Berkeley, CA, USA, 2019; pp. 419–453.
28. Moisan, F.; Gonzalez, C. Security under Uncertainty: Adaptive Attackers Are More Challenging to Human Defenders than Random Attackers. *Front. Psychol.* **2017**, *8*, 982. [CrossRef] [PubMed]
29. Nguyen, T.H.; Kar, D.; Brown, M.; Sinha, A.; Jiang, A.X.; Tambe, M. Towards a Science of Security Games. In *Mathematics & Statistics*; Springer: Berlin/Heidelberg, Germany, 2016; Volume 6.
30. També, M. *Security and Game Theory: Algorithms, Deployed Systems, Lessons Learned*; Cambridge University: Cambridge, UK, 2011.
31. Zhang, H.; Zheng, K.; Wang, X.; Lou, S.; Wu, B. Efficient Strategy Selection for Moving Target Defense Under Multiple Attacks. *IEEE Access* **2019**, *7*, 65982–65995. [CrossRef]
32. Connell, W.; Menasce, D.A.; Albanese, M. Performance Modeling of Moving Target Defenses with Reconfiguration Limits. *IEEE Trans. Dependable Secur. Comput.* **2018**, *99*, 1. [CrossRef]
33. Lei, C.; Ma, D.-H.; Zhang, H.-Q. Optimal Strategy Selection for Moving Target Defense Based on Markov Game. *IEEE Access* **2017**, *5*, 156–169. [CrossRef]
34. Sharma, D.P.; Cho, J.-H.; Moore, T.J.; Nelson, F.F.; Lim, H.; Kim, D.S. Random Host and Service Multiplexing for Moving Target Defense in Software-Defined Networks. In Proceedings of the IEEE International Conference on Communications (ICC), Shanghai, China, 26–28 May 2019.
35. Lim, K.; Tuladhar, K.M.; Kim, H. Detecting Location Spoofing using ADAS sensors in VANETs. In Proceedings of the IEEE Consumer Communications & Networking Conference (CCNC), Las Vegas, NV, USA, 11–14 January 2019.
36. Kim, H.; Ben-Othman, J. A Collision-free Surveillance System using Smart UAVs in Multi Domain IoT. *IEEE Commun. Lett.* **2018**, *22*, 2587–2590. [CrossRef]
37. Khraisat, A.; Gondal, I.; Vamplew, P.; Kamruzzaman, J.; Alazab, A. A Novel Ensemble of Hybrid Intrusion Detection System for Detecting Internet of Things Attacks. *Electronics* **2019**, *8*, 1210. [CrossRef]

© 2019 by the author. Licensee MDPI, Basel, Switzerland. This article is an open access article distributed under the terms and conditions of the Creative Commons Attribution (CC BY) license (http://creativecommons.org/licenses/by/4.0/).

Article

InSight2: A Modular Visual Analysis Platform for Network Situational Awareness in Large-Scale Networks

Hansaka Angel Dias Edirisinghe Kodituwakku [1,*], Alex Keller [2] and Jens Gregor [1]

[1] Department of Electrical Engineering and Computer Science, The University of Tennessee, 1520 Middle Dr, Knoxville, TN 37996, USA; jgregor@utk.edu
[2] School of Engineering, Stanford University, 450 Serra Mall, Stanford, CA 94305, USA; axkeller@stanford.edu
* Correspondence: angelk@utk.edu

Received: 16 September 2020; Accepted: 13 October 2020; Published: 21 October 2020

Abstract: The complexity and throughput of computer networks are rapidly increasing as a result of the proliferation of interconnected devices, data-driven applications, and remote working. Providing situational awareness for computer networks requires monitoring and analysis of network data to understand normal activity and identify abnormal activity. A scalable platform to process and visualize data in real time for large-scale networks enables security analysts and researchers to not only monitor and study network flow data but also experiment and develop novel analytics. In this paper, we introduce InSight2, an open-source platform for manipulating both streaming and archived network flow data in real time that aims to address the issues of existing solutions such as scalability, extendability, and flexibility. Case-studies are provided that demonstrate applications in monitoring network activity, identifying network attacks and compromised hosts and anomaly detection.

Keywords: visual analytics; cybersecurity awareness; incident response; anomaly detection

1. Introduction

One of the prominent issues security analysts and researchers face when analyzing network data, whether archived or real-time streaming flow data, is finding tools that can extract, enrich, index, filter, process, and visualize the large-scale network data. For exploratory visual analysis in threat hunting and forensic study, a tool that allows processing of network flows filtered by a complex pipeline is important to find threats and events for proper incident response and decision-making. Flow data enriched with Open Source Intelligence (OSINT) as well as proprietary information provide valuable information for the analysis. Intuitive visualizations can help the human analysts not only understand the typical behaviors but also detect anomalies and further investigate them.

Furthermore, when generating datasets to develop novel analytics researchers also have to implement the frameworks to manipulate flow data and generate visualizations since general purpose data analysis tools are not designed to connect to network sockets to read different formats of streaming flow data and process flows at gigabit speeds. Existing works focus on providing visualizations and analytics for specific networks and applications and may not be ideal for real-time visual flow data analysis for large-scale networks. By learning past behavioral patterns, future states of the network can be predicted such as bandwidth utilization patterns to detect anomalies. These functional requirements for situational awareness are critical for identifying incidents and threats, investigating anomalies and making decisions [1–4]. A flexible platform that can provide a framework for researchers to read and manipulate flow data and augment them with contextual information such as geolocation and known threat labels can improve and streamline analytics development.

1.1. Motivation

InSight2 is conceived as a platform specifically designed for flow data analysis that creates a synergy between researchers who develop analytics and analysts that make use of them. This led to the development of InSight2 with the goals to monitor networks in real time for cybersecurity awareness, visually analyze archived flow data for incident response, augment and extract a subset of flow data for the development of novel analytics such as anomaly detection and deploy said analytics back in the system for continued growth of its capabilities. These goals require a common efficient flow processing back-end and a software architecture that overcome issues with existing solutions.

Network flow data are used for the analysis in this work as most networks collect and store some form of flow data. A network flow summarizes the attributes of a communication between two nodes in a computer network, such as size of the data transferred during a given session, source and destination identifiers, protocol used, etc. Capturing and processing network flows instead of network packets has many advantages. Since flow records do not contain any payload data, they are dramatically smaller in size and offer better user privacy as well as efficient storage and faster processing. In most cases, network packet processing techniques, such as deep packet inspection, are becoming less effective due to the widespread use of end-to-end encryption and privacy concerns. Enriching flow data with attributes such as geographic location, Domain Name Service (DNS) hostname, domain name, known malicious status, Autonomous System (AS) number, etc. provides vital information for real-time monitoring and analysis. Enriched flow data can be filtered, sorted and aggregated to generate real-time interactive visualizations to provide comprehensive network visibility. Visualizations based on port, protocol, geographic location, and custom attributes allow for detection of anomalies more effectively [5,6]. It is essential for modern networks to have an operational tool to visualize this data in real time to obtain situational awareness.

While most cybersecurity incidents are caused by intentional malicious activities, some firewall misconfigurations, software bugs, human errors, etc. can degrade the security posture of a network and leave it vulnerable for attackers. Comprehensive data about the network usage down to each device including attributes, such as traffic characteristics, communication with known threats, and use of restricted ports and protocols, can uncover risk factors proactively. Researchers and network operators can gain a vast amount of knowledge through exploratory visual analytics on datasets such as penetration testing data and Capture the Flag (CTF) competition data.

Development of flow analytics requires access to comprehensive datasets. Publicly available benchmark datasets such as KDD-99 may not be suitable in certain cases [7]. Due to the lack of an open platform to collect, enrich, organize, and manipulate flow data to develop flow analytics, each research group has to develop their own code to perform these functions. An open platform that provides such capabilities while facilitating the implementation of the developed analytics as plugin modules and see their output using a unified visualization framework, has the potential to lower the barrier to entry as well streamline the development for researchers.

1.2. Related Work

In this section, we examine current state-of-the-art of network situational awareness in large-scale networks and identify their shortcomings and challenges when adapting them to analyze flow datasets, streaming flow data, and developing analytics. To reduce the amount of data to be processed for large-scale networks, preprocessing steps such as filtering and sampling have been suggested in the literature [8]. These lossy preprocessing steps may, however, eliminate important data. In fact, it has been shown that anomaly detection algorithms can degrade when data are sampled [9,10]. We have analyzed 14.6TB of network traces from the Global Ring Network for Advanced Applications Development (GLORIAD) project [11,12]. Operational from 2012 to 2015, GLORIAD was one of the largest Research and Education (R&E) networks of its time, connecting researchers and scientists at a global scale. We found the GLORIAD traffic to consist of many small flows less than 100 kB in size with a relatively small number of larger 'elephant flows' accounting for the majority of data transmitted.

Only by amalgamating the many small flows can a complete picture of the traffic be obtained. Even though accuracy with sampling may be adequate for operational measurements, detection of malicious activity such as slow port scan traffic may go undetected. We developed a novel multi-threaded flow processing back-end to process all flow data that scales well on modern CPU architectures to overcome this problem.

A number of purpose-built flow data tools have been developed for deployment on specific networks. The GLORIAD team based at the University of Tennessee had previously developed a tool called InSight for their internal operational measurement purposes. InSight sampled Argus flow data [13] and filtered out flows smaller than 100 kB. InSight2 was inspired by this work, and archived flow data from the GLORIAD R&E network was used for testing prior to deployment on live university networks. Other than the name and lineage, InSight2 does not have any relation with InSight as the code bases and capabilities are completely different. Details are provided in Section 2.

Network monitoring using In-band Network Telemetry (INT) is one of the recent developments in network monitoring [14]. However, it requires the maximum number of hops in the INT reports to be six or less and an INTCollector to be placed on every sink switch which takes the switch resources away from its core functions and may be infeasible in certain situations. Another network operations related tool, NetSage [15,16], focuses on visualizing the R&E network infrastructure of the National Science Foundation. NetSage filters out flows smaller than 100 MB. The current implementation consists of a web portal for visualizations of the traffic measurements. Internet2 provides the Deepfield Analytics Service (DAS) [17] for visualization and analysis of cloud and network data and Network Diagnostic Tool [18] for network diagnostics. DAS is only available to Internet2 members. ESNet that is part of the U.S. Department of Energy has a visualization tool for displaying network bandwidth utilization [19]. SiLK [20] is a collection of Unix command-line tools for querying and analyzing converted NetFlow records. It requires the conversion of flow data to SiLK data-structure and additional software are needed for their visualization such as Analysis Pipieline [21]. NVisionIP [22], VisFlowConnect [23], and NfSen [24] provide visualization capabilities for NetFlow data generated by Cisco routers, but none of the projects appear to be active. Tstat visualizes traffic patterns at the network and transport levels [25] and is used by NetSage. However, Tstat also appears to be inactive. Commercial products are considered outside of the scope of our work.

1.3. Outline

The paper is organized as follows. Section 2 describes the software architecture from the modular scalable processing back-end and the visualization front-end to the deployment mechanism. We also explore the design decisions that address the issues with existing systems for network situational awareness. Section 3 presents case-studies relating to real-time situational awareness, incident response and anomaly detection. Section 4 provides the conclusions.

2. System Architecture

The core of InSight2 is a novel and flexible multi-threaded system architecture. Multi-threaded software can increase the throughput by delegating tasks to separate processor cores making the software more scalable. It also employs an optimized data flow control mechanism, uses a redundancy based indexed schemaless distributed database and a search engine, and visualizations to present the information using a web-based front-end. Compared to relational databases, schemaless indexed databases can significantly reduce the time needed to process queries (to the point of supporting real-time). They also allow custom attributes to be added without having to recreate the data tables. These functionalities are required for the generation of in-depth visualizations and running analysis tasks on complex filter pipelines.

InSight2 incorporates up-to-date security features to protect from unauthorized access such as server-side authentication and encryption using Transport Layer Security (TLS). Furthermore, its platform nature allows researchers to collect, enrich, organize, and manipulate flow data from real and virtual environments, such as Software Defined Networks (SDN), to aid flow analysis research. The modular architecture allows such analytics to be implemented within InSight2 as modules extending its functionality. This modular nature also allows the sharing of these modules for extendability.

We studied the proof-of-concept InSight tool in-depth developed by GLORIAD at the University of Tennessee. It was designed to be an internal tool for the visualization of archived Argus flow data. The code base consisted of Perl 5 scripts that were invoked as Unix cronjobs at set intervals. Each script would carry out a specific task such as reading archived Argus data, extracting elephant flows, adding geolocation information, host attributes, etc. A ZeroMQ publisher-subscriber queue was used to share data between scripts. MySQL and SQLite databases were used for intermediate data storage of host attributes. During this study, we discovered issues with scaling, due to the use of single-threaded enrichment back-end and the use of SQL databases to hold contextual information and temporary flow records. It had issues with usability due to the lack of a proper installation mechanism and a run-time data-flow structure. It also had timing alignment issues due to being a collection of independent scripts that passed data between each other asynchronously. It was not extensible due to the lack of a core enrichment engine nor a mechanism to add modules to further process the enriched data to bring out more insights from the data.

InSight2 was inspired by InSight but has been built from the ground-up addressing the above issues. InSight2 consists of a core enrichment engine written in Python 3 along with peripheral modules that allow further management, processing, and analysis of the flow data. It uses an indexed schemaless database structure and does not depend on intermediate databases to hold temporary results. These design decisions enable better maintainability of the code-base as well as minimize the bottlenecks present in InSight architecture. The system runs in containerized environments, which simplifies the installation, streamlines updating, and eliminates dependency conflicts with the host operating system while allowing it to run on a wide range of hardware and virtualized environments, including standard commodity computers, Virtual Machines (VM) and SDNs. InSight2 uses the open-source Elasticsearch database, Kibana visualizations [26], and Docker containers but does not rely on any commercial software.

2.1. Overview

The flow data are processed through InSight2 as shown in Figure 1. Network packets are converted to network flows and are enriched using different information sources such as OSINT or proprietary information. They are stored in the database which has a search-engine functionality capable of on-the-fly data manipulation without storing an intermediate copy of the results. They are used to build the real-time visualizations and are further processed by the flow analysis plugins. Final results of the flow analysis plugins are added back to the database so that they can also be visualized in the same manner. Each component of the system function as modules as described below, providing data ingestion, database maintenance, and further analysis.

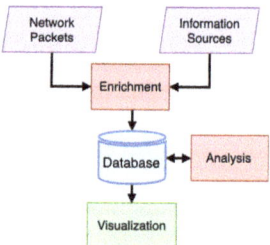

Figure 1. Data-flow overview.

Furthermore, the web-based Graphical User Interface (GUI) hosts the dashboards that allow for visual filtering of the displayed data in real-time. They react in unison according to the filters specified for the particular dashboard. Visual filtering includes limiting the scope of the displayed data based on the parameters such as time-range, IP/port range, geographic location, etc. Once the filters are defined the filtered data can be exported in CSV format to facilitate the use of third-party data analysis tools for further study. This facilitates extraction of critical enriched data as well as the outputs of the analysis modules. Below, we discuss the system modules in detail with reference to Figure 2, the system architecture.

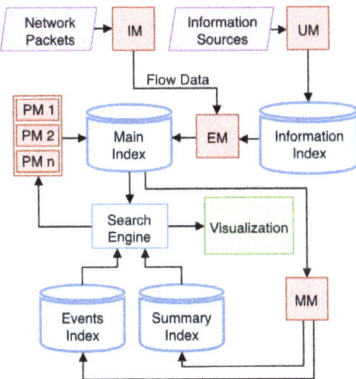

Figure 2. System architecture.

2.2. Input Module

The Input Module (IM) carries out flow data generation, storage, and presentation to the other modules. IM uses the Argus client 'rabins' to process six widely used flow standards in real time: NetFlow, sFlow, JFlow, IPFIX, Flow-tools, and Argus itself. IM also supports ingesting data from raw network streams such as network mirrors, which also are converted into Argus format. Furthermore, generated flow data can be archived for future enrichment and archived flow data can be re-enriched with new information such as updated threat-lists and host attributes at any time for forensic study.

Network packets are converted into bi-directional network flows. They are subjected to time-series binning which allows real-time stream block processing to synchronize the enrichment process. In this method, a network flow record is not only created when a FIN or RST flag is seen in the packet header but also at the end of each time bin. This method is lossless and retains all the information in every network flow. The flow data created at this step are enriched in-memory for efficiency and then are sent to the database for indexing. It facilitates the generation of real-time visualizations when a flow does not terminate at or before the time boundary. Even though this fragments a flow record at the boundary, they are re-aligned based on the timestamp for accurate flow reconstruction for analysis.

2.3. Enrichment Module

The Enrichment Module (EM) augments the raw flow data with contextual information and is crucial in visualizing relationships between end-points in the network. End-points consist of user-devices in the network and critical infrastructure such as network equipment and servers. These connections, and statistics about them, can be filtered to obtain detailed information, as explored in detail in Section 3, providing network administrators and security analysts the capability to evaluate the data from various perspectives for situational awareness. The contextual information can be intrinsic, such as the hostname, the institution associated with the end-point, physical location, type and the owner of the end-point, or extrinsic, such as whether the IP address has been involved with botnet activity, compromised hosts, distribution of malware, as well as the geographical location if it is a public IP address. Intrinsic information is maintained by the network administrators while extrinsic information such as OSINT is kept up-to-date by different organizations that specialize in a particular task such as tracking botnet activity on the Internet.

EM is multi-threaded, so that it scales well with the processor speed as well as the core count to handle any incoming flow rate to deliver real-time visualizations. EM synchronizes with the bin duration provided by IM to capture its output at the time boundaries and the flow records are passed in memory. Once the records are extracted, it delegates the data into each core of the CPU by invoking process threads. Each thread augments the flow records with the contextual information and queues them for indexing into the database. When indexed, they are made available in the Main Index, which is used for generating visualizations as well as providing ground level data for further analysis by analytics modules.

Each database is divided into multiple shards and stored across multiple distributed servers. Shards are duplicated into backup shards providing data redundancy and faster access.

2.4. Updater Module

The Updater Module (UM) keeps the contextual information up-to-date by tracking changes to contextual data, both intrinsic and extrinsic, so that flow data enrichment is always carried out with the most up-to-date information. This allows on-demand enrichment of past flow data with new contextual information for forensic study.

2.5. Maintenance Module

Storage of enriched flow data is limited by the storage capacity of the host system. The Maintenance Module (MM) routinely prunes the Main Index by deleting older indexes when the system begins to run out of space. As a part of the pruning process, MM extracts and stores security incidents and flagged events in a separate index named Events Index. Enriched data are furthermore summarized to retain high level information about the performance of the network in the Summary Index. In the long term, MM contributes to retaining high level summarized information while retaining highly granular information in the short term.

2.6. Plug-in Modules

Plug-in Modules (PM) extend the core functionality of InSight2 provided by the aforementioned core modules. The Main Index, Summary Index, and the Events Index are all made accessible to the plug-in modules to enable flow analysis. Novel analytics are implemented and incorporated as modules, extending the capability of the system. Dashboards specific to the output of each plug-in visualize the results. When a plug-in module is created, a dashboard is created that reads data from the relevant index.

2.7. Front-End Functionality and Security

Figure 3 illustrates web-based GUI and security features of InSight2. The GUI consists of various tabs that are created per use case and each tab has one or more dashboards. A Dashboard consists of a set of modular visualizations and can be configured to show a particular aspect of the data. For example it can be, a bar chart of the traffic of top ten countries sorted in descending order along with a line chart showing total traffic variation over time. Information shown in the dashboard can be filtered visually from the GUI limiting the displayed data to a given scope—for example, filtering all Secure Shell (SSH) traffic from outside the network to a particular host in the network during the past hour. When this filter is applied to the previous scenario it will result in a bar chart showing the top incoming SSH traffic from outside the network to that particular host during the past hour sorted in descending order by the country as well as in the line plot spanning over the last hour. New dashboards can be created to group frequently used visualizations together or by a specialized purpose. Dashboards can be saved for later use as well as for reporting purposes. Data displayed in the dashboard after being filtered according to the filter pipeline can be exported and saved to the disk in CSV format to create new datasets. Furthermore, when accessing data in time ranges past current retention, an option to re-enrich the data is provided.

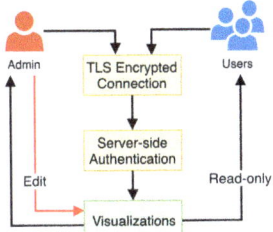

Figure 3. Web-interface security.

The dashboards are capable of detecting the user agent of the web-browser, either mobile or desktop, and serving the appropriate format of the dashboard that fits the available screen real-estate. The visualizations are created on-demand within the web-browser. The web-interface is secured with modern authentication and encryption standards to restrict access only to authorized users. It uses TLS (HTTPS) encryption to thwart man-in-the-middle (MITM) attacks. Based on the user type, relevant dashboards are loaded when each user logs in, controlling access to information. This also allows dashboards to be modified only by authorized users, making it possible to delegate monitoring and generate reports to less privileged users. It also facilitates collaborating with remote analyst teams for network forensics.

2.8. Deployment Mechanism

Virtualization allows InSight2 to be installed on a wide range of operating systems, hardware, and virtual environments such as SDN [27,28]. Using InSight2 with SDNs is not tested as of writing of this paper, but the Ingestion Module is designed to connect to either a supported network flow source such as Argus with the default port 561 over the network or convert network packets from a network mirror port to Argus format. Modular structure of the system provides better maintainability as well as deployability of the system. They can be installed in either hardware computers or VMs. InSight2 modules use application level abstraction that packages code and dependencies together, and shares the operating system (OS) kernel. InSight2 packages each of its modules as a separate Docker image and connects them using the Docker network. This is an abstract network that enables each image to be installed on either a single host or across multiple hosts as they communicate using network sockets. Figure 4 illustrates the connectivity between the containers.

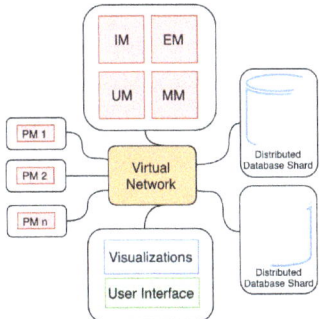

Figure 4. Deployment mechanism.

The database is distributed using shards which can be installed in separate containers. Duplicate shards aid in mitigating data loss in case of container failure or reachability issues. Furthermore, duplicate shards are searched in parallel during a search query to produce faster results.

3. Case Studies

In this section, we illustrate how we used InSight2 for situational awareness in three different scenarios: real-time situational awareness, incident response, and anomaly detection. Network packets are converted into Argus flow data which contain information extracted from the packet headers as well as measurements computed for the packets associated with each flow including number of bytes transmitted, start and end times, etc. The flow data are enriched using information available from other sources—for example, the mapping of IP addresses to geolocations using MaxMind GeoIP databases [29]. The enriched flow data are stored in a searchable indexed database. Plug-in modules add data analysis capabilities. Data are visualized using intuitive dashboards that combine different metrics by common needs such as measurement and security. InSight2 is currently deployed at Stanford University, Queen's University, Canada, and the University of Tennessee, Knoxville with plans underway for further deployments at additional universities and research institutes. The data for the case studies discussed here were obtained from the deployment at Stanford University. There, InSight2 is connected to a 10 Gbps mirror port in one of the School of Engineering (SoE) primary datacenters connecting over 200 physical servers and approximately 100 virtual machines, handling both administrative applications and research computation. Up-stream connectivity to the campus backbone is provided via 2 × 10 Gbps (Link Aggregation Control Protocol) LACP connection.

3.1. Real-Time Situational Awareness

Flow data are enriched in real time and the resulting visualizations are used for network situational awareness. Dashboards group a set of web-based visualizations that show information about the status of the network for current or past time periods. Each dashboard represents a particular aspect of the network traffic. Figure 5 illustrates the first page seen when the web-based GUI is loaded. Sensitive information such as IP addresses and country names are pixelated to preserve privacy. This dashboard gives a high-level view of the network traffic using gauges that display average network bandwidth utilization, packet loss, packet retransmissions, and producer-consumer ratio (PCR). Number of bytes transmitted as a function of time is shown along with four histograms displaying the top fifteen IP addresses and protocols for flows that originate from both source and destination. Tag clouds show the top ten countries sorted by the amount of traffic generated, while the segmented pie-chart provides organization name, project, and department information. Geographic heat-maps provide a geo-spatial breakdown of flows. Figure 5 shows a second dashboard that gives a more detailed view of the network traffic including connectivity between organizations, traffic heat-map, and composition by country, average number of hops, TCP handshake times, average packet sizes, etc. The connectivity map shows the top three organizations with the highest amount of traffic utilization and what other top organizations they communicated with. This is useful to identify the "top talkers" in the network. The sent and received traffic per country is plotted in a matrix on a time axis which allows for understanding traffic breakdown per country in the given time-frame. PCR is plotted in more detail on the time axis that visually shows the ingress and egress traffic ratio. TCP connection times, packet drops, and packet retransmissions indicate potential network congestion. Other visualizations include packet size and number of packets sent by country. Data enrichment capabilities and the ability to manipulate the data in real time from within the web interface enable interactive visualizations that do not require writing manual queries to the database.

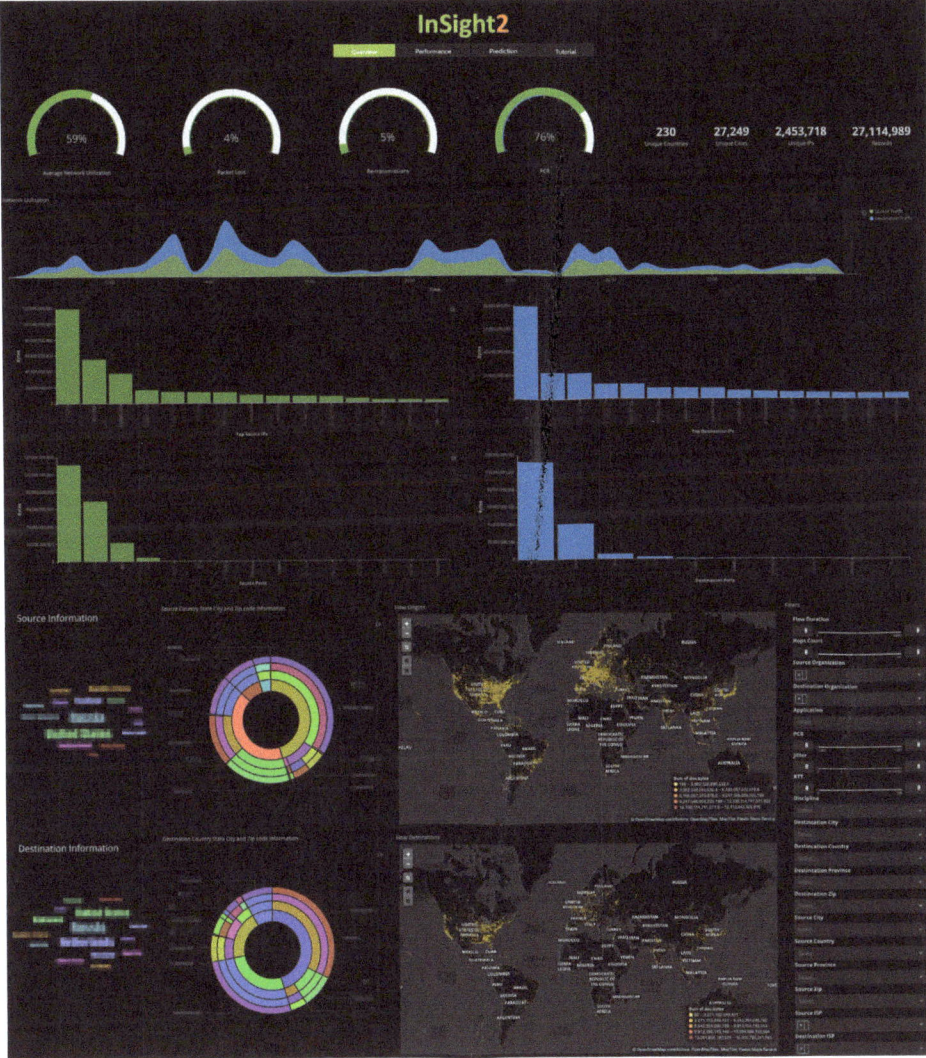

Figure 5. Network traffic overview dashboard.

Visualizations discussed in Figures 5 and 6 were built to understand the various aspects of the particular traffic under inspection. Filters can be applied to limit the scope within that data. For example, by applying outgoing port to port 80, the visualizations will adjust on-the-fly to show only the flows with HTTP requests that are exiting the network. This allows for understanding how the network handles web traffic as the flows are bi-directional. Furthermore, existing visualizations can be modified. For example, the country can be replaced with city for more granular visualizations. Finally, new dashboards can be created and saved to be re-used at any time.

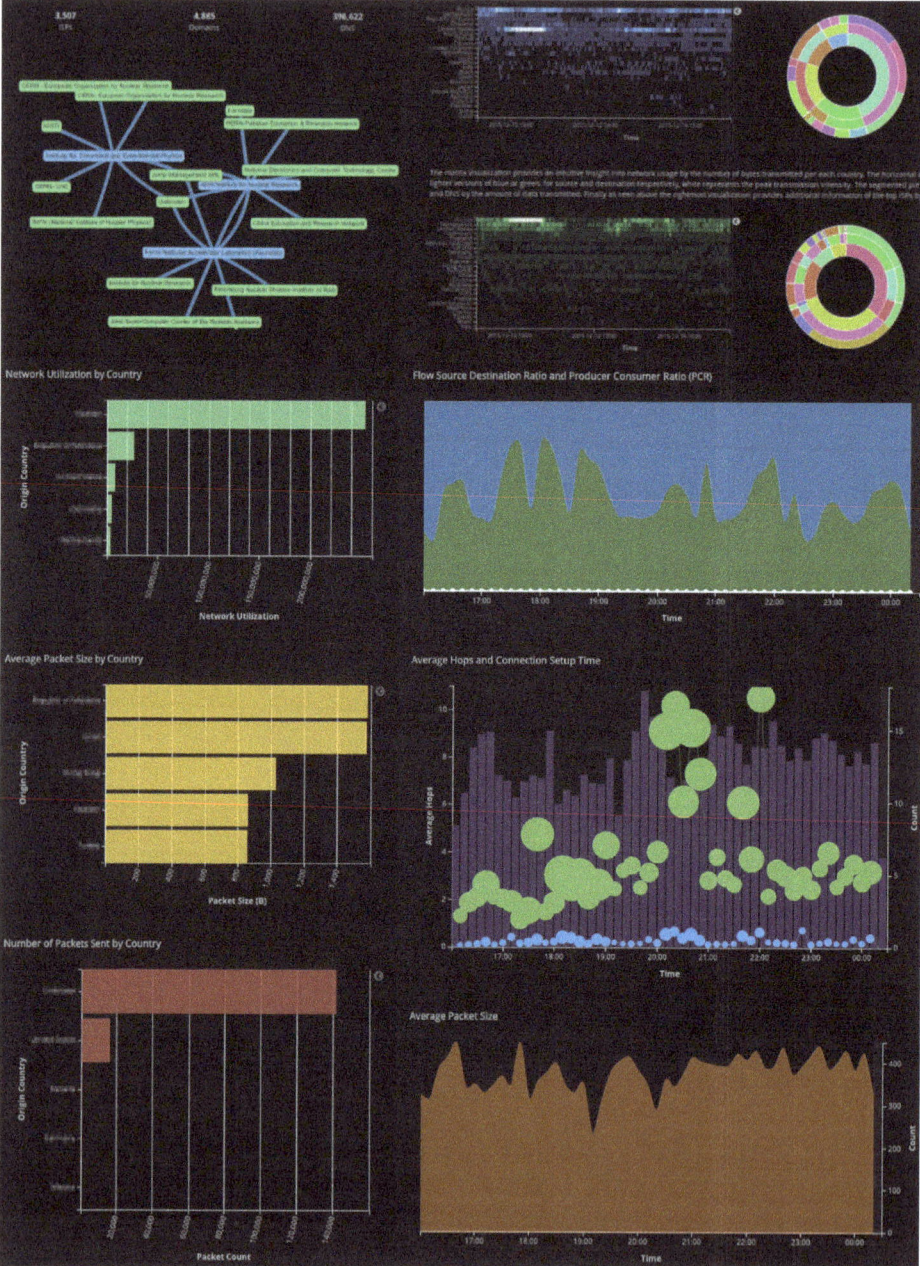

Figure 6. Network performance dashboard.

An important factor when exploring flow data is having the ability to filter and sort data according to one or more criteria, such as isolating all traffic to certain source or destination hosts or countries, ports or protocols, flows of given sizes, Internet Service Providers (ISPs) with the most packet losses, hosts that communicate with suspicious IP addresses, etc. This has been a core objective in the development of InSight2. For example, all visualizations within a dashboard can be focused on traffic originating from the host currently generating the most traffic simply by selecting it via the dashboard. More sophisticated and granular filtering is possible via the advanced controls in the dashboard. This method allows for selecting traffic based on any one or more features associated with each flow including augmented features added during enrichment. For example, the network operator can create a filter that isolates all traffic to SSH traffic on port 22. In such cases, visualizations made for displaying source and destination hosts are sorted by the number of flows in descending order. Additionally, other sections of the dashboard will show where they are coming from geographically and topologically along with other attributes such as whether they have been previously tagged as malicious. A workflow such as this allows the network administrator to find potential threats that try to infiltrate critical systems in the network, and learn more information about them.

Furthermore, in conjunction with numerical and categorical filters, geographic areas can be easily isolated as well, either by clicking on countries in the tag cloud or by drawing a region of interest on the global map. To filter information based on an organization, project, or department of interest, the user simply clicks on the desired section on the segmented pie chart. The dashboards can be customized by modifying, adding, removing, relocating, and resizing visualizations to meet the needs of each deployment site.

3.2. Incident Response

We used flow data from Western Regional Collegiate Cyber Defense Competition 2019 (WRCCDC) [30] for visual analysis of the attacks and to find compromised hosts. Here, we discuss the process of knowledge inference that can be used in incident response and decision-making from the point of view of the participating Stanford University team. A simple forensic study workflow is followed in order to find compromised IPs in the network as shown in Figure 7.

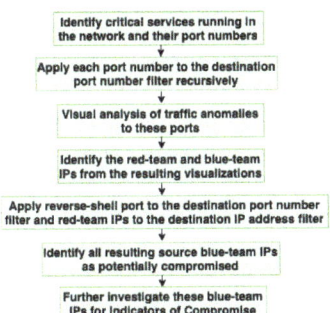

Figure 7. A simple forensic study workflow.

The WRCCDC competition consisted of eight blue teams whose objective was to defend network services such as web and email servers. The blue teams operated on subnets 10.47.x.0/24, where x denotes team numbers 1 through 8. A red team operating on a wide network range of 10.128.0.0/9 attacked the blue team servers using various techniques. A service check engine predominantly operating at 10.0.0.111 (and various other IPs) periodically checked the status of the blue team services to see if attacks had successfully disabled their critical network applications. Archived network traces are analyzed in this subsection using InSight2 to explore the attack sequence to demonstrate the network forensics capability of the platform.

Figure 8 shows the initial dashboard created for the analysis which contains the top IP addresses of both attackers and defenders before applying any filters. This visualization provides a starting point for the analysis when determining which services were attacked. We can discern some details of the attacks based on the destination port numbers and protocols. For example, port 3389 is used for RDP (Remote Desktop Protocol) connections in Windows systems, port 5900 is used by VNC (Virtual Network Computing), port 22 for SSH, port 80/443 for web services, etc. Analyzing the source and destination IP addresses associated with these ports, we observed ports 137 (NetBIOS) and 445 (SMB) were under sustained attack by the red team. After a successful attack we see a connection initiated from the victim to port 4444 back to the attacker. Being the default port for the Metasploit [31] reverse shell, we can reasonably conclude the attacker has taken control of the target system and is engaged in post-exploitation techniques. This is shown in the Figure 9, where the reverse shell traffic is filtered to show the timeline, statistics of associated IP addresses and port numbers, as well as blue and red team IPs.

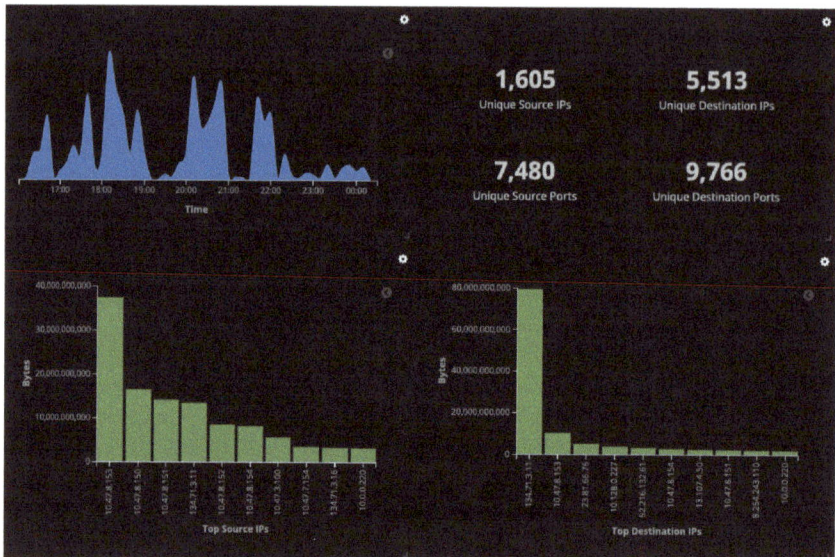

Figure 8. Attack statistics before filtering.

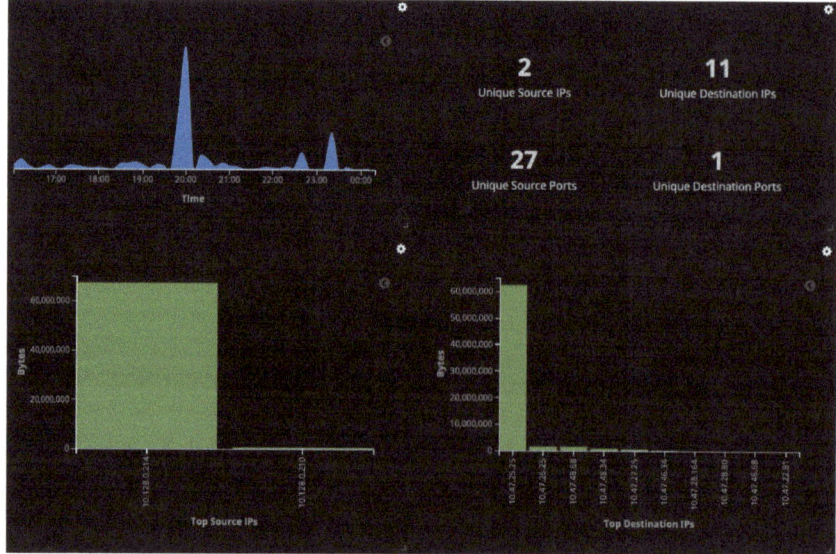

Figure 9. Attack statistics after filtering.

This case study illustrates how security analysts and incident response teams can find the end-points associated with network attacks, gauge what attacks are being launched, and isolate the command and control traffic to and from the compromised host(s).

3.3. Anomaly Detection

Plug-in modules that extend the functionality of the system reuse the data ingestion, enrichment, indexing and visualizing capabilities of InSight2 to do further analysis. With reference to Figure 2, a plug-in module reads data from the Main Index, the Events Index, and/or the Summary Index and writes its results to its own index. Dashboards tied with the analytics module visualize the output of that index.

Flow analysis is critical to identify anomalies automatically [32,33]. To illustrate the modular analytics capability of InSight2, we describe Markov chain prediction of network bandwidth utilization to detect anomalies. A Markov chain is a probabilistic finite-state machine for which future transitions depend only on the present state and not the states visited before reaching it. States here are defined by a simple three-level discretization of the network bandwidth utilization: low, medium, and high. The state transition matrix which makes up the Markov chain is inferred from observed transition frequencies when going from each state to every other state. The expected number of transitions needed before going from one state to another and the standard deviation associated therewith can be calculated from the transition matrix. Analyzing the bandwidth utilization data of the Stanford University for one week shows the daily mean time before high bandwidth usage is replaced by low bandwidth usage is about six hours with a standard deviation of about 1.3 h. Once the module was deployed on InSight2, we were able to detect a statistical anomaly where almost 16 h of high bandwidth usage was reported as shown in Figure 10. This simple module demonstrates an example of automating the analysis to detect anomalies.

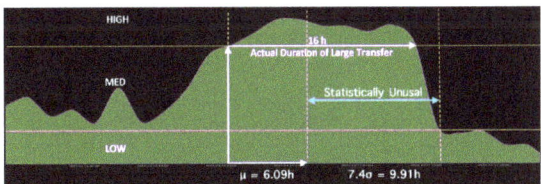

Figure 10. Network bandwidth utilization prediction.

Plug-in modules such as this can be developed and shared by the community of InSight2 users. They are language agnostic since they query, process, and insert data to the ES database without requiring any message passing between the core modules.

4. Conclusions

InSight2 is open-source software [34] that aims to fulfill the need for a platform for real-time situational awareness, incident response, and development and deployment of flow analytics in large-scale networks. We discuss the novel software architecture that addresses the issues with current solutions such as scalability, extendability, and flexibility. We discuss three case studies from one of its deployments at Stanford University on real-time situational awareness, anomaly detection and incident response using WRCCDC dataset. In these situations, we use real-time graphs to understand its behavior under normal conditions, infer knowledge from visual analysis for incident response that can be used to make decisions to improve the security, and detect abnormal behavior using automated anomaly detection.

The intuitive web-based front-end allows analysts and researchers to explore the data quickly, reducing the effort and time needed to monitor network traffic compared to querying databases followed by manual visualization of each result. A novel system architecture is developed for efficient flow data enrichment leveraging modern multi-core CPUs. The modular architecture allows users to develop their own extensions to the platform. Finally, a container-based installation mechanism allows for ease of deployment and seamless maintenance of the software.

Author Contributions: Conceptualization, H.A.D.E.K.; Methodology, H.A.D.E.K., J.G.; Project administration, J.G.; Software architecture, H.A.D.E.K.; Supervision, J.G.; Validation, J.G., A.K.; Software implementation, H.A.D.E.K., Software deployment, H.A.D.E.K., A.K.; Writing—original draft, H.A.D.E.K.; Writing—review and editing, J.G., A.K.; All authors have read and agreed to the published version of the manuscript.

Funding: This material is based upon work supported by the National Science Foundation under Grant No. IRNC-1450959.

Conflicts of Interest: The authors declare no conflicts of interest.

References

1. Onwubiko, C. Functional requirements of situational awareness in computer network security. In Proceedings of the IEEE International Conference on Intelligence and Security Informatics (ISI 2009), Richardson, TX, USA, 8–12 June 2009; pp. 209–213.
2. Husák, M.; Jirsík, T.; Yang, S.J. SoK: contemporary issues and challenges to enable cyber situational awareness for network security. In Proceedings of the 15th International Conference on Availability, Reliability and Security (ARES 2020), Dublin, Ireland, 25–28 August 2020; pp. 1–10.
3. Li, J.; Yi, X.; Wei, S. A Study of Network Security Situational Awareness in Internet of Things. In Proceedings of the 2020 International Wireless Communications and Mobile Computing (IWCMC), Byblos, Lebanon, 3–29 June 2020; pp. 1624–1629.
4. Gutzwiller, R.; Dykstra, J.; Payne, B. Gaps and Opportunities in Situational Awareness for Cybersecurity. In *Digital Threats: Research and Practice*; ACM: New York, NY, USA, 2020; pp. 1–6.
5. Zhang, T.; Liao, Q.; Shi, L. Bridging the gap of network management and anomaly detection through interactive visualization. In Proceedings of the 2014 IEEE Pacific Visualization Symposium, Yokohama, Japan, 4–7 March 2014; pp. 253–257.
6. Franke, U.; Brynielsson, J. Cyber situational awareness–a systematic review of the literature. *Comput. Secur.* **2014**, *46*, 18–31. [CrossRef]
7. Divekar, A.; Parekh, M.; Savla, V.; Mishra, R.; Shirole, M. Benchmarking datasets for anomaly-based network intrusion detection: KDD CUP 99 alternatives. In Proceedings of the 2018 IEEE 3rd International Conference on Computing, Communication and Security (ICCCS), Kathmandu, Nepal, 25–28 October 2018; pp. 1–8.
8. Silva, J.M.; Carvalho, P.; Lima, S.R. A Modular Traffic Sampling Architecture: Bringing Versatility and Efficiency to Massive Traffic Analysis. *J. Netw. Syst. Manag.* **2017**, *25*, 643–648. Available online: https://www.overleaf.com/project/5f8fbb2378dfed00019bbb5b (accessed on 15 September 2020). [CrossRef]
9. Mai, J.; Chuah, C.N.; Sridharan, A.; Ye, T.; Zang, H. Is sampled data sufficient for anomaly detection? In Proceedings of the 6th ACM SIGCOMM conference on Internet measurement, Rio de Janeiro, Brazil, 25–27 October 2006; pp. 165–176.
10. Carela-Español, V.; Barlet-Ros, P.; Cabellos-Aparicio, A.; Solé-Pareta, J. Analysis of the impact of sampling on NetFlow traffic classification. *J. Comput. Netw.* **2011**, *55*, 1083–1099. [CrossRef]
11. Cole, G.; Bulashova, N. GLORIAD: A ring around the Northern Hemisphere for science and education connecting North America, Russia, China, Korea and Netherlands with advanced network services. In Proceedings of the TERANA Networking Conference, Pznan, Poland, 19–21 May 2005.
12. Cole, G.; Jun, L.; Sobieski, J.; Kim, D.; Riley, D. NSF Award: GLORIAD, Award number: IRNC-0963058. 2010. Available online: https://www.nsf.gov/awardsearch/showAward?AWD_ID=0963058 (accessed on 15 September 2020).
13. QoSient Argus. Available online: https://qosient.com/argus/ (accessed on 15 September 2020).
14. Hyun, J.; Van Tu, N.; Yoo, J.H.; Hong, J.W. Real-time and fine-grained network monitoring using in-band network telemetry. *Int. J. Netw. Manag.* **2019**, *29*, e2080. [CrossRef]
15. Gonzalez, A.; Leigh, J.; Peisert, S.; Tierney, B.; Lee, A.; Schopf, J.M. NetSage: Open Privacy-Aware Network Measurement, Analysis, In addition, Visualization Service. 2016. Available online: https://escholarship.org/uc/item/5rz6t3q4 (accessed on 15 September 2020).

16. Gonzalez, A.; Leigh, J.; Peisert, S.; Tierney, B.; Balas, E.; Radulovic, P.; Schopf, J.M. Big Data and Analysis of Data Transfers for International Research Networks Using NetSage. In Proceedings of the 2017 IEEE International Congress on Big Data (BigData Congress), Boston, MA, USA, 11–14 December 2017; pp. 344–351.
17. Deep Field Analytics. Available online: https://www.internet2.edu/media/medialibrary/2014/07/01/IS-deepfield-analytics.pdf (accessed on 15 September 2020).
18. Network Diagnostic Tool. Available online: http://software.internet2.edu/ndt/ndt-cookbook.pdf (accessed on 15 September 2020).
19. US Department of Energy Energy Sciences Network. Available online: https://my.es.net/Network (accessed on 15 September 2020).
20. System for Internet Level Knowledge (SiLK). Available online: https://tools.netsa.cert.org/silk/ (accessed on 15 September 2020).
21. Analysis Pipeline. Available online: https://tools.netsa.cert.org/analysis-pipeline5/ (accessed on 15 September 2020).
22. Lakkaraju, K.; Yurcik, W.; Lee, A.J. NVisionIP: Netflow visualizations of system state for security situational awareness. In Proceedings of the 2004 ACM workshop on Visualization and data mining for computer security, Washington, DC, USA, 29 October 2004; pp. 65–72.
23. Yin, X.; Yurcik, W.; Treaster, M.; Li, Y.; Lakkaraju, K. VisFlowConnect: Netflow visualizations of link relationships for security situational awareness. In Proceedings of the 2004 ACM workshop on Visualization and Data Mining for Computer Security, Washington, DC, USA, 29 October 2004; pp. 26–34.
24. Haag, P. Watch your Flows with NfSen and NFDUMP. In Proceedings of the 50th RIPE Meeting, Stockholm, Sweden, 2–6 May 2005.
25. Mellia, M.; Carpani, A.; Cigno, R.L. Measuring IP and TCP behavior on Edge Nodes. In Proceedings of the Global Telecommunications Conference, Taipei, Taiwan, 17–21 November 2002; pp. 2533–2537.
26. Elastic Stack and Product Documentation. Available online: https://www.elastic.co/guide/index.html (accessed on 15 September 2020).
27. Sun, G.; Xu, Z.; Yu, H.; Chen, X.; Chang, V.; Vasilakos, A.V. Low-latency and resource-efficient service function chaining orchestration in network function virtualization. *IEEE Internet Things J.* **2020**, *7*, 5760–5772. [CrossRef]
28. Sun, G.; Zhou, R.; Sun, J.; Yu, H.; Vasilakos, A.V. Energy-Efficient Provisioning for Service Function Chains to Support Delay-Sensitive Applications in Network Function Virtualization. *IEEE Internet Things J.* **2020**, *7*, 6116–6131. [CrossRef]
29. MaxMind, LLC. GeoIP Database. Available online: http://www.maxmind.com (accessed on 15 September 2020).
30. Western Regional Collegiate Cyber Defence Competition. Available online: http://archive.wrccdc.org/pcaps/2019/regionals/ (accessed on 15 September 2020).
31. Holik, F.; Horalek, J.; Marik, O.; Neradova, S.; Zitta, S. Effective penetration testing with Metasploit framework and methodologies. In Proceedings of the 2014 IEEE 15th International Symposium on Computational Intelligence and Informatics (CINTI), Budapest, Hungary, 19–21 November 2014; pp. 237–242. [CrossRef]
32. Zhang, J.; Chen, C.; Xiang, Y.; Zhou, W.; Vasilakos, A.V. An effective network traffic classification method with unknown flow detection. *IEEE Trans. Netw. Service Manag.* **2013**, *10*, 133–147. [CrossRef]
33. Fadlullah, Z.M.; Taleb, T.; Vasilakos, A.V.; Guizani, M.; Kato, N. DTRAB: Combating against attacks on encrypted protocols through traffic-feature analysis. *IEEE ACM Trans. Netw.* **2010**, *18*, 1234–1247. [CrossRef]
34. InSight2 Installation and Configuration Repository. Available online: https://github.com/angelkdev/InSight2 (accessed on 15 September 2020).

Publisher's Note: MDPI stays neutral with regard to jurisdictional claims in published maps and institutional affiliations.

© 2020 by the authors. Licensee MDPI, Basel, Switzerland. This article is an open access article distributed under the terms and conditions of the Creative Commons Attribution (CC BY) license (http://creativecommons.org/licenses/by/4.0/).

Article

C^3-Sex: A Conversational Agent to Detect Online Sex Offenders

John Ibañez Rodríguez [1], Santiago Rocha Durán [1], Daniel Díaz-López [2,*], Javier Pastor-Galindo [3] and Félix Gómez Mármol [3]

[1] Faculty of Computer Engineering, Escuela Colombiana de Ingeniería Julio Garavito, AK.45 No.205-59, Bogotá 111166, Colombia; john.ibanez@mail.escuelaing.edu.co (J.I.R.); santiago.rocha@mail.escuelaing.edu.co (S.R.D.)
[2] School of Engineering, Science and Technology, Universidad del Rosario, Carrera 6 No. 12 C-16, Bogotá 111711, Colombia
[3] Faculty of Computer Science, Campus de Espinardo, University of Murcia, 30100 Murcia, Spain; javierpg@um.es (J.P.-G.); felixgm@um.es (F.G.M.)
* Correspondence: danielo.diaz@urosario.edu.co

Received: 8 August 2020; Accepted: 9 October 2020; Published: 27 October 2020

Abstract: Prevention of cybercrime is one of the missions of Law Enforcement Agencies (LEA) aiming to protect and guarantee sovereignty in the cyberspace. In this regard, online sex crimes are among the principal ones to prevent, especially those where a child is abused. The paper at hand proposes C^3-Sex, a smart chatbot that uses Natural Language Processing (NLP) to interact with suspects in order to profile their interest regarding online child sexual abuse. This solution is based on our Artificial Conversational Entity (ACE) that connects to different online chat services to start a conversation. The ACE is designed using generative and rule-based models in charge of generating the posts and replies that constitute the conversation from the chatbot side. The proposed solution also includes a module to analyze the conversations performed by the chatbot and calculate a set of 25 features that describes the suspect's behavior. After 50 days of experiments, the chatbot generated a dataset with 7199 profiling vectors with the features associated to each suspect. Afterward, we applied an unsupervised method to describe the results that differentiate three groups, which we categorize as indifferent, interested, and pervert. Exhaustive analysis is conducted to validate the applicability and advantages of our solution.

Keywords: chatbot; online child sexual abuse; criminal profiling; natural language processing; law enforcement agencies

1. Introduction

Human or drug trafficking, sexual abuse, child bullying, animal cruelty or terrorism are examples of realities that should not exist. Unfortunately, they are still present in our societies causing irreparable harm to innocent beings. Another problem that is also detected today and affects minors is the production, possession, and distribution of online child sexual abuse content, an offence clearly defined by the Luxembourg Guidelines (http://luxembourgguidelines.org) and supported by 18 organizations interested in the protection of children from sexual exploitation and sexual abuse.

During the last decades, this scourge has been magnified by the proliferation of forums, online communities, and social networks [1,2]. The content derived from child sexual abuse is now more accessible than before and, in turn, more present than ever [3]. In this sense, the Child Sexual Exploitation database of INTERPOL registered almost three million images and videos containing child sexual abuse content to the date of July 2020. This database has been useful to identify 23,100 victims worldwide (more than 3800 in 2019), and to chase 10,579 criminals.

The current era of the information facilitates the formation of criminal groups that share and spread sexual content. As a result, many sex offenders can directly see or download such illegal multimedia that would once have been much harder to come by. According to [4], 75% of all exploitation cases were linked to the possession and distribution of child sexual content, 18% was related to child sex trafficking, and 10% to production itself.

Law Enforcement Agencies (LEAs) are making a great effort to pursue these criminals and seize child sexual content [5]. Analyzing P2P networks to measure the existing child sexual abuse content [6], blocking the access to specific illegal webpages [7] or automatically detecting suspect material with visual recognition techniques [8] are some of the dimensions that have been considered to uncover pedophiles and protect the rights of children. Nevertheless, despite those praiseworthy trials of chasing sexual offenders, there is still work to be done to have a sophisticated and effective system to avoid online child abuse [9]. Due to the problem severity, new approaches scaling to such magnitude are needed.

Fortunately, Artificial Intelligence (AI) is being developed by leaps and bounds and constitutes a new pillar on which to build more refined solutions. This paradigm is being implemented, for example, to assist in decision-making, to extract patterns in datasets, or to predict future events. Nonetheless, considering that child abuse content is mainly distributed through the immersion of the Internet and especially frequent in online communities, we are particularly interested in tools adaptable to World Wide Web scenarios.

Natural Language Processing (NLP) is one of the most used AI techniques for social media applications and aims to analyze and understand human language, or even to replicate it with empathetic responses. One of the most common NLP applications is the Artificial Conversational Entities (ACE), also known as chatbots, which do not require human intervention for maintaining a conversation with a user. On the other hand, it faces some challenges such as response blandness (tendency to generate uninformative responses), speaker consistency (possibility of making contradictory interventions), word repetitions, lack of grounding (responses out of context), empathy incorporation or explanation capacity [10]. According to the Hype Cycle for emerging technologies by Gartner [11], conversational AI platforms remain in the phases of "innovation trigger" and "peak of inflated expectations", meaning that they are currently getting substantial attention from the industry.

The paper at hand proposes C^3-Sex, an AI-based chatbot that automatically interacts with Omegle (https://www.omegle.com) users to analyze their behavior around the topic of child illegal content. The analysis of messages from a suspect that follow the tone of the conversation produces a vector with features that allow to describe the suspect behavior. This tool could be potentially useful for Law Enforcement Agencies (LEAs) who work in chasing and putting an end to child sexual abuse. In particular, this chatbot is an enhanced version of the one presented in [12], indeed including more sophisticated techniques, support for mobile apps and a more refined behavior.

The remainder of the paper is structured as follows. Section 3 describes some remarkable related works found in the literature. In Section 4, the key goals and components of C^3-Sex are introduced, while the main aspects of the data science lifecycle and the achieved proposal are presented in Section 5. Section 6 discusses the different features that support the profiling of a suspect based on his/her interaction with the C^3-Sex. Then, in Section 7 we perform an exhaustive evaluation of the proposal and analyze the obtained results. Finally, Section 8 contains some highlights derived from the work done and mentions some future research directions.

2. Background

A chatbot is generally built upon the following elements, as shown in Figure 1:

- Interaction channel, the front-end interface with the end-user in the form of an email agent, instant messaging service, web page or mobile app.
- Natural Language Processor (NLP), the component in charge of understanding the human peer.

- Natural Language Generator (NLG), responsible for responding to human interventions.
- Knowledge-based data, the element that brings the contextual information of the chatbot.
- Business logic, which defines how to interact with the end-user.
- Machine learning models, those parts of the system that enable automatic operating and simulates human behavior.

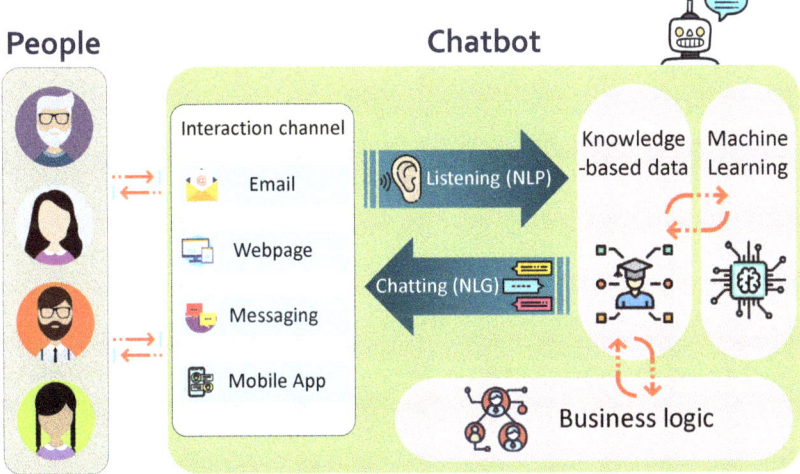

Figure 1. Anatomy of a chatbot.

Chatbots are used in a variety of fields for different purposes. The most common ones are support bots, which assist customer requests related to the delivery of a service or use of a product; financial bots, which resolve inquiries about financial services; or information bots that help new users on a page or disoriented people around a topic. Recently, they are also being applied to innovate in cybersecurity through the form of training bots or guide bots. The formers are a productive tool to educate end-users [13] and cyber analysts [14] in security awareness and incident response, whereas the latter informs the end-users about terms of security and privacy, such as Artemis [15].

In our context, a chatbot could be trained to pretend being an adult person interested in acquiring child sexual abuse material and then be deployed on the web to seek perverts, interact with them, and infer a possible possession of illicit content in an automatic and scalable manner. The software-driven agent(s) can automatically log in to communities, social networks, and forums to interact with large numbers of users. It is worth mentioning that LEAs would not be as effective in doing this manually, having the whole Internet at their disposal and few resources. As far as we know, the use of chatbots to profile suspects in an active way of child sexual content has been investigated little. Only a few approaches [16,17] employ them to emulate a victim, such as a child or a teenager.

3. State of the Art

Several scientific works have been conducted so far in the field of chatbots. Thus, for instance, Gapanyuk et al. [18] propose a hybrid chatbot model composed of a question–answering module and a knowledge-based scheme. The question-answering module contains a list of pairs of questions and answers so, when a user asks a question that matches one of the lists, the corresponding answer is returned to the user. This work's main contribution is the implementation of a rule-based system that is encapsulated in a meta-graph as multiple agents.

Most early works about conversation systems are generally based on knowledge and designed for specific domains. These knowledge-based approaches require no data to be built, but instead, they require much more manual effort and expert knowledge to build the model, which is usually expensive. Thus, [19] proposes a deep learning hybrid chatbot model, which is generative-based. This proposal is composed of 22 response models, including retrieval-based neural networks, generation-based neural networks, knowledge-based question–answering systems, and template-based systems. In addition, it develops a reinforcement learning module based on estimating a Markov decision process.

The integration of an emotional module to chatbots is one way to engage users, i.e., to give the conversational system the ability to be friendly and kind depending on the user's current emotional state. Following this approach, Reference [20] builds a complex embedded conversational agent system, capable of doing high-quality natural language processing and sophisticated manipulation of emotions based on the Plutchik Model [21]. This chatbot analyzes and synthesizes the actual emotional status and the emotional expression depicted on the user messages so that a response can be generated in a customized way.

With the assumption that linguistic style can be an indicator of temperament, a chatbot with an explicit personality was proposed in [22]. The objective of this chatbot is to generate responses that are coherent to a pre-specified disposition or profile. The proposal uses generic conversation data from social media to generate profile-coherent responses, representing a specific reply profile suitable for a received user post.

Heller et al. [23] describe another related work, where a chatbot named Freudbot was constructed using the open-source architecture of Artificial Intelligence Markup Language (AIML). This chatbot aimed to improve the student–content interaction in online learning ecosystems. Explicitly, this chatbot technology is promising as a teaching and learning tool in online education.

In turn, Sabbagh et al. [14] present the HI^{2}P TOOL, which is focused on encouraging an information security culture and awareness between users. This tool incorporates different types of learning methods and topics like incidents, response, and security policies. The interaction with the user is based on the ALICE chatbot using the AIML, making the solutions efficient and straightforward.

Another case of a chatbot used for security training is presented in [13], where the chatbot Sally can interact with some groups of employees in a company with different education or experience on security. Sally was able to provide security training, which was evidenced by a growing knowledge of the target users.

Furthermore, the work presented in [24] investigates the behavior of people when they are aware that they are interacting with chatbots. The results show that the conversation can become pure and composed of short messages in such a situation, even if it can be extended in time. Conversely, conversations with a human can become complex and composed of long messages, but shorter in time. Additionally, the same research found that language skills, such as vocabulary and expression, are easily transferred to a machine.

Emotional Chatting Machine (ECM) [25] is a proposal with a machine learning approach that considers the emotional state of the conversation to generate appropriate responses in content (relevant and grammatical) and in emotion (emotionally consistent).

Particularly related to the topic of sexual harassment and online child sexual abuse, Zambrano et al. [16] present BotHook, a chatbot to identify cyber pedophiles and catch cybercriminals. In this work, a module of attraction of pedophile interests and characterization was developed. Likewise, the work introduced in [17] discusses the efficiency of current methods of cyber perverts detection and proposes some futuristic methods such as metadata and content data analysis of VoIP communications, as well as the application of fully automated chatbots for undercover operations.

In the same direction, a system to detect online child grooming is proposed in [26], which uses Bag of Words (BoW) to select words in the context of grooming, and Fuzzy-Rough Feature Selection (FRFS) for the selection of the most important features. Finally, a fuzzy twin Support Vector Machine was used for text classification using two training datasets: one from Perverted Justice (http://www.perverted-justice.com/) and another one from PAN13 https://pan.webis.de/data.html#pan13-author-profiling.

Likewise, an alternative method for detecting grooming conversations is proposed in [27] where a group of 17 characteristics associated with grooming conversations are identified and then used for text classification. Such proposed method exposes a low-computational cost and an accuracy (96.8%) close to the one obtained with more computational-demanding classifiers like Support Vector Machine (SVM) (98.6%) or K-nearest neighbor (KNN) (97.8%).

The authors of [28] propose a classifier for the detection of online predators, which employs Convolutional Neural Networks. Such proposal gets a classification performance (F1-score) that is 1.7% better than the one obtained with an SVM approach. Two datasets were used for the training of the model: PAN-2012 (https://pan.webis.de/data.html#pan12-sexual-predator-identification) and conversations gathered by the Sûreté du Quebec (Police for the Canadian province of Quebec).

A model that employs user reactions coming from different social networks to detect cyberbullying incidents is proposed in [29]. Such proposal argues that not all text with profanity addressed to a person may be actually considered as proof of bullying, as it depends on the reaction of the person. A dataset composed of 2200 posts were manually labeled as bullying or not-bullying and used to train a Support Vector Machine classifier.

A conversational agent pretending to be a child is proposed in [30], which aims to prevent cybercrimes associated with pedophilia, online child sexual abuse and sexual exploitation. This conversational agent uses game theory to manage seven chatter-bots that address a conversation strategy with the aim of identifying pedophile tendencies without making the chatter to suspect about the agent.

Grooming Attack Recognition System (GARS) is a system that calculates dynamically the risk of cybergrooming that a child is exposed to along a conversation [31]. The risk is calculated using fuzzy logic controllers, which consider the exchanged dialogs, the personality of the interlocutors, the conversations history of the child and the time that the child profile has been exposed on the Internet.

A study of different computational approaches to forecast social tension in social media (Twitter) is defined in [32], which is accompanied by a comparison of approaches based on precision, recall, F-measure, and accuracy. The considered approaches are tension analysis engine, which is their own proposal based on the conversation analysis method MCA (Membership, categorization, analysis) [33], machine learning approach with Naive Bayes classifier and sentiment analysis with the SentiStrength tool [34]. Conversational analysis and syntactic and lexicon-based text mining rules showed a better performance than machine learning approaches.

The proposal designed in [35] tries to detect whether an adult is pretending to be a child as part of an online grooming abuse. The proposal identifies a person as an adult or a child based on the writing style, and then it determines if a child is a fake child or not. That paper suggests that it is challenging to separate children from adults in informal texts (blogs or chat logs). Such proposal uses a set of 735 features gathered from the literature, which are used to build models based on algorithms like Adaboost, SVM, and Naive Bayes.

In rows 1–8 of Table 1 we found different literature efforts concerning the development of chatbots (conversational agents) that mainly uses rule-based and generative-based models, aimed to resolve different industry requirements or to improve the quality of bot conversation. Only works at rows 9–10 propose solutions that integrate conversational agents and classificators to detect cyber pedophiles. On the other hand, rows 11–18 contain different works that propose classificator models that employ different artificial intelligence techniques to detect sexual crimes, cyber bullying and harassment conversation. In the paper at hand, we propose a chatbot to face child abuse with a different approach that has not been considered before: in essence, our chatbot C^3-Sex emulates an individual interested in acquiring child sexual abuse material and then evaluates the responses and behavior to profile the suspect using 25 different features, which altogether provide essential information to LEA in the hunting of perverts.

Table 1. Comparative table of the analyzed related works.

	Related Work	Year	Type	Main Component(s)	Aim
1	Gapanyuk et al. [18]	2018	Conversational model	Question and Answering model using cosine distance, metagraph model	Develop a bot that provides information about goods
2	Serban et al. [19]	2017	Conversational model	Recurrent, sequence-to-sequence and latent variable NN	Develop a social bot
3	Tatai et al. [20]	2003	Conversational model	Emotional state generator, Plutchik emotional model	Develop an emotion sensitive social bot
4	Qian et al. [22]	2018	Conversational model	Identity coherent conversation machine	Develop a social bot with a predefined personality
5	Heller et al. [23]	2005	Conversational model	AIML Knowledge base	Develop a bot to support on-line education
6	Sabbagh et al. [14]	2012	Conversational model	AIML Knowledge base	Develop a bot for education in security
7	Kowalski et al. [13]	2013	Conversational model	Question and Answering model	Develop a bot for security training
8	Zhou et al. [25]	2018	Conversational model	Sequence-to-sequence model with emotion category embedding	Develop an emotion sensitive social bot
9	Zambrano et al. [16]	2017	Conversational model	Sequence-to-sequence model (Planned but not implemented)	Detect cyber pedophile
10	Laorder et al. [30]	2013	Conversational model	Question-answering based on AIML and game theory	Detect Paedophile, online child sexual abuse and Sexual exploitation
11	Anderson et al. [26]	2019	Classificator model	Fuzzy-Rough feature selection (FRFS) + Fuzzy Twin SVM	Detect grooming
12	Gunawan et al. [27]	2018	Classificator model	SVM and KNN	Detect grooming
13	Ebrahimi et al. [28]	2016	Classificator model	Convolutional Neural Network	Detect juvenile abuse, Sexual solicitation
14	Dadvar et al. [29]	2012	Classificator model	SVM	Detect cyberbullying
15	Michalopoulos et al. [31]	2014	Classificator model	Fuzzy logic controllers to evaluate risk	Detect grooming
16	Burnap et al. [32]	2015	Classificator model	Own tension analysis algorithm based on MCA (Membership, categorization, analysis), Naive Bayes and SentiStrength	Detect social tension
17	Meyer et al. [35]	2015	Classificator model	Adaboost, SVM and Naive Bayes	Detect grooming

4. Goals and Taxonomy of C^3-Sex

The implementation of an Artificial Conversational Agent (ACE) encompasses many challenges [10]. It is an arduous task bringing along a substantial technical complexity. On the other hand, it is crucial to define the context in which a chatbot will operate and the objectives it pursues.

The primary purpose of this agent is to chase pedophiles on the web and the detection of sexual content. To this end, we intend to introduce the software-controlled instrument into suspicious chat rooms to interact with other users. Therefore, for the proper functioning of the tool, C^3-Sex should comply with the following properties and directives:

1. Appropriateness: C^3-Sex should manage those situations in which the conversation is out of scope. It should be programmed to bring the subject back into our field of interest, that is, minors' sexual content. To this end, C^3-Sex has to emulate a human behavior accurately, being pragmatic in interpreting the suspect's intent, and responding accordingly. It is essential to be consistent with the created answer, depending on whether it is being asked, ordered, or affirmed. In essence, our software-controlled agent should not be revealed.
2. Platform flexibility: The ACE has to change scenarios from chat rooms to other environments dynamically. To date, C^3-Sex operates in Omegle rooms and is able to smoothly migrate at any time to Snapchat or Telegram sessions. The latter is a requirement directly extracted from our specific experience, where Omegle's suspects used to ask for a change of application. In this sense, Snapchat and Telegram are frequently used for sexting as it lets users transfer media files without limitations and without saving the conversation.
3. Illegal content holders hunting: C^3-Sex should exhibit the behavior of a human interested in acquiring online child sexual abuse content in order to pinpoint suspects possessing illegal content (such as images or videos) who are willing to share it with others.
4. Illegal content bidder hunting: The chatbot should also exhibit the behavior of a human interested in distributing online child sexual abuse content to identify suspects eager to obtain and consume this kind of illegal content.
5. Suspect profiling: The solution should produce a vector of features that characterize the conversation maintained between the chatbot and the suspect to help in the profiling of the latter.

C^3-Sex combines two main parts to fulfill these goals and achieve a functional conversational model, namely: the interactive module and the analysis module. The interactive module holds some interaction interfaces, a knowledge-based system, and a machine learning generative model. The analysis module, in turn, includes an emotional classifier and an opinion classifier. As a result of both modules execution, the chatbot finally outputs a vector of 25 features that characterize the ended conversation. The components and the overall functioning workflow of C^3-Sex can be observed in Figure 2 and they are further described next.

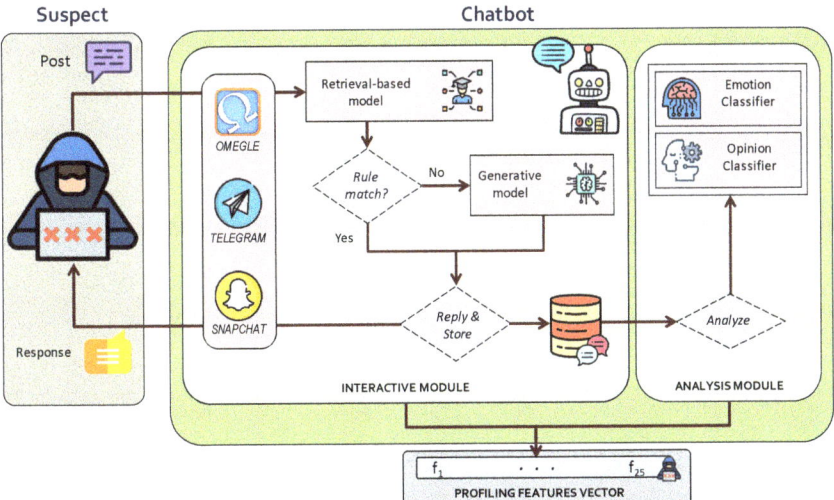

Figure 2. C^3-Sex overview depicting the workflow of posts and replies.

4.1. Interactive Module

The interactive module is active throughout the conversation with the suspects. Concretely, it is in charge of emulating a human behavior and consequently storing information of the suspect. This part is directly linked with the aforementioned properties 1, 2, 3, and 4. The main elements making this module work are explained below:

- Interaction channel: This is the front-end of the chatbot enabling the communication link with the users. This channel can dynamically switch to the different available interfaces. Thus, the default initial Omegle interface can migrate to Snapchat or Telegram interfaces. It is worth mentioning that this change is transparent to the rest of the system, which in turn continues running uninterruptedly. That is to say, the conversation continues naturally despite the change of platform.
- Retrieval-based model: It contains the context's information and leads the conversation when the theme of child content is in the air. In this model, the knowledge is expressed using a set of rules in AIML (Artificial Intelligence Markup Language). Therefore, the response to a question is selected according to the match with one rule. Next, Listing 1 shows an example of an AIML rule within our context, where the *pattern* is checked against the received post, and *template* contains the associated response to be sent if the post actually matches the *pattern*.

  This model is executed whenever the chatbot receives a post from the suspect. When a post matches a defined rule, it means that the conversation has reached the context of interest. As a result, the corresponding reply is provided to continue the proper flow of the interactions. More details are discussed in depth in Section 4.

  However, and despite the effective management of controlled posts by the retrieval-based model, this component is unfortunately not suitable for out-of-context situations. The knowledge base is not infinite, and there may be cases where no answer can be found (there is not a matching rule for a received post). As C^3-Sex should be able to send valid responses, even in the absence of matching rules, a second model comes into play to alleviate such a shortcoming.

- Generative model: This component is activated whenever the model mentioned above does not know how to respond in the first instance. In this case, the generative model can reply to requests out of our scope of online child sexual abuse content, allowing C^3-Sex to still send valid responses and not get blocked when the conversation goes sideways.

  This model uses memory to maintain the learned patterns, and when a question (post) comes from the chat, the algorithm replies with the patterns saved in memory. The specific configuration of the model is addressed in Section 5. Note that responses from a generative model may be more natural than those from a retrieval-based model, but may contain grammatical failures or sometimes have no sense. Some examples can be appreciated in Table 2.

Listing 1. Example of Artificial Intelligence Markup Language (AIML) Rule.

```
<aiml version="1.0">
  <category>
    <pattern>* DO * WANNA * WITH ME
    </pattern>
    <template>actually im looking
    for some hot content, do you
    have?</template>
  </category>
  ...
</aiml>
```

Table 2. Example of replies from the generative model of C^3-Sex.

Post	Reply
I need some good content	I will make this one more!
I want weird porn material	I don't think so. Weird?
What are you doing now?	I am chatting with you.
I want to offer you something, however it is a little weird and illegal	Is it weird that I want to be the case?

As seen before, the three elements (interaction channel, retrieval-based model and generative model) are strongly linked to the chatbot functionality. Additionally, it is crucial to store the conversation maintained by the interactive module in a database so that the next module can sucessfully operate.

4.2. Analysis Module

The analysis module is activated once the conversation has finished, that is, when the suspect leaves the chatroom/app or C^3-Sex receives (but does not download) media content. This module is responsible for characterizing the conversation. Specifically, there are two models which process and analyze the stored messages:

- Emotional classification model: It assesses the underlying emotions within a given conversation. This model classifies each suspect's response into one of the following six emotions: anger, disgust, fear, joy, sadness, and surprise. The emotional classifier makes use of a Support Vector Machine (SVM) as classification and supervised learning paradigm (see Section 5).
- Opinion classification model: It extracts the opinion inclined towards (or against) specific topics. In this case, a message is categorized as a positive claim or as a negative statement. The opinion classification model is an excellent complement to the emotional classification model for comparing results and guaranteeing consistency. The opinion classifier uses a Multinomial Naïve Bayes algorithm, as specified in Section 5.

Once the conversation managed by the interactive module has finished, and the consequent emotional and opinion metrics have been obtained, the chatbot returns a vector of features that describes suspect's behavior. These features are detailed in Section 6. Nonetheless, before studying the profiling vector, we will detail exactly how each of the mentioned AI models has been built and developed.

5. Artificial Intelligence Models in C^3

Every AI-based model of our proposed chatbot has followed a generic data science life cycle to enable a custom, automatic, and intelligent Artificial Conversational Entity. Specifically, each model has gone through the following phases [36]:

1. Business understanding: Definition of the data context where the solution will be deployed and executed.
2. Data acquisition: Step focused on obtaining data that will be used and applied in the modeling phase.
3. Modeling: Specific technologies, algorithms, or paradigms are selected to achieve the defined goals. The selected models should be configured and trained to adapt them to the use case (defined in step 1) through useful information (collected in step 2).
4. Deployment: The AI elements are jointly embedded in the system to enhance the effectiveness and efficiency of the tool. The modeled tool (developed in step 3) is launched in real scenarios as a validated product that works according to its mission. This phase involves continuous updating and maintenance of the application.

Table 3 shows this data science life cycle for the C^3-Sex models.

Table 3. Data science life cycle for C^3-Sex models.

Model	Business Understanding	Data Acquisition	Modeling	Deployment
Retrieval-based model	Child illegal content	Context information	100 AIML rules	Context responses
Generative model	Child illegal content	Cornell Movies (https://www.cs.cornell.edu/~cristian/Cornell_Movie-Dialogs_Corpus.html)	LSTM-NN	Out-of-context responses
Emotion classifier	Pedophiles	SemEval [37]	Support Vector Machine	Sentiment extraction
Opinion classifier	Pedophiles	Opinion dataset [38]	Multinomial Naive Bayes	Opinion extraction

5.1. Business Understanding

Social media has brought many advantages to our society, but it has also led to the generation of comfortable spaces for dishonest, beguiler, and deceitful people [39]. Apart from social networks [40], other services have been successful in recent years, such as the Deep Web or anonymous chats due to their private nature. These virtual environments mask users interested in certain crimes, such as child sexual abusers, protecting them from being arrested.

Given that C^3-Sex is designed to explicitly detect criminals related to child sexual content, it is interesting to deploy it in those websites or scenarios where it is most likely to interact with suspects. In this regard, our context is defined by the Omegle platform where there are chatrooms under the 'sex' topic, with the possibility of migrating to other anonymous applications such as Telegram or Snapchat. Therefore, our software-driven agent should operate correctly in holding conversations in those platforms, considering their jargon, common patterns, and specific properties.

The retrieval-based model and the generative model should be prepared to manage conversations around sex and minors, whereas the emotion classifier and opinion classifier should work adequately to categorize the behavior and mood of potential pedophiles. The combination of the four models should guarantee:

- Maintenance of a fluid conversation in an anonymous chatroom that is primarily about sex, even if a change of platform is required.
- Analysis of the stored messages to characterize the suspect's interest in obtaining or distributing illegal child content.
- Profile the suspects depending on their level of interest towards online child sexual abuse content.

5.2. Data Acquisition

AI models should be trained with datasets to configure their operation. Given those datasets, AI models, and particularly machine learning algorithms, learn from observations and extract patterns. Once the model is configured, validated, and ready, the data of our context feeds those technologies to produce the desired outputs.

The retrieval-based model is used for knowledge representation and is built upon a series of rules focused on maintaining a conversation pivoting around the topic of online child sexual abuse. It is worth noting that this model is built with handmade rules, so it is not explicitly trained. However, a previous data acquisition is essential before creating rules for the context.

The more sophisticated the rules, the better the agent's simulation. In this regard, it is vital to understand the use case and analyze conversations in order to create custom, effective and valid rules.

Secondly, the generative model has been trained with the Cornell Movie Dialogs dataset [41]. This dataset contains conversations between characters from more than 600 movies. In this sense, the model gets configured with a naturalness attractive for the bot to respond in casual or jovial situations, for which a very elaborated response is not expected.

The emotion classifier is a model trained with the SemEval dataset [37]. This dataset contains news headlines from major newspapers (e.g., The New York Times, CNN, BBC News), each of them labeled through a manual annotation with one of the following six emotions: anger, disgust, fear, joy, sadness, and surprise. Therefore, this classifier is able to categorize a sentence or text in one of those six possible feelings.

In turn, the opinion classifier was trained with a dataset [38] containing reviews of movies, restaurants, and other products labeled as positive (1) or negative (0). Thanks to this classifier, we can categorize chat interactions.

5.3. Modeling

Once the data has been selected, found, or collected, the AI models should be configured, trained, and validated for the use case. To this extent, it is necessary to define the hyperparameters of the models (their configuration) to determine their behavior. Subsequently, they are trained with labeled datasets so that they learn to predict or classify. In the production environment, each model responds following the patterns learned from the training phase.

The retrieval-based model has been build with 100 AIML rules. These rules are used for different phases of the interaction: the formation of the friendly relationship (focused on capturing suspect's interest), the establishment of the sexual relationship (which already conveys an interest in the exchange of sexual material) and the assessment of risk (to gain the trust of the suspect). The old version of C^3-Sex [12] used a set of 60 AIML rules, so it has been augmented from a deeper review of experienced conversations between the C^3-Sex and suspects. Particularly, 40 additional AIML rules that consider new terms used in sexual-related conversations were created, giving the C^3-Sex the capacity to answer properly in more situations. Also, this new set of AIML rules hinders an easy identification of C^3-Sex as a chatbot by the suspects. Some examples of new AIML rules are shown in Listing 2.

The generative model responds according to the message sent by the suspect. This model was built using a Long Short Term Memory (LSTM) Neural Network (NN) [42] and trained with the Cornell Movie Dialogs dataset, where each instance of data is preprocessed by eliminating additional blank spaces, numbers and special characters, and treated as a Post-Reply message. Thus, to train this model, 50 epochs (top-down analysis of the datasets) and approximately four days of continuous execution were necessary. In addition, different manual validations were performed over the chatbot to analyze the error generated in the training of the LSTM-NN, avoiding both overfitting and underfitting.

Listing 2. Example of added AIML Rules.

```
<aiml version="1.0">
    <category>
        <pattern>* FEEL *</pattern>
        <template>cool, u know whats more exciting? pics haha</template>
    </category>
    ...
    <category>
        <pattern>* link to * child *</pattern>
        <template>
        i d like but i prefer in private add me snap, my nickname
        is p\_ramirezxxx
        <template>
    </category>
    ...
    <category>
        <pattern> * bored * bit horny </pattern>
        <template>
        I m horney and bored, do u wanna shared some content
        </template>
    </category>
</aiml>
```

The emotion classifier is based on the supervised method of Support Vector Machine (SVM). Given the SemEval dataset of sentences labeled with associated emotions (among the six possible), SVM is trained to seek those hyperplanes that separate the groups of sentences of each class. Consequently, the emotion classifier learns to interpret the emotion of future inputs with a rate of 0.5, which indicates how much the model changes in response to the estimated error each time that the model weights are updated [43]. The model can map a message within a specific hyperplane, that is, within an emotion.

The opinion classifier uses a Multinomial Naïve Bayes solution [44] with a simple preprocessing (steaming, removing stopwords, among other techniques) and an alpha of 1. It calculates the sentiment of texts considering the joint probabilities between the appearing words and existing sentiments (in our case, negative or positive). The representation of messages is based on a matrix with the words as columns.

5.4. Deployment

The deployment of C^3-Sex (https://github.com/CrkJohn/Snapchat, https://github.com/Santiago-Rocha/PGR) may be done through a connection with an online chat service where it can be possible to interact with different suspects. The four designed models should work together within the same system as a unique entity. In this manner, suspects directly interact with C^3-Sex through the interaction channel, remaining unaware that the communication is actually conducted with its AI-powered software modules.

In the interactive module, the communication with Omegle is done through a Chrome driver handled by functions of the Python library Selenium (https://pypi.org/project/selenium), which allows manipulating the DOM tree of the Omegle interface. On the other hand, the Snapchat management was implemented in the Snapbot (https://github.com/CrkJohn/Snapchat) module with the usage of Appium (http://appium.io/), Selenium, and Android Studio (https://developer.android.com/studio). Appium and Selenium allow us to manipulate the application as if automation tests or unit tests were conducted, and Android Studio allows us to know the identifiers of each component like login button, text entry, or send button. Therefore, we could achieve smooth and error-free navigation on the Snapchat. A physical mobile device was employed to install the Snapbot module and manage Snapchat conversations.

The implementation of the retrieval-based model was based on AIML (Artificial Intelligence Mark-up Language). Such model makes use of the library python-aiml (https://pypi.org/project/

python-aiml/), which serves as an inference engine to read XML files in the AIML way, which compose the entire knowledge base of the retrieval-based conversational module. Additionally, the generative model of C^3-Sex was implemented using Python 3 with a variety of libraries: (i) TensorFlow (https://www.tensorflow.org) as Machine Learning library used to build the recurrent neural network (RNN) which simulates the generative conversational agent, (ii) Sklearn (https://pypi.org/project/scikit-learn/) for the adoption of ML algorithms, such as the Bayesian network in the opinion classification model, and (iii) Pandas (https://pandas.pydata.org), a data analysis library used to manage and read data structures, such as CSV and dataframes.

Concerning the analysis module, the opinion classifier has been developed with Pandas Section 5.4 and the emotion classifier has been developed with R (https://www.r-project.org/).

The chatbot should be coherent, cohesive, and tolerant against unforeseen situations. To this end, C^3-Sex interventions are handled by the generative and retrieval-based models, generating trust with the suspect and subtly giving direction to the conversation to finally guess the possession of or expectation about possessing illegal content. In those cases where there is an imminent delivery of multimedia, a change of online chat service should be suggested in the conversation since Omegle does not allow to exchange images or videos. In fact, the change of platform must not affect the operation of the modules.

At the final step, the chatbot should be able to analyze the suspect's mood in his/her messages and his/her opinion on specific topics. For this purpose, C^3-Sex uses the emotion and opinion classification models to analyze the conversation once it has ended. This analysis of the suspect's behavior in conjunction with other metrics will be used to definitively build the profiling vector of the suspect.

6. Profiling Vector of the Suspect

Once a conversation is completed in both Omegle and Snapchat platforms, the entire record of such conversation is thoroughlyanalyzed to extract metrics about the the suspect's interest in child sexual abuse content. Finally, the profiling vector is completed with 25 features extracted and calculated from the interaction with the suspect. These features are shown in Table 4.

Features f_{1-6} refer to metrics of time for the conversation maintained between the C^3-Sex and the suspects. These metrics are important to identify how much a suspect is interested in a conversation. This due to the conversation time could be considered as an indicator of the interest of the suspect in the conversation topic. Features f_{8-11} aims to identify the activity of the suspect, and the subsequent activation of the generative or retrieval models to generate a response from the C^3-Sex to the suspect. Analyzing the number of responses generated by each model is critical to understand the behavior of the C^3-Sex.

If the proportion of replies from the retrieval model f_9 is bigger than the proportion of replies from the generative model f_{10}, for a total suspect posts (f_8), it is possible to deduce that the suspect has an affinity with child sexual abuse content, as the retrieval model only gets activated when a message coming from the suspect matches an AIML rule built specifically for the child sexual abuse context.

Features f_{12-13} refer to the amount of links to external sites that the suspect shares, which is useful to quantify the grade of confidence that the suspect has gained, as the existence of more external URLs implies a bigger willing to share content.

In turn, features f_{14-19} and f_{24} allow to determine how pleasant or enjoyable a conversation was for the suspect according to the emotions identified in the suspect posts by the emotion classifier. Positive emotions (Joy, Surprise) imply that the suspect is comfortable with the conversation topic.

Features f_{20-23} represent the general opinion of the suspect about the conversation topic maintained with the C^3-Sex . This also helps to identify the affinity of the suspect with a specific topic.

At last, f_{25} allows to identify the response time between a message of the C^3-Sex and a reply from the suspect, which is also an important aspect to determine the interest of the suspect in a conversation topic, due to the fact that a shorter response time may imply a greater interest.

Table 4. Suspect profiling features.

Feature	Name	Domain	Description
f_1	Initial timestamp in Snapchat	Time	Full timestamp (year, month, day, hour, minute, second) of the start of the conversation in Snapchat
f_2	Initial timestamp in Omegle	Time	Full timestamp (year, month, day, hour, minute, second) of the start of the conversation in Omegle
f_3	Final timestamp in Snapchat	Time	Full timestamp (year, month, day, hour, minute, second) of the end of the conversation in Snapchat
f_4	Final timestamp in Omegle	Time	Full timestamp (year, month, day, hour, minute, second) of the end of the conversation in Omegle
f_5	Snapchat duration	$\mathbb{Q}^+$	Total time on the Snapchat website in seconds, subtraction between metrics f_1 and f_3
f_6	Omegle duration	$\mathbb{Q}^+$	Total time on Omegle in seconds, subtraction between metrics f_2 and f_4
f_7	Conversation length	$\mathbb{N}^+$	Number of interactions by suspect and bot.
f_8	Suspect posts	$\mathbb{N}^+$	Number of times that the suspect interacted with our bot
f_9	Suspect posts about sex	[0–1]	Proportion of times that child sexual abuse rules (from the retrieval-based model) matched the total of suspect posts. (f_8)
f_{10}	Suspect posts not about sex	[0–1]	Proportion of times that the generative model responded because the retrieval model could not respond over the total of suspect posts (f_8)
f_{11}	Received bytes	$\mathbb{N}^+$	Total number of bytes of text received by the suspect
f_{12}	URLs from Snapchat	$\mathbb{N}^+$	Number of hyperlinks received by the suspect in Snapchat
f_{13}	URLs from Omegle	$\mathbb{N}^+$	Number of hyperlinks received by the suspect at Omegle
f_{14}	Anger	[0–1]	Rate between the number of times that the anger emotion was identified in the suspect posts and f_8
f_{15}	Disgust	[0–1]	Rate between the number of times that the disgust emotion was identified in the suspect posts and f_8
f_{16}	Fear	[0–1]	Rate between the number of times that the fear emotion was identified in the suspect posts and f_8
f_{17}	Sadness	[0–1]	Rate between the number of times that the sadness emotion was identified in the suspect posts and f_8
f_{18}	Joy	[0–1]	Rate between the number of times that the joy emotion was identified in the suspect posts and f_8
f_{19}	Surprise	[0–1]	Rate between the number of times that the surprise emotion was identified in the suspect posts and f_8
f_{20}	Negativity	[0–1]	Rate between the number of times that a negative post was identified by the opinion model and f_8
f_{21}	Positivity	[0–1]	Rate between the number of times that a positive post was identified by the opinion model and f_8
f_{22}	Neutrality	[0–1]	Rate between the number of times that a neutral post was identified by the opinion model and f_8
f_{23}	Opinion about sex	[0–1]	Average of the opinions found in the suspect's posts about sex, where 0 is a negative opinion and 1 is a positive opinion. Not neutral opinions are considered.
f_{24}	Emotion about sex	[0–1]	Average of the emotions found in the suspect's posts about sex, where 0 is a negative emotion (anger, disgust, fear, sadness) and 1 is a positive emotion (joy, surprise)
f_{25}	Response speed	$\mathbb{Q}^+$	Average time between C^3-Sex messages and suspect responses

7. Experiments

Our proposal's suitability has been validated through different experiments that aim to verify the chatbot's effectiveness in (i) holding a conversation with a human, (ii) characterizing the suspect's behavior with features, and (iii) automatically contacting a big number of users. Figure 3 shows the design of the experiments.

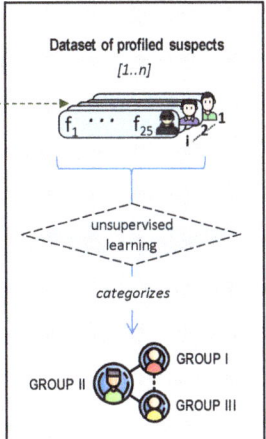

Figure 3. Experiments launched with C^3-Sex.

The settings of the experiment are described in Section 7.1, while the analysis of results is carried out in Section 7.2.

7.1. Experimental Setup

Experiments were conducted by running an instance of a C^3-Sex that connects to the online chat platform Omegle. Omegle was chosen because it allows to easily start communication with a peer in some place of the word around a common interesting topic. For these experiments, the interesting topic that was configured was "sex". In this way, the C^3-Sex may start a conversation with a suspect, who has also typed "sex" as interesting topic during the access to the Omegle chat service.

In the experiments, C^3-Sex simulated the behavior of a person interested in online child sexual abuse material, building trust for the suspects to transfer content and allowing profiling them according to the previously defined metrics based on the conversation performed.

The chatbot was executed for 50 days to contact and profile suspects through Omegle automatically. The workflow of each interaction of the C^3-Sex with a suspect is detailed in the left of Figure 3:

1. C^3-Sex joins into an Omegle chat room typing sex as an interesting topic.
2. Suspect i also logs in to the chat room.
3. The conversation starts between the user and the chatbot. In case the suspect is willing to exchange multimedia content, the C^3-Sex suggests to use Snapchat to exchange the multimedia content.
4. At some point, suspect i leaves the chat room.
5. Accordingly, C^3-Sex closes the conversation.
6. Finally, C^3-Sex analyzes the conversation using the profiling metrics and AI models (emotional classification and opinion classification model), calculates the features, and saves the associated profiling vector in a dataset for further analysis. Then, the C^3-Sex returns to step 1 to start a new conversation with another suspect.

As a result of 50 days of experimentation, the chatbot has built a dataset of n profiled suspects composed by the associated n vectors of features. In the following section, we will study in-depth the results of the experiments, executed for 50 days (from 19-04-2020 to 07-06-2020), time in which the C^3-Sex was able to contact 7199 suspects.

7.2. Classification of Suspects

After completing the 50 days of activity, we had a database containing the information about 7199 suspects, including the 25 features from f_1 to f_{25} calculated for each of them. The operation of the chatbot has finished, and the investigator should inspect the dataset to detect pedophiles. In our case, the high complexity of analyzing these unlabeled users by hand led us to propose an unsupervised learning method to provide a first aggregated description of the results. This analytical procedure may support the competent authorities with the manual exploration and identification of sex offenders.

Since our ultimate goal is to detect perverts, we individually pursued those users who demonstrated suspicious patterns in obtaining or transmitting sexual content. With this aim, a possible solution was to differentiate various groups within our dataset, which shared common ways of thinking, behaving, and acting, that is, with similar metrics. Thanks to our subsequent expert analysis of the groups that emerged, we could infer different types of suspects according to their features. One or more groups could share aspects related to pedophilia.

In this context, we chose the K-Means clustering method [45] due to its simplicity and efficiency, linear computational complexity, and convergence in detecting clusters. Our dataset did not present particularities to force us to use a specific algorithm in this sense, and K-Means maintains an excellent accuracy–complexity ratio.

K-Means algorithm classified each instance of the dataset in the most similar cluster among the existing K groups. In particular, this method is based on the distances between instances' features, so the cluster assigned was the one in which the centroid was closest to the instance in the feature multidimensional space. Due to this algorithm's nature, it operates better as there are fewer dimensions (features).

In our specific use case, the original set of 25 features would have produced a large multidimensional space and subsequent inaccurate clusters. For this reason, we performed a feature selection process in which we tested several combinations among the 25 features, always standardizing them and reducing the dimensionality with a Principal Component Analysis (PCA, limited to 95% of explained variance). As a result, we obtained that the configuration with lower values of distortion was formed by the features f_5 (Snapchat duration), f_6 (Omegle duration), f_9 (Suspect posts about sex), f_{23} (Opinion about sex), f_{24} (Emotion about sex) and f_{25} (Speed of response). Fortunately, those metrics aggregate other (discarded) features and are, from our intuition, truly relevant for the search of perverts.

Finally, considering only the selected features, we calculated the optimal number of clusters K with the distortion score elbow method [45]. Assuming that a larger number of clusters always groups more precisely (in its maximum case, each instance would belong to its own cluster), the idea is to find the lowest possible K that guarantees the best balance between the average distance of the cluster points to its centroid (distortion) and the total number of clusters. Figure 4a shows that $K = 3$ is the optimal number of groups in our dataset to divide the suspects faithfully. On the other side, Figure 4b represents the inter distances between the cluster centers as a result of a Multidimensional Scaling (MDS). The clusters' size is proportional to the number of elements they contain, in particular 1489, 1673 and 4037 instances.

Based on the formed groups, in the following sections, we characterize and compare each group of suspects according to their features.

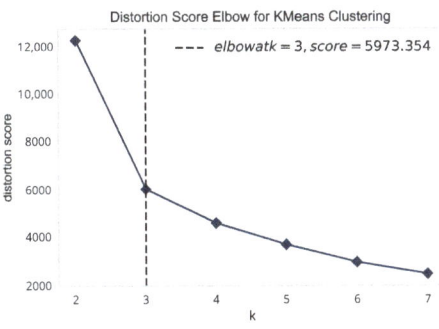 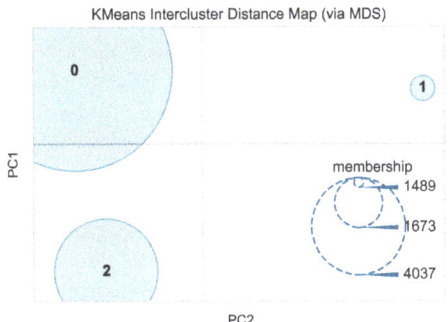

(**a**) Distortion scores per number of possible clusters   (**b**) Clusters with the best accuracy and performance

Figure 4. Application of K-Means to generate groups of suspects.

7.3. Comparative of Detected Groups

Once we detected groups of similar suspects, we compared them according to the selected features. To provide a clear and homogeneous view of the differences, we applied a quantile-based scaler to the features (where values go from 0 to 1). Note that this transformation is applied to each feature independently and maps the original values to a uniform distribution using quantiles information. Another type of transformation could significantly offer bad visualizations for comparison, alter the data distribution, or suffer a lot from outliers. Conversely, the quantile transform respects the essence of the variability of the data and maintains the so valuable outliers. Despite some distribution distortion, the conversion facilitated a great comparison. Note that we did not want to remove anomalies because those specific antipatterns might potentially expose pedophiles.

In this regard, Figure 5 shows the high-level comparative of each group according to the scaled features. For this particular experiment we have deducted that the three found groups refer to the categories of indifferent, interested, and pervert as we will explain next. These assigned names for each cluster come from the analysis of the features of the included conversations, but it is not a judgment statement, and must just be considered as a way to qualitative describe the results.

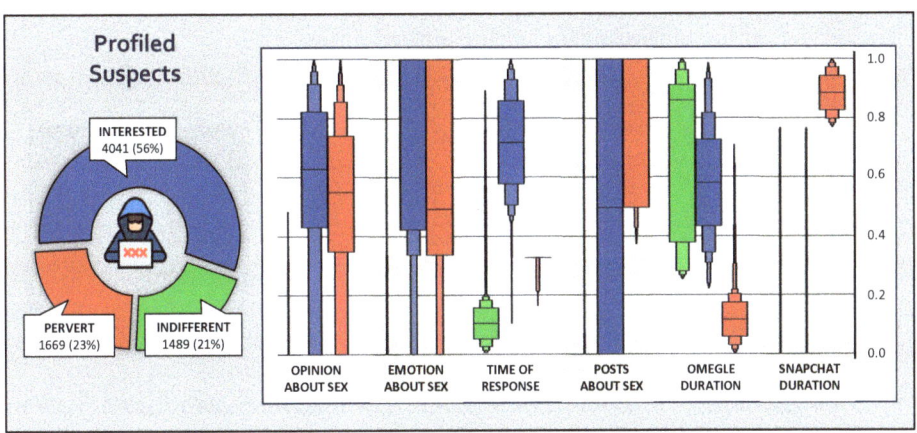

Figure 5. Comparative of the different types of profiled suspects.

7.3.1. Indifferent

We have considered the green group, the cluster with 1489 elements, to be the indifferent users. In terms of online child sexual abuse, they have a bad opinion about it, demonstrate bad emotions when dealing with child abuse topics, and do not send sexual messages. Moreover, they are also the quickest to respond, which could demonstrate an apparent devotion to rejecting the transmission of compromised content. Another proof of the latter is that they spend all the conversation on Omegle, the platform that does not allow sending multimedia, and when the change to Snapchat is made, they leave.

7.3.2. Interested

The cluster denoted in blue, composed of 4041 suspects, demonstrates a higher interest. In general, they have a favorable opinion about child abuse proposals and present positive emotions in their sexual interactions. Nevertheless, in this group, there are users with different frequencies in posting sexual content. In this sense, one could say that these suspects might be interested in reading or talking about sex, without welcoming the transmission of content. Therefore, they are not fast to respond, which might indicate skepticism when talking about sex, although they do not reject it. Also, they focus their activity in Omegle, where no photos or videos are allowed.

7.3.3. Pervert

Finally, the red-colored group is formed by 1669 suspects and presents a similar interest in child sex content as interested users. However, the main difference with the rest of the groups is their predisposition for the transmission of sexual content. They demonstrate a receptive attitude to having a conversation about sex as the interested group, but they take little time to respond, they spend a short time in Omegle and register a high activity in Snapchat, the platform to which users move to exchange multimedia.

Therefore, having described the different types of profiled suspects, in the following section, we analyze in-depth the values of the different metrics by groups to precisely characterize their behavior.

7.4. Behavior of Profiled Suspects

Considering the principal differences among the distinguished groups, it is worth analyzing each group's specific behavior in detail. Figure 6 compares the distribution of values per detected group for the selected features. Note that it is a more accurate perspective than Figure 5, which served us to categorize each group.

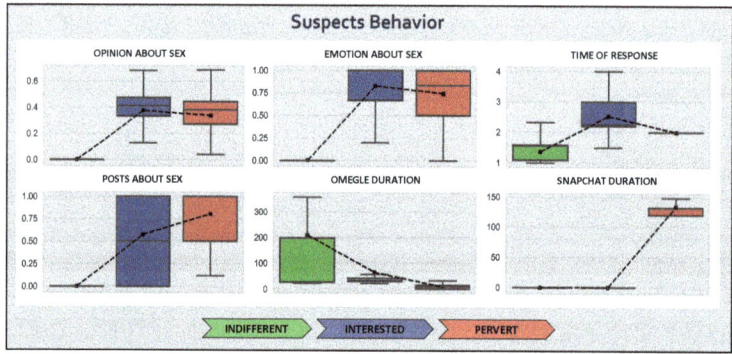

Figure 6. Behavior of profiled groups.

First of all, interested and pervert suspects maintain a positive tendency around 0.4, reaching even a 0.7 of positivity, in child abuse interactions. On the other hand, as indifferent users practically do not send messages about child sexual aspects, it demonstrates a totally negative opinion against minor contents.

Moreover, interested and pervert suspects experience very positive emotions within sexual messages, more than 0.75 of emotional positivism for the former in most cases, and more than 0.50 in the latter. On the contrary, indifferent users also show strong negative emotions since they do not post about sex.

In terms of response time, indifferent users generally take less than 2 s to respond. On the other hand, interested users typically require 2 and 3 s to respond, eventually reaching 4 s. Surprisingly, perverts spend exactly two seconds, on average, to interact with our chatbot. This non-variable value makes us suspect that these users, apart from being interested in sex, were actually programmed to interact automatically, that is, they were probably bots controlled by software. Nevertheless, these numbers demonstrate that all users using Omegle are interested in chatting and go away from keyboard.

As mentioned before, indifferent users do no usually send messages with sexual content. Regarding interested users, there is a continuous uniform distribution at all levels, from suspects who barely post about sex to others who mention sexual topics within almost 100% of their messages. Additionally, most perverts send sexual messages in over 50% of their interactions.

Finally, the differences in chatroom duration are significantly disparate. In the Omegle conversation, indifferent users spend up to 200 s (3–4 min) in most cases, while interested and pervert suspects do not usually reach 50 s in the best cases. When moving to Snapchat, both the indifferent and interested users definitely leave the conversation. Yet, perverts remain online in Snapchat for a considerable period of 2 to 4 min in predisposition to broadcast or receive multimedia.

With these results, C^3-Sex is proven to be a powerful tool for profiling suspects based on premeditated metrics. The following sections show how long the chatbot has been running in order to profile 7199 users effectively.

7.5. Experiments Overview and Limitations

C^3-Sex is an automatic and software-controlled conversational agent deployed on the web that interacts with users autonomously. Figure 7 represents the weekly temporal activity of our chatbot in the experiments, showing its high capacity to analyze the network. On average, C^3-Sex can interact with 900 suspects weekly, a value that is already high compared to a manual intervention replicated by a human. In particular, in the seventh week, the chatbot chatted with more than 500 users, and in weeks 2, 3, and 5, it surpassed the incredible number of 1500 suspects. Moreover, in the second week C^3-Sex was able to maintain contact with nearly 2500 network users.

The diagram helps us to evaluate the chatbot very positively, which was able to stay online throughout the eight weeks of the experiment, with a total of 7199 users contacted. Therefore, this tool has is a valuable support for profiling people on a large scale and with descriptive statistics of behavior.

However, C^3-Sex also has some limitations. Although it profiles suspects through a wide range of features, it is unfeasible for the current version of the C^3-Sex to infer the potential intention of the suspect or the probability of being a pedophile. Such determination is the job of the cyber intelligence analyst who employs C^3-Sex and who must analyze the generated clusters to determine the type of users allocated in each one. A coming step in the evolution of C^3-Sex could be to use the 7199 conversations, which were categorized with an unsupervised method in Section 7 as belonging to indifferent, interested or pervert suspects, to train a supervised model that may be able to make an automatic categorization of the new suspects that interact with C^3-Sex. On the other hand, artificial intelligence models included in Table 3 could be trained with more data and thus achieve a more intelligent chatbot when interacting and analyzing.

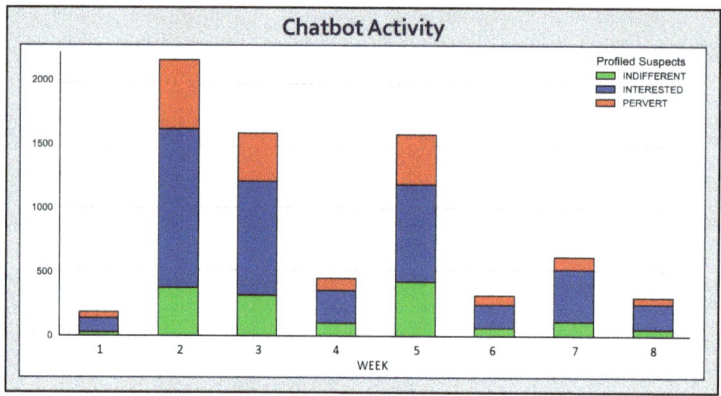

Figure 7. Chatbot activity.

8. Conclusions and Future Work

With the aim of humbly contributing to the honorable task of prosecuting sexual crimes, specifically online child sexual abuse, our conversational agent C^3-Sex has been proposed along with this paper. C^3-Sex is composed of four models, namely: Retrieval-based, Generative, Emotional classification, and Opinion classification. Altogether, these models constitute a solution able to keep conversations with suspects and profile them with up to 25 features related to the conversation's content and the behavior in the chatroom. The proposed analytic classification task reveals that these profiling features segregate different kinds of users and behaviors. The final goal of C^3-Sex is to profile users and expose potential holders and bidders of illegal content related to online child sexual abuse, who can later be investigated by a Law Enforcement Agency.

As future work, we are willing to investigate and compile real chats of sex offenders to implement a supervised model and provide a robust categorization mechanism. The latter would also allow the validation of the tool with a specific numeric accuracy. In fact, in the next release, the chatbot could include the percentage of the dangerousness of the other peer, having predictive functions on-the-fly as the conversation goes on. We also plan to improve the models that compose our chatbot, so a more human-like interaction between the chatbot and the suspects can be performed, reducing the probability that the suspect can unveil C^3-Sex. This should be achieved by generating more specific AIML rules for the retrieval model and training the generative model with a dataset associated with a context of sexual conversations. Additionally, in the future, we expect to address other types of sexual crimes related to children, like grooming, sexual exploitation, sexting, sextortion, sex scam, or sex trafficking. Some of these new types of sexual crimes would require C^3-Sex to keep more complex conversations for a longer time.

Author Contributions: Conceptualization, D.D.-L.; methodology, D.D.-L. and F.G.M.; software, J.I.R. and S.R.D.; validation, J.P.-G.; formal analysis, J.I.R., S.R.D. and D.D.-L.; investigation, J.I.R., S.R.D. and J.P.-G.; resources, D.D.-L.; data curation, J.I.R., S.R.D. and Javier Pastor-Galindo; writing—original draft preparation, J.I.R., S.R.D. and D.D.-L.; writing—review and editing, J.P.-G. and F.G.M.; visualization, J.P.-G. and F.G.M.; supervision, D.D.-L. and F.G.M.; project administration, D.D.-L. and F.G.M.; funding acquisition, D.D.-L. and F.G.M. All authors have read and agreed to the published version of the manuscript.

Funding: This work has been partially supported by the Colombian School of Engineering Julio Garavito (Colombia), by the Escuela de Ingeniería, Ciencia y Tecnología and the Dirección de Investigación e Innovación at the Universidad del Rosario (Colombia), by an FPU predoctoral contract (FPU18/00304) granted by the Spanish Ministry of Science, Innovation and Universities, as well as by a Ramón y Cajal research contract (RYC-2015-18210) granted by the MINECO (Spain) and co-funded by the European Social Fund.

Conflicts of Interest: The authors declare no conflict of interest.

Abbreviations

The following abbreviations are used in this manuscript:

ACE	Artificial Conversational Entity
AI	Artificial Intelligence
AIML	Artificial Intelligence Markup Language
ECM	Emotional Chatting Machine
KNN	K-Nearest Neighbor
LEA	Law Enforcement Agency
LSTM-NN	Long Short Term Memory Neural Network
MCA	Membership, Categorization, Analysis
MDS	Multidimensional Scaling
NLG	Natural Language Generator
NLP	Natural Language Processor
P2P	Peer to Peer
PCA	Principal Component Analysis
SVM	Support Vector Machine
VoIP	Voice over IP

References

1. Díaz López, D.O.; Dólera Tormo, G.; Gómez Mármol, F.; Alcaraz Calero, J.M.; Martínez Pérez, G. Live Digital, Remember Digital: State of the Art and Research Challenges. *Comput. Electr. Eng.* **2014**, *40*, 109–120. [CrossRef]
2. Pina Ros, S.; Pina Canelles, A.; Gil Pérez, M.; Gómez Mármol, F.; Martínez Pérez, G. Chasing offensive conducts in social networks: A reputation-based practical approach for Frisber. *ACM Trans. Internet Technol.* **2015**, *15*, 1–20. [CrossRef]
3. Taylor, M.; Quayle, E. *Child Pornography: An Internet Crime*; Psychology Press: London, UK, 2003.
4. Adams, W.; Flynn, A. *Federal Prosecution of Commercial Sexual Exploitation of Children Cases, 2004–2013*; US Department of Justice, Office of Justice Programs, Bureau of Justice: Washington, DC, USA, 2017.
5. Krone, T. International Police Operations Against Online Child Pornography. In *Trends and Issues in Crime and Criminal Justice*; Australian Institute of Criminology: Canberra, Australia, 2005; Volume 2005.
6. Wolak, J.; Liberatore, M.; Levine, B.N. Measuring a year of child pornography trafficking by U.S. computers on a peer-to-peer network. *Child Abus. Negl.* **2014**, *38*, 347–356. [CrossRef] [PubMed]
7. McIntyre, T.J. Blocking child pornography on the Internet: European Union developments. *Int. Rev. Law Comput. Technol.* **2010**, *24*, 209–221. [CrossRef]
8. Ulges, A.; Stahl, A. Automatic detection of child pornography using color visual words. In Proceedings of the 2011 IEEE International Conference on Multimedia and Expo, Barcelona, Spain, 11–15 July 2011; pp. 1–6. [CrossRef]
9. Gottfried, E.D.; Shier, E.K.; Mulay, A.L. Child Pornography and Online Sexual Solicitation. *Curr. Psychiatry Rep.* **2020**, *22*, 10. [CrossRef] [PubMed]
10. Gao, J.; Galley, M.; Li, L. Neural Approaches to Conversational AI. *Found. Trends Inf. Retr.* **2019**, *13*, 127–298. [CrossRef]
11. Walker, M. Hype Cycle for Emerging Technologies, 2018. In *2018 Hype Cycles: Riding the Innovation Wave*; Gartner: Stamford, CT, USA, 2018.
12. Murcia Triviño, J.; Moreno Rodríguez, S.S.; Díaz López, D.O.; Gómez Mármol, F. C3-Sex: A Chatbot to Chase Cyber perverts. In Proceedings of the 4th IEEE Cyber Science and Technology Congress, Fukuoka, Japan, 5–8 August 2019; pp. 50–57. [CrossRef]
13. Kowalski, S.; Pavlovska, K.; Goldstein, M. Two Case Studies in Using Chatbots for Security Training. In *Information Assurance and Security Education and Training*; Dodge, R.C., Futcher, L., Eds.; Springer: Berlin/Heidelberg, Germany, 2013; pp. 265–272.

14. Sabbagh, B.A.; Ameen, M.; Wätterstam, T.; Kowalski, S. A prototype For HI2Ping information security culture and awareness training. In Proceedings of the 2012 International Conference on E-Learning and E-Technologies in Education (ICEEE), Lodz, Poland, 24–26 September 2012; pp. 32–36. [CrossRef]
15. Filar, B.; Seymour, R.; Park, M. Ask Me Anything: A Conversational Interface to Augment Information Security Workers. In Proceedings of the SOUPS, Santa Clara, CA, USA, 12–14 July 2017.
16. Zambrano, P.; Sanchez, M.; Torres, J.; Fuertes, W. BotHook: An option against Cyberpedophilia. In Proceedings of the 2017 1st Cyber Security in Networking Conference (CSNet), Rio de Janeiro, Brazil, 18–20 October 2017; pp. 1–3. [CrossRef]
17. Açar, K.V. Webcam Child Prostitution: An Exploration of Current and Futuristic Methods of Detection. *Int. J. Cyber Criminol.* **2017**, *11*, 98–109. [CrossRef]
18. Gapanyuk, Y.; Chernobrovkin, S.; Leontiev, A.; Latkin, I.; Belyanova, M.; Morozenkov, O. The Hybrid Chatbot System Combining Q&A and Knowledge-base Approaches. In Proceedings of the 7th International Conference on Analysis of Images, Social Networks and Texts (AIST 2018), Moscow, Russia, 5–7 July 2018; pp. 42–53.
19. Serban, I.V.; Sankar, C.; Germain, M.; Zhang, S.; Lin, Z.; Subramanian, S.; Kim, T.; Pieper, M.; Chandar, S.; Ke, N.R.; et al. A deep reinforcement learning chatbot. *arXiv* **2017**, arXiv:1709.02349.
20. Tatai, G.; Csordás, A.; Kiss, Á.; Laufer, L.; Szaló, A. The chatbot who loved me. In Proceedings of the ECA Workshop of AAMAS, Budapest, Hungary, 10–15 May 2003.
21. Plutchik, R. The nature of emotions: Human emotions have deep evolutionary roots, a fact that may explain their complexity and provide tools for clinical practice. *Am. Sci.* **2001**, *89*, 344–350. [CrossRef]
22. Qian, Q.; Huang, M.; Zhao, H.; Xu, J.; Zhu, X. Assigning Personality/Profile to a Chatting Machine for Coherent Conversation Generation. In Proceedings of the IJCAI, Stockholm, Sweden, 13–19 July 2018; pp. 4279–4285.
23. Heller, B.; Proctor, M.; Mah, D.; Jewell, L.; Cheung, B. Freudbot: An investigation of chatbot technology in distance education. In Proceedings of the EdMedia: World Conference on Educational Media and Technology. Association for the Advancement of Computing in Education (AACE), Montreal, QC, Canada, 27 June 2005; pp. 3913–3918.
24. Hill, J.; Ford, W.R.; Farreras, I.G. Real conversations with artificial intelligence: A comparison between human—human online conversations and human–chatbot conversations. *Comput. Hum. Behav.* **2015**, *49*, 245–250. [CrossRef]
25. Zhou, H.; Huang, M.; Zhang, T.; Zhu, X.; Liu, B. Emotional chatting machine: Emotional conversation generation with internal and external memory. In Proceedings of the Thirty-Second AAAI Conference on Artificial Intelligence, New Orleans, LA, USA, 2–7 February 2018.
26. Anderson, P.; Zuo, Z.; Yang, L.; Qu, Y. An Intelligent Online Grooming Detection System Using AI Technologies. In Proceedings of the 2019 IEEE International Conference on Fuzzy Systems (FUZZ-IEEE), New Orleans, LA, USA, 23–26 June 2019; pp. 1–6. [CrossRef]
27. Gunawan, F.E.; Ashianti, L.; Sekishita, N. A simple classifier for detecting online child grooming conversation. *Telkomnika Telecommun. Comput. Electron. Control* **2018**, *16*, 1239–1248. [CrossRef]
28. Ebrahimi, M. Automatic Identification of Online Predators in Chat Logs by Anomaly Detection and Deep Learning. Master's Thesis, Concordia University, Montreal, QC, Canada, 2016.
29. Dadvar, M.; de Jong, F. Cyberbullying Detection: A Step toward a Safer Internet Yard. In Proceedings of the 21st International Conference on World Wide Web, Association for Computing Machinery, New York, NY, USA, 16 April 2012; pp. 121–126. [CrossRef]
30. Laorden, C.; Galán-García, P.; Santos, I.; Sanz, B.; Hidalgo, J.M.G.; Bringas, P.G. Negobot: A Conversational Agent Based on Game Theory for the Detection of Paedophile Behaviour. In *International Joint Conference CISIS'12-ICEUTE'12-SOCO'12 Special Sessions*; Herrero, Á., Snášel, V., Abraham, A., Zelinka, I., Baruque, B., Quintián, H., Calvo, J.L., Sedano, J., Corchado, E., Eds.; Springer: Berlin/Heidelberg, Germany, 2013; pp. 261–270.
31. Michalopoulos, D.; Mavridis, I.; Jankovic, M. GARS: Real-time system for identification, assessment and control of cyber grooming attacks. *Comput. Secur.* **2014**, *42*, 177–190. [CrossRef]
32. Burnap, P.; Rana, O.F.; Avis, N.; Williams, M.; Housley, W.; Edwards, A.; Morgan, J.; Sloan, L. Detecting tension in online communities with computational Twitter analysis. *Technol. Forecast. Soc. Chang.* **2015**, *95*, 96–108. [CrossRef]

33. Thelwall, M.; Buckley, K.; Paltoglou, G.; Cai, D.; Kappas, A. Sentiment strength detection in short informal text. *J. Am. Soc. Inf. Sci. Technol.* **2010**, *61*, 2544–2558. [CrossRef]
34. Sacks, H.; Jefferson, G.; Schegloff, E. *Lectures on Conversation*; Wiley-Blackwell: Hoboken, NJ, USA, 2010.
35. Meyer, M. Machine Learning to Detect Online Grooming. Master's Thesis, Department of Information Technology, Uppsala University, Uppsala, Sweden, 2015.
36. Ericson, G.; Rohm, W.; Martens, J.; Sharkey, K.; Casey, C.; Harvey, B.; Nevil, T.; Gilley, S.; Schonning, N. Team Data Science Process Documentation. *Retrieved April* **2017**, *11*, 2019.
37. Strapparava, C.; Mihalcea, R. Semeval-2007 task 14: Affective text. In Proceedings of the Fourth International Workshop on Semantic Evaluations (SemEval-2007), Prague, Czech Republic, 23–24 June 2007; pp. 70–74.
38. Kotzias, D.; Denil, M.; De Freitas, N.; Smyth, P. From group to individual labels using deep features. In Proceedings of the 21th ACM SIGKDD International Conference on Knowledge Discovery and Data Mining, San Jose, CA, USA, 12–15 August 2015; pp. 597–606.
39. Gómez Mármol, F.; Gil Pérez, M.; Martínez Pérez, G. Reporting Offensive Content in Social Networks: Toward a Reputation-based Assessment Approach. *IEEE Internet Comput.* **2014**, *18*, 32–40. [CrossRef]
40. Pastor Galindo, J.; Nespoli, P.; Gómez Mármol, F.; Martínez Pérez, G. The not yet exploited goldmine of OSINT: Opportunities, open challenges and future trends. *IEEE Access* **2020**, *8*, 10282–10304. [CrossRef]
41. Danescu-Niculescu-Mizil, C.; Lee, L. Chameleons in imagined conversations: A new approach to understanding coordination of linguistic style in dialogs. In Proceedings of the Workshop on Cognitive Modeling and Computational Linguistics, Portland, OR, USA, 23 June 2011.
42. Hochreiter, S.; Schmidhuber, J. Long Short-Term Memory. *Neural Comput.* **1997**, *9*, 1735–1780. [CrossRef]
43. Brownlee, J. Understand the Impact of Learning Rate on Neural Network Performance. 2019. Available online: https://machinelearningmastery.com/understand-the-dynamics-of-learning-rate-on-deep-learning-neural-networks (accessed on 12 December 2019).
44. Gupte, A.; Joshi, S.; Gadgul, P.; Kadam, A.; Gupte, A. Comparative study of classification algorithms used in sentiment analysis. *Int. J. Comput. Sci. Inf. Technol.* **2014**, *5*, 6261–6264.
45. Kodinariya, T.M.; Makwana, P.R. Review on determining number of Cluster in K-Means Clustering. *Int. J.* **2013**, *1*, 90–95.

Publisher's Note: MDPI stays neutral with regard to jurisdictional claims in published maps and institutional affiliations.

© 2020 by the authors. Licensee MDPI, Basel, Switzerland. This article is an open access article distributed under the terms and conditions of the Creative Commons Attribution (CC BY) license (http://creativecommons.org/licenses/by/4.0/).

Article

Practical Implementation of Privacy Preserving Clustering Methods Using a Partially Homomorphic Encryption Algorithm

Ferhat Ozgur Catak [1,*], Ismail Aydin [2], Ogerta Elezaj [1] and Sule Yildirim-Yayilgan [1]

1. Department of Information Security and Communication Technology, NTNU Norwegian University of Science and Technology, 2815 Gjøvik, Norway; ogerta.elezaj@ntnu.no (O.E.); sule.yildirim@ntnu.no (S.Y.-Y.)
2. Cyber Security Engineering, Istanbul Sehir University, 34865 Istanbul, Turkey; ismail.aydin@tubitak.gov.tr
* Correspondence: ferhat.o.catak@ntnu.no

Received: 31 December 2019; Accepted: 24 January 2020; Published: 31 January 2020

Abstract: The protection and processing of sensitive data in big data systems are common problems as the increase in data size increases the need for high processing power. Protection of the sensitive data on a system that contains multiple connections with different privacy policies, also brings the need to use proper cryptographic key exchange methods for each party, as extra work. Homomorphic encryption methods can perform similar arithmetic operations on encrypted data in the same way as a plain format of the data. Thus, these methods provide data privacy, as data are processed in the encrypted domain, without the need for a plain form and this allows outsourcing of the computations to cloud systems. This also brings simplicity on key exchange sessions for all sides. In this paper, we propose novel privacy preserving clustering methods, alongside homomorphic encryption schemes that can run on a common high performance computation platform, such as a cloud system. As a result, the parties of this system will not need to possess high processing power because the most power demanding tasks would be done on any cloud system provider. Our system offers a privacy preserving distance matrix calculation for several clustering algorithms. Considering both encrypted and plain forms of the same data for different key and data lengths, our privacy preserving training method's performance results are obtained for four different data clustering algorithms, while considering six different evaluation metrics.

Keywords: cryptography; clustering; homomorphic encryption; machine learning

1. Introduction

In recent years, there is an increasing demand for outsourced cloud systems that allow tenants to rapidly handle sensitive data that are collected from systems, including military systems, health care systems, or banking systems. Additionally, public big data systems are needed to analyze data without breach data privacy.

Machine learning [1] is a new era that can gather valuable information to make decisions in an efficient way using several public cloud systems. Occasionally, cloud platforms contain data batches with different privacy policies; however, they still have to be analyzed in a mutual way. Consider the case of different medical institutes that want to jointly build a disease diagnosis model using a machine learning algorithm. In this case, privacy policies and General Data Protection Regulation (GDPR) prevent these medical institutes from sharing with each other [2,3]. In this case, traditional machine learning methods cannot be applied. Sensitive data cannot be distributed publicly every time, due to the different privacy policies that different parties have.

1.1. Current Solutions

How can sensitive data be distributed and computed between the different computer cluster nodes without losing their privacy? Generally, in order to find a correct answer to this problem, symmetric and asymmetric (with public-private key cryptography) encryption algorithms [4] are applied to the data. The power of cryptographic algorithms consists of key pairs built on the power of privacy and randomness.

The power of cryptographic algorithms consists of key pairs built on the privacy of cryptographic keys and the strength of randomness.

In classical asymmetric key cryptographic methods, the secret key must be shared with the cloud service provider in order to perform arithmetic operations on the data. In this case, the data owner should trust to the cloud service provider. However, the need for disclosure of confidential information arises and existing laws prevent it. In addition, many asymmetric key encryption methods are not probabilistic, thus they are only one representation of a value. For example, with a secret key $1234567890abcdef$, the result of 0 will always be $B1B38CACC1F6A6BEB068BEC2C0643185$. The data sets we use today are sparse and therefore consist of a large number of 0s. If the value 0 has only one encrypted form, it can be easily estimated by an attacker. In order to protect privacy, the encryption methods must also be probabilistic.

When a privacy-preserving technique is implemented using classical cryptographic algorithms, the computation nodes are required to have appropriate cryptographic key or keys from public/private key pairs, and the keys have to be exchanged using a secure communication channel. The biggest problem in this type of system is that the cloud server that will perform the computation has keys that can access the private data in plain form [5].

Considering privacy issues related to data handling for machine learning algorithms, there are two primary approaches;

First, using the characteristic features of datasets for the generalization and suppression for anonymization. After that, the anonymized version of data can be distributed [6] to other computation parties to execute a data analytics algorithm.

Secondly, using cryptographically secure multi-party computation algorithms to build cryptographic protocols that can compute the same result with a plain form of data using an encrypted version of the data. This approach is applicable mostly when the relationship between data sharing nodes is symmetrical. A symmetrical relationship indicates that if a dataset is partitioned and distributed to different parties, then the model that is built by a machine learning algorithm and applied to the dataset is the same. Thus, the final result of the algorithm execution shows that all parties learn the same model based on a shared dataset.

The main difference between these two methods consists of the fact that in the first approach (the anonymization approach), the computation parties do not execute machine learning algorithms on the original data, which belong to themselves, and the data owner itself does not get back an output of the computation.

The objective of the method presented in this paper is to allow a user to create different cluster models using homomorphic encryption algorithm without accessing the plain version of the data. Accordingly, the data owner also would not know anything about the data clustering model. The algorithm performance was measured in different size and the clustering performance with different metrics is investigated using various metrics.

1.2. Contribution

Currently, the data, which need to be preserved privately, are part of a large volume and may differ in a wide range of ranges. There are several approaches in the literature for providing privacy for sensitive data. Classical cryptographic methods are not sufficient in each case for privacy issues with a data system. When a process needs to run on a sensitive database with the protection of classical cryptographic systems, there will be a need for a proper key for deciphering. After that, the needed process can be executed on the decrypted version of the data. Thus, the privacy of the data is violated and there is a threat of data disclosure. Considering their competence level, it is evident that this procedure cannot be used in every situation. Of concern to this data privacy breach problem, there is always a need for a system that has to protect the confidentiality of this data when a computation is executed.

In our previous work, [7], we applied partially homomorphic encryption methods to build a probabilistic classifier using the extreme learning machine algorithm. Within the scope of this previous study, we have examined how to create classification models using homomorphic encryption algorithms. We created the privacy-protected version of the ELM algorithm, which constructs a classification model by creating a linear equation.

In this research, we have designed a system that uses Paillier Cryptography for clustering data systems without violating their privacy. This system would use the data in an efficient way while preserving the privacy of data.

The contributions of this paper are twofold. First:

- The Paillier cryptosystem encryption-based clustering model building protocol is proposed for preserving privacy and thus clustering model training is performed. Secondly:
- The computation of the distance metric matrix of four different clustering algorithms is distributed to independent parties, thus minimizing the overall computational time.

The remainder of the paper is organized as follows. Section 2 presents the related work on using cryptographic models and technologies for privacy preserving machine learning and data analysis. Section 3 describes data clustering, cluster model evaluation metrics, and homomorphic encryption. Section 4 describes our proposed privacy-preserving clustering learning model. Section 5 evaluates the proposed learning model. Finally, Section 6 concludes this paper.

2. Related Work

In this section, we describe the general overview of literature related to privacy-preserving machine learning models.

In [8], the authors suggested a privacy enhanced version of the Iterative Dichotomiser 3 (ID3) algorithm by using secure two party computation. They have stated that their learning model requires relatively less communication stages and bandwidth. In this method, the classification algorithm is combined with a decision tree algorithm, while the privacy of data is preserved as different users work. The results of each party are merged by using different cryptographic protocols. If there are more than two parties involved in the computation, their protocol cannot be applied to build a classification model. The ID3 algorithm can be used for only discrete values datasets; however, most of the today dataset contains only continuous variables.

In [9], the authors examined the balance between learnability from data and privacy, while developing a privacy preserving algorithm for sensitive data. They focused on the privacy preserving version of the logistic regression classification algorithm. Limiting the sensitivity due to distortion is calculated when a noise-adding feature is implemented to the regularized version of the logical regression classification algorithm. A privacy-preserving version of the regularized logistic regression algorithm is built, solving a perturbed optimization problem. Their approach uses a differential privacy technique to build a classifier model in a privacy-preserving environment. The differential privacy model can be applied to statistical databases only.

In [10], the authors proposed data processing methods that are aim to merge different security requirements on different platforms. They calculated the mathematical representation of the distributed data to its original form and tried to accurately approximate the true values of the original data from distributed data. They added some perturbation to the original dataset using a randomizing function.

Xu K., Yue H. et al. considered the conventional methods of cryptography for different nodes that do not share open data in respect of adequation. For this concern, the authors applied to minimize the data that requires being computed, by using the data locality feature of the Apache Hadoop system to protecting the privacy [11]. Their approach is based on the map-reduce paradigm, which is a quite old, big data analysis technique. In this work, the authors use a random matrix to preserve the privacy of the dataset content.

In [12], the authors inspected the overheads of the privacy and transmission of partitioned data in supervised and non-supervised contexts. They concluded that is better to transmit the hyper-parameters of the generative learning models that are built on local data nodes to a central database, instead of sharing the original data. Their paper showed that through reproducing artificial samples from the original data probability distributions using Markov Chain Monte Carlo method, it is mathematically possible to represent all the data with a mean model. In their approach, the authors do not share a perturbated matrix, they instead produce artificial samples from the underlying distributions using the Markov Chain Monte Carlo technique. Producing artificial data can sometimes cause errors on the model. The resulting model may not perform adequately on the test dataset.

In [13], the authors used artificial neural networks to gather information and construct models from various datasets. Their aim was to design a practical system that allows different parties to jointly build an accurate neural network model without sharing their private data. The authors concluded that their approach has a strong privacy-preserving component compared to any existing method due to the minimal data sharing between parties, which is actually a small number of neural network hyper-parameters. In this paper, the authors proposed a distributed version of the stochastic gradient descent (SGD) optimization algorithm. Their classification model evaluation results are almost the same to the original model evaluation results. However, there are many other optimization techniques that have more efficiency over time.

In [14], the authors proposed a system that allows users an on point representative model for sensitive data, which also protects its privacy. The system calculates the frequencies of specific values or a group of values. In this phase, the model protects the data privacy. The authors stated that there is no information sharing except the frequency of data values. The proposed model works only in a specific scenario where each customer (or party) sends only one flow of communication to the central miner, and no customer sends data to other customers.

In [15], the authors developed a system that guarantees the data from a single party data source will not be disclosed. In order to protect the data privacy, Shamir's secret sharing is used with the distributed version of the decision tree learning, ID3 algorithm. The ID3 algorithm can be used for only discrete values datasets.

In [16], the authors used a multi key-fully homomorphic encryption alongside a hybrid structure, which combined double decryption. The authors stated that these two privacy-preserving algorithms are acceptable to use with different types of deep learning algorithms over encrypted data. Each party in computation chooses its key pairs and encrypts its own data. Encrypted data is sent to a cloud system and the computation over the data is executed by these two systems.

In [17], the authors developed a privacy preserving version of the Bayes classifier learning method for horizontally partitioned data and proposed two protocols. One of these protocols is a secure two-party protocol and the other one is a secure multi-party protocol. The secure multi-party protocol is used between owners of sensitive data and a semi trusted server for sending messages that have to be classified. The two party protocol is used for sensitive data between a user and a semi-trusted server for broadcasting the classification results made on the server. In this work it is assumed that these two protocols are trusted and can preserve privacy. Their approach can be applied only into a Bayesian classifier.

In [18], the authors proposed a parallel computing approach that can perform with high performance on the computationally intensive works of data mining. In this study they considered a system built on the idea of the existence of a cluster and a grid resource in one tier, and in an another tier the system would simply abstract the communication for algorithm development. Their framework is applied only for classification algorithms.

3. Preliminaries

In this section, we briefly introduce clustering methods, homomorphic encryption, and the Paillier Cryptosystem used for our proposed privacy-preserving cluster learning model. We also introduce the evaluation metrics used to analyze the experimental results.

3.1. Data Clustering

Clustering can be described as combining the same instances of a dataset into the same group depending on special features of the data. For the evaluation criteria of a clustering model, the elements that are similar to each other should be in the same group as much as possible.

There are four different approaches for clustering algorithms: centroid-based, distribution-based, density-based, and connectivity-based.

Centroid-Based Clustering: Centroid-based clustering [19] presents a cluster by a central vector, which does not have to be an instance of the dataset. The required number of clusters should be predetermined and this is the main drawback of this type of algorithm (such as the K-Means algorithm). After deciding the number of clusters, then the central vectors for each cluster are computed by different distance metrics from the clusters to find the nearest instance to form clusters.

Distribution-Based Clustering: In this type of algorithm, instances of training data are clustered due to their statistical characteristics. Clusters can be defined as instances belonging most likely to the same distribution. One well-known mixture model for this method is a *Gaussian mixture* [20]. A dataset at the start is modeled with a *Gaussian distribution* arbitrarily and then the parameters are optimized to fit the dataset. It is an iterative algorithm, thus the output will converge into an optimum model.

Density-Based clustering: In this type of algorithm, clusters are defined, taking into consideration the area of high density of instances in the input dataset. Density-based clustering algorithms use different criteria for defining the density. One of the well-known criteria is called *density reachability* [21]. This logic works for searching for instances in a dataset that are within a certain threshold value, and adding those instances into the same cluster.

Connectivity-based clustering: This clustering method collects instances of an input dataset that are more interconnected to nearby instances, than farther away instances, to build a cluster model. In this approach, clusters are built based on their distance metrics, thus, a cluster can be defined by the maximum distance metric needed to group instances. Different clusters will be built at different distances, as a result, this approach can be visualized with a *dendrogram*, because these type of clustering methods yield a hierarchical model. As a result, a certain cluster also can be merged with another cluster.

There are a variety of different clustering algorithms. In our work, we analysed the *K-Means*, *Hierarchical*, *Spectral*, and *Birch* clustering algorithms.

The *K-Means algorithm* builds a cluster model considering *distance metrics* between input instances. The algorithm is applied generally when the data has a flat geometry, creating only a few clusters, and the number of clusters is evenly sized.

The *Hierarchical algorithm* builds a cluster model, taking into consideration the *pairwise distance metrics* between input instances. This algorithm is applicable generally when creating a large amount of clusters, and when there are possible connectivity constraints to building clusters.

The *Spectral algorithm* builds a cluster model taking into account the nearest-neighbors [22]. This algorithm is applicable generally when the input data has a non-flat geometry, creating only a few clusters and the number of clusters is even sized.

The *Birch algorithm* builds a cluster model taking into consideration the *Euclidean distance* between input instances. This algorithm is applicable generally when the data is large and data reduction, in respect to outlier removal, is needed.

3.2. Evaluation Metrics

In this paper, our proposed system is evaluated in consideration of six different metrics.

Homogeneity: This metric can only be fulfilled if the members of each single class are assigned into a distinct cluster in terms of homogeneity [23]. Thus, each cluster consists of only instances of a single class. The class distribution within each cluster should be to only a single class. This metric gets a value between 0 and 1. If clustering is done in the most ideal way, than the homogeneity value will be equal to 1. The homogeneity value can be evaluated to understand the quality of the clustering of elements belonging to the same class in the same cluster. The ratio of current homogeneity value and the ideal value can be expressed as $H(C|K)$, and this expression is equal to 0 when there is a perfect clustering. The value of $H(C|K)$ is dependent on the size of the dataset, therefore, instead of using this value, the normalized version is used and it is;

$$\frac{(H(C|K))}{(H(C))}. \quad (1)$$

Thus, the output value with 1 is desirable, and 0 is the undesirable state, the homogeneity can be defined as:

$$h = \begin{cases} 1 & if\ H(C,K) = 0 \\ 1 - \frac{H(C|K)}{H(C)} & else \end{cases} \quad (2)$$

where

$$H(C|K) = -\sum_{k=1}^{|K|}\sum_{c=1}^{|C|} \frac{a_{ck}}{N} \log \frac{a_{ck}}{\sum_{c=1}^{|C|} a_{ck}}$$
$$H(C) = -\sum_{c=1}^{|C|} \frac{\sum_{k=1}^{|K|} a_{ck}}{n} \log \frac{\sum_{k=1}^{|K|} a_{ck}}{n}. \quad (3)$$

Completeness is a symmetrical version of the homogeneity metric, expressing the balance of instances belonging to the same class, inside the same cluster. If the instances belonging to the same class are represented in the same cluster as the result of the clustering of the data, this metric, which takes the ideal value in this situation, is regarded as 1. If the grouping is far from the ideal state, the value of this metric will be 0. In order to satisfy this criterion, each of the clusters should be comprised of elements that belong to only one class. The distribution of cluster labeling inside each single class label is used to analyze the completeness metric. In ideal conditions, the value will be 0, $H(K|C) = 0$. The worst case conditions occur when each class is labeled by each cluster. In this case, the value will be 0, $H(K) = 0$. Completeness can be defined as:

$$c = \begin{cases} 1 & if\ H(K,C) = 0 \\ 1 - \frac{H(K|C)}{H(K)} & else \end{cases} \quad (4)$$

where

$$H(K|C) = -\sum_{c=1}^{|C|}\sum_{k=1}^{|K|} \frac{a_{ck}}{N} \log \frac{a_{ck}}{\sum_{k=1}^{|K|} a_{ck}}$$
$$H(K) = -\sum_{k=1}^{|K|} \frac{\sum_{c=1}^{|C|} a_{ck}}{n} \log \frac{\sum_{c=1}^{|C|} a_{ck}}{n}. \quad (5)$$

V-Measure is a balance metric between homogeneity and completeness criteria. *V-Measure* is calculated by using the harmonic mean of the homogeneity and completeness values of the cluster model. This metric value is between 0 and 1. As described above in previous homogeneity and completeness sections, these two metrics have working logics that are opposite to each other.

An increase in the homogeneity value results in a decrease in the completeness value, and vice versa. The *V-Measure* is shown as:

$$v - Measure = \frac{(2 * homogenity * completeness)}{(homogenity + completeness)}. \quad (6)$$

Adjusted Rand Index: The *Rand Index* metric is a measure of similarities between two data clustering labels, and should be calculated in order to find the *Adjusted Rand Index* (ARI). While the *Rand Index* may vary between 0 and 1, the ARI can also have negative values. ARI metric is shown as:

$$ARI = \frac{(RI - Expected\ RI)}{(max(RI) - Expected\ RI)}. \quad (7)$$

ARI becomes 0 when the clustering is done randomly and independent of the number of clusters; however, if the clusters are similar or identical, then the metric value becomes 1. This metric is also symmetrical.

$$(adjusted\_rand\_score(a,b) == adjusted\_rand\_score(b,a)). \quad (8)$$

Adjusted Mutual Information For this metric, as with *ARI*, mutual information describes how much information is shared between different clusters. Thus, *adjusted mutual information* can be considered as a similarity metric. In our paper, adjusted mutual information (AMI) measures the number of mutual instances between different clusters. This metric value becomes 1 when the clusters are completely identical, and when the clusters are independent from each other this metric value is equal to 0. Thus, there is no information shared between clusters. AMI is the adjusted shape of mutual information. The mutual information value is shown as:

$$AMI(U,V) = \frac{MI(U,V) - E(MI(U,V))}{max(H(U), H(V)) - E(MI(U,V))}. \quad (9)$$

This metric is also symmetrical, as with the ARI metric.

The *Silhouette Coefficient* metric is calculated by using both intra-cluster distance and mean nearest-cluster distance and is the distance between an instance and the nearest cluster that the instance that is not a part of each of the instances in a dataset. The silhouette coefficient is shown as:

$$Silhouette\ Coeff. = \frac{(b-a)}{max(a,b)} \quad (10)$$

where b is the distance value between an instance and the nearest cluster (which the instance does not belong to) and a is the mean value of distances within the cluster that the instance is a part of. This metric obtains the mean value of all silhouette coefficient values for the instances in the dataset.

The Silhouette coefficient can achieve values between -1 and 1. When the Silhouette coefficient is closer to -1, then it is more likely that the instance is in the wrong cluster. If this metric is considered in all instances in the dataset, the more the value moves closer to -1. In this case, the clustering is not accurate and the instances are more likely assigned to wrong clusters. On the contrary, if the metric value gets closer to 1, then, the clustering model is more accurate. When the metric value is near 0, then the clusters are overlapped.

3.3. Homomorphic Encryption

Homomorphic encryption allows for computing with encrypted data and achieves the same results with the plain version of the data. The most important feature of this type of cryptographic scheme is to preserve the privacy of the sensitive data [24] as they allow work on the encrypted data instead of its plain form. Homomorphic encryption schemes can also be used in connecting different kinds of services without putting at risk the exposure of sensitive data. Homomorphical encryption systems can be divided into two different groups; partially homomorphic algorithms and fully homomorphic algorithms. In our research, we applied the partially additive homomorphic Paillier Cryptosystem.

One can say that a public-key encryption scheme is additively homomorphic if, given two encrypted messages, such as $[\![a]\!]$ and $[\![b]\!]$, there exists a public-key summation operation $\oplus$ such that $[\![a]\!] \oplus [\![b]\!]$ is an encryption of the plaintext of $a + b$. The formal definition is that an encryption scheme is additively homomorphic if for any private key, public key (sk, pk), the plaintext space $\mathcal{P} = \mathbb{Z}_N$ for $a, b \in \mathbb{Z}_N$.

$$Enc_{pk}(a + b \bmod N) = Enc_{pk}(x) \times Enc_{pk}(y)$$
$$Enc_{pk}(x \cdot y \bmod N) = Enc_{pk}(x)^y. \tag{11}$$

3.3.1. Paillier Cryptosystem

The Paillier cryptosystem [25] is an asymmetric, probabilistic, and public key cryptosystem. It preserves privacy [26] depending on the complexity of the problem of computing the *n-th* residue classes. A Paillier cryptosystem is a partially homomorphic system, which means that the encryption of M_1 and M_2 plain datasets with a K public key gives the same result as the encryption of multiplication of the same two datasets ($M1 + M2$), using the same K public key. This encryption algorithm works by performing the two main jobs in an order. The first one is the key generation and the second one is the encryption/decryption of the dataset.

The Paillier cryptographic system has partial homomorphic properties, which makes this algorithm more appropriate against data exposure and it is used in several fields that have sensitive data, such as medicine, finance, and the military. These partial homomorphic properties are:

- The addition of two encrypted messages: The result of adding two encrypted messages that are exactly the same with the result of the plain version of adding these two messages.
- Multiplication of an encrypted message with a non-encrypted message: Multiplying an encrypted message with a number N is same with multiplying the plain form of that message with the same message N and encrypting it.

Let us give a set of possible plaintext messages M and a set of secret and public key pairs $K = pk \times sk$, where pk is the public key and sk is the secret key of the cryptosytem. Then the Paillier homomorphic encryption cryptosystem satisfies the following property for any two plaintext messages m_1 and m_2 and a constant value a:

$$Dec_{sk}\left(Enc_{pk}(m_1) \times Enc_{pk}(m_1)\right) = m_1 + m_2 \tag{12}$$

$$Dec_{sk}\left(Enc_{pk}(m_1)^a\right) = a \times m_1. \tag{13}$$

One of the main properties of the Paillier cryptosystem is that it is based on a probabilistic encryption algorithm. Large-scale data sets are sparse matrices, in which most of the elements are zero. In order to prevent the guessing of elements of the input data set, the Paillier cryptosystem has probabilistic encryption that does not encrypt two equal plaintexts with the same encryption key into the same ciphertext.

3.3.2. Floating Point Numbers

As the Paillier encryption systems works only on integer values, the proposed protocols are only capable of handling integers. This is considered to be an obstacle in applying these algorithms, as mostly of the real datasets contain continuous values. Nonetheless, in the case of an input dataset with real numbers in the protocol, we need to map the floating point input data vectors into the discrete domain with a conversion function (i.e., scaling).

The number of digits in the fractional part of the data, which is defined as a floating point number, can be very large and if these numbers are to be used, then it would definitely require more processing power and processing time. For this reason, in this work only the first five fractional digits have been used, and, due to this fact, the fractional digits of input data more than these digits are not used. These may be possibly generated as a result of computational work and are rounded into five digits. Using this approach potentially creates minimal data losses and deviations of computation. It is seen that this effect on the calculation result depends on the grouping algorithm used, but overall it does not affect the effectiveness of the algorithms.

4. System Model

4.1. Development Environment

In our experiments, we used Python 64-bit 3.3 to implement the mentioned algorithms to encrypt the training datasets using Paillier cryptography. We used scikit-learn, scipy python libraries to build privacy-preserving versions of four different clustering algorithms, which have been described in Section 3.

Each algorithm was run on a computer with an Intel Core i7-6700HQ CPU @ 2.60GHz octa core processor (4 real and +4 pseudo) along with 16 GB of RAM. Parallel computing has not been used in every experiment but where an algorithm allows the use of all processor cores while running, that property has been used.

4.2. Sequence Diagram

In this section, we explain our privacy preserving clustering training approach with several sequence diagrams. This system not only aims to preserve privacy but also aims to analyze and handle the data efficiently in terms of execution time and processing power. The model has two distinct phases; the client computation and the model building.

4.2.1. Client Computation

In this research, the client has the plain version of the data. The main task of the client side is to handle and to encrypt the sensitive data when it is needed. The client generates a public/private Paillier key pair by establishing a key exchange session with cloud service providers, then sends its public key and the encrypted form of the training data.

The cloud service provider also does not store the plain version of the sensitive data but has a huge computation infrastructure to compute intensive data analysis. On the cloud servers, the computations are executed to get an encrypted distance metric matrix using Pailler crypto-system encrypted data that will be used by the clustering algorithms. Cloud servers perform all computations for the tenant without violating their privacy and it sends the computation results (distance metric matrix). The tenant, then uses the encrypted distance metric matrix in order to build a cluster model. As illustrated in Figure 1, in this system, the tenant side does not require the plain data to use, but instead the cluster information can be derived from the encrypted distance metric matrix.

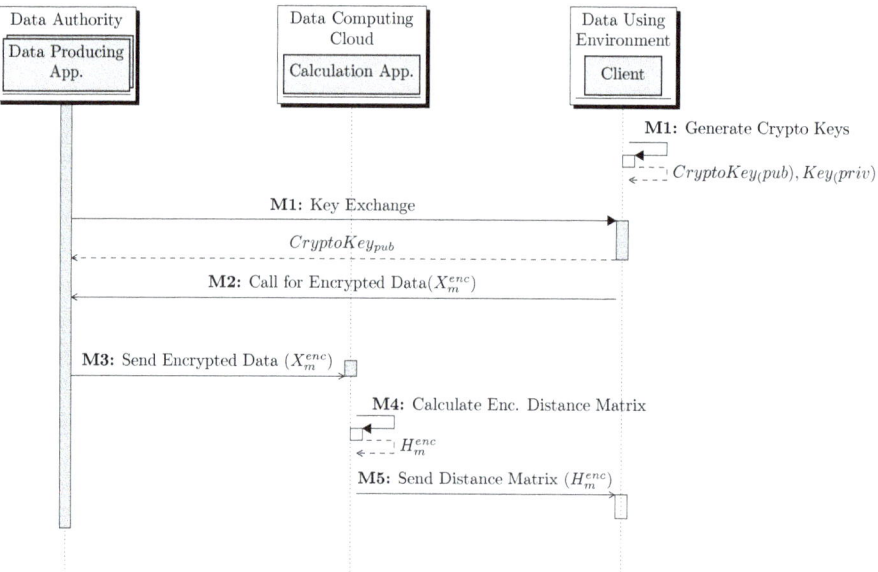

Figure 1. Sequence diagram for the client side.

The pseudocode for the *Client Computation* phase of the proposed system is shown in Algorithm 1. In lines 2–3, the client creates public/private crypto keys, keeps the private key internally, and sends the public key to the cloud service provider. In lines 4–5, the client encrypts the training dataset and sends it to the cloud service provider. At this stage, the cloud service provider has access to the public key and to the data to perform the calculation. In lines 6–7, the cloud server performs distance calculations on the encrypted data using the metric information given as parameters. It sends the encrypted distance matrix back to the client.

Algorithm 1 Client initialization and distance computation.

1: **Inputs:**
   Dataset, $\mathcal{X}$, Crypto key length: l, metric type: m
2: $(Key_{pub}, Key_{priv}) \leftarrow KeyGen(l)$ ▷ Key Generation
3: $res \leftarrow sendCryptoKey(Key_{pub})$ ▷ send public key to cloud service provider
4: $[\![\mathcal{X}]\!] \leftarrow encrypt(Key_{priv}, \mathcal{X})$ ▷ encrypt dataset
5: $sendEncryptedDataSet([\![\mathcal{X}]\!])$ ▷ send encrypted dataset to cloud service provider
6: $[\![\mathcal{H}]\!] \leftarrow distanceMatrix([\![\mathcal{X}]\!], m)$ ▷ Cloud service provider computes the distance matrix of each input instance according to metric parameter
7: $res \leftarrow sendDistanceMatrix([\![\mathcal{H}]\!])$ ▷ Cloud service provider sends the encrypted distance matrix to client

---

4.2.2. Model Building at the Client Side

At this stage, the client has an encrypted distance matrix. The first action of the client is to decrypt this matrix using the private key Key_{priv}. Since the clustering algorithms used in this study are distance based, the clustering model can be constructed by using the distance matrix as a parameter, explained in detail in Algorithm 2. As illustrated in Figure 2, the client receives the encrypted distance matrix and decrypts it using the private key.

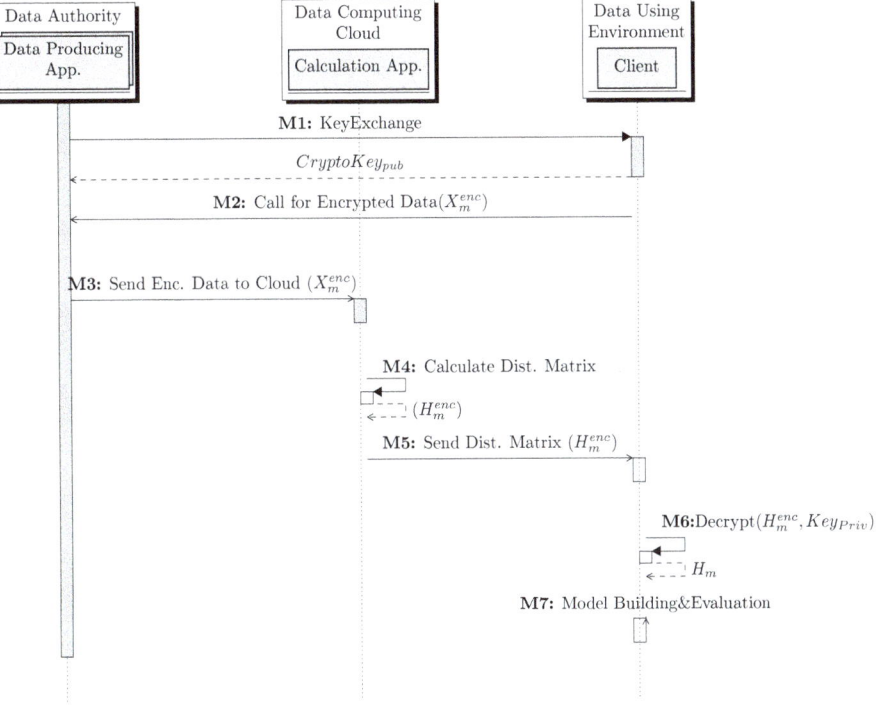

Figure 2. Sequence diagram for model building at the client side.

Algorithm 2 Cluster model building.

1: **Inputs:**
   Encrypted distance matrix: $[\![\mathcal{H}]\!]$, Crypto keys: Key_{pub}, Key_{priv}, Cluster
   algorithm: $\mathcal{C}$
2: $\mathcal{H} \leftarrow decryptDistanceMatrix([\![\mathcal{H}]\!])$ ▷ Client decrypt the distance matrix
3: $m \leftarrow clusteringAlgorithm(\mathcal{C}, \mathcal{H})$ ▷ Run clustering algrithm with plain domain distance matrix
4: **Outputs:**
   Clustering model m

---

5. Experimental results

In this research, the clustering methods that have been described in Section 3.1 were examined by running each of them on 10 different datasets (from 500 to 5000 instances) using five different crypto key lengths. The data length of each dataset is also the name of the dataset (dataset 500, dataset 1000 etc.) and the used key lengths vary between 64 bits and 1024 bits.

All algorithms were run on a Python environment and all algorithm codes have been modified to create the same number of clusters (20 *clusters*) from the given data.

5.1. Plaintext Results

The experimental results for the plain domain computations are shown in Table 1. As no key is used, there is only one column of the table for the score of each evaluation metric. In the case of plain data computations, the same metric scores have been obtained except for the silhouette coefficient. The change is not large enough to change the metric score, thus it can be neglected. The evaluation metrics obtained the maximum scores in plain data computation (except *silhouette coefficient*), as the used data are *artificial* and not *real*.

According to Table 1, all clustering performance metrics are the same for all algorithms. The K-means clustering algorithm takes more time to build a model for 1000–3000 row datasets, Hierarchical clustering needs more time for 4000–5000 row datasets, and the Spectral algorithm needs a lower time for training for all datasets.

Table 1. Plain domain evaluation of metric scores.

Length	Algorithm	Homogenity	Completeness	V-Measure	Adj. Rand Ind.	Adj. Mut. Inf	Silhouette Coeff.	Time (s)
1000	K-means	1.0	1.0	1.0	1.0	1.0	1.0	36.95
	Hierarchical	1.0	1.0	1.0	1.0	1.0	1.0	14.89
	Spectral	1.0	1.0	1.0	1.0	1.0	1.0	8.62
	Birch	1.0	1.0	1.0	1.0	1.0	1.0	15.07
2000	K-means	1.0	1.0	1.0	1.0	1.0	1.0	163.58
	Hierarchical	1.0	1.0	1.0	1.0	1.0	1.0	123.05
	Spectral	1.0	1.0	1.0	1.0	1.0	1.0	26.47
	Birch	1.0	1.0	1.0	1.0	1.0	1.0	96.95
3000	K-means	1.0	1.0	1.0	1.0	1.0	1.0	413.85
	Hierarchical	1.0	1.0	1.0	1.0	1.0	1.0	410.76
	Spectral	1.0	1.0	1.0	1.0	1.0	1.0	53.83
	Birch	1.0	1.0	1.0	1.0	1.0	1.0	347.99
4000	K-means	1.0	1.0	1.0	1.0	1.0	1.0	789.06
	Hierarchical	1.0	1.0	1.0	1.0	1.0	1.0	964.55
	Spectral	1.0	1.0	1.0	1.0	1.0	1.0	98.60
	Birch	1.0	1.0	1.0	1.0	1.0	1.0	795.86
5000	K-means	1.0	1.0	1.0	1.0	1.0	1.0	1292.19
	Hierarchical	1.0	1.0	1.0	1.0	1.0	1.0	1592.06
	Spectral	1.0	1.0	1.0	1.0	1.0	1.0	148.25
	Birch	1.0	1.0	1.0	1.0	1.0	1.0	1509.15

5.2. Encrypted Domain Results

The experimental results of the encrypted domain computations are shown in Tables 2–5. Based on these results, we conclude that there is no need to use 64 bit and 128 bit keys, as, these keys did not effect the results of this work and, for this reason, the scores of 64 and 128 bit keys are not shown on the table.

According to Table 2, all clustering performance metrics are different for all algorithms. Generally Birch clustering algorithm takes more time to build a model, Spectral clustering needs needs lower time for training for all datasets. Generally Spectral clustering algorithm performance results are better than the other algorithms.

Table 2. Encrypted domain evaluation of metric scores for 256 bit key length.

Length	Algorithm	Homogenity	Completeness	V-Measure	Adj. Rand Ind.	Adj. Mut. Inf	Silhouette Coeff.	Time (s)
1000	K-Means	0.741	0.724	0.732	0.488	0.705	**0.350**	**228.41**
	Hierarchical	0.708	0.661	0.684	0.422	0.638	0.272	**167.56**
	Spectral	**0.842**	**0.848**	**0.845**	**0.694**	**0.832**	0.291	219.22
	Birch	0.735	0.610	0.667	0.427	0.588	0.292	175.84
2000	K-Means	0.811	0.793	0.802	**0.720**	0.786	**0.352**	849.30
	Hierarchical	0.722	0.706	0.714	0.504	0.696	0.306	725.57
	Spectral	**0.822**	**0.829**	**0.825**	0.676	**0.816**	0.273	**646.24**
	Birch	0.706	0.569	0.630	0.364	0.557	0.282	**1275.94**
3000	K-Means	0.743	0.727	0.735	0.543	0.721	**0.338**	2056.95
	Hierarchical	0.703	0.675	0.689	0.428	0.669	0.295	2163.89
	Spectral	**0.837**	**0.837**	**0.835**	**0.710**	**0.830**	0.277	**1427.10**
	Birch	0.719	0.520	0.603	0.296	0.511	0.267	**3687.20**
4000	K-Means	**0.795**	**0.781**	0.788	0.671	**0.788**	0.345	3194.87
	Hierarchical	0.641	0.688	0.664	0.384	0.636	0.264	**3254.51**
	Spectral	0.777	**0.780**	0.779	0.579	0.774	0.281	3012.27
	Birch	0.721	0.533	0.613	0.363	0.527	0.271	3222.27
5000	K-Means	0.766	0.751	0.759	0.604	0.748	**0.340**	6257.98
	Hierarchical	0.654	0.686	0.670	0.405	0.649	0.245	5737.90
	Spectral	**0.793**	**0.799**	**0.796**	**0.625**	**0.791**	0.288	**4031.74**
	Birch	0.723	0.581	0.644	0.373	0.577	0.245	**7191.10**

According to Table 3, generally the Spectral clustering algorithm performance results are better than the other algorithms.

Table 3. Encrypted domain evaluation of metric scores for 512 bit key length.

Length	Algorithm	Homogenity	Completeness	V-Measure	Adj. Rand Ind.	Adj. Mut. Inf	Silhouette Coeff.	Time (s)
1000	K-Means	0.741	0.724	0.732	0.488	0.705	**0.350**	725.91
	Hierarchical	0.739	0.679	0.708	0.453	0.657	0.296	579.17
	Spectral	**0.815**	**0.819**	**0.817**	**0.646**	**0.803**	0.284	770.70
	Birch	0.735	0.610	0.667	0.427	0.588	0.292	580.24
2000	K-Means	0.811	0.793	0.802	**0.720**	0.786	**0.352**	2994.48
	Hierarchical	0.747	0.728	0.737	0.540	0.719	0.298	2329.03
	Spectral	**0.833**	**0.840**	**0.836**	0.707	**0.828**	0.270	**2298.56**
	Birch	0.706	0.569	0.630	0.364	0.557	0.282	4679.90
3000	K-Means	0.743	0.727	0.735	0.543	0.721	**0.338**	5541.25
	Hierarchical	0.699	0.670	0.684	0.454	0.663	0.292	**6722.31**
	Spectral	**0.817**	**0.821**	**0.819**	**0.678**	**0.813**	0.269	**5333.42**
	Birch	0.719	0.520	0.603	0.296	0.511	0.267	5497.33
4000	K-Means	**0.795**	**0.781**	0.788	0.671	**0.788**	0.345	10,375.71
	Hierarchical	0.659	0.703	0.680	0.407	0.654	0.264	**9928.27**
	Spectral	0.766	0.770	0.768	0.556	0.762	0.283	**11,371.57**
	Birch	0.721	0.533	0.613	0.363	0.527	0.271	9963.32
5000	K-Means	0.766	0.751	0/59	0.604	0.748	**0.340**	15,636.90
	Hierarchical	0.634	0.659	0.646	0.360	0.629	0.245	15,691.37
	Spectral	**0.806**	**0.813**	**0.809**	**0.648**	**0.804**	0.280	**14,343.40**
	Birch	0.723	0.581	0.644	0.373	0.577	0.245	**16,862.45**

According to Table 4, generally, the Spectral clustering algorithms need a lower training time. The Spectral clustering algorithm performance results are better than the other algorithms.

Table 4. Encrypted domain evaluation of metric scores for 1024 bit key length.

Length	Algorithm	Homogenity	Completeness	V-Measure	Adj. Rand Ind.	Adj. Mut. Inf	Silhouette Coeff.	Time (s)
1000	K-Means	0.741	0.724	0.732	0.488	0.705	**0.350**	4186.54
	Hierarchical	0.727	0.671	0.698	0.429	0.649	0.277	3534.10
	Spectral	**0.823**	**0.822**	**0.822**	**0.644**	**0.810**	0.295	3458.16
	Birch	0.735	0.610	0.667	0.427	0.588	0.292	3977.07
2000	K-Means	0.811	0.793	0.802	0.720	0.786	**0.352**	15,613.31
	Hierarchical	0.706	0.717	0.712	0.490	0.696	0.299	14,633.00
	Spectral	**0.841**	**0.848**	**0.845**	**0.730**	**0.836**	0.271	13,878.68
	Birch	0.706	0.569	0.630	0.364	0.557	0.282	14,160.69
3000	K-Means	0.743	0.727	0.735	0.543	0.721	**0.338**	34,827.04
	Hierarchical	0.673	0.702	0.687	0.455	0.666	0.271	34,352.03
	Spectral	**0.814**	**0.820**	**0.817**	**0.688**	**0.810**	0.266	31,492.53
	Birch	0.719	0.520	0.603	0.296	0.511	0.267	39,601.76
4000	K-Means	0.795	**0.781**	**0.788**	**0.671**	**0.788**	**0.345**	56,810.99
	Hierarchical	0.646	0.689	0.667	0.389	0.640	0.264	58,259.58
	Spectral	**0.777**	0.780	0.778	0.576	0.773	0.283	56,022.39
	Birch	0.721	0.533	0.613	0.363	0.527	0.271	56,957.61
5000	K-Means	0.766	0.751	0.759	0.604	0.748	**0.340**	93,672.79
	Hierarchical	0.651	0.680	0.665	0.401	0.646	0.245	90,506.86
	Spectral	**0.792**	**0.798**	**0.795**	**0.621**	**0.790**	0.288	96,517.34
	Birch	0.723	0.581	0.644	0.373	0.577	0.245	90,333.51

According to Table 5, generally, the Spectral clustering algorithm performance results are better than the other algorithms.

Table 5. Encrypted domain evaluation of metric scores for 2048 bit key length.

Length	Algorithm	Homogenity	Completeness	V-Measure	Adj. Rand Ind.	Adj. Mut. Inf	Silhouette Coeff.	Time (s)
1000	K-Means	0.746	0.729	0.739	0.490	0.708	**0.351**	28,773.89
	Hierarchical	0.729	0.675	0.705	0.435	0.649	0.290	24,757.46
	Spectral	**0.835**	**0.839**	**0.836**	**0.663**	**0.819**	0.295	**19,632.64**
	Birch	0.744	0.618	0.677	0.435	0.592	0.301	29,590.15
2000	K-Means	0.813	0.799	0.808	0.724	0.79	**0.355**	99,402.12
	Hierarchical	0.735	0.717	0.731	0.516	0.703	0.311	104,967.10
	Spectral	**0.833**	**0.842**	**0.839**	**0.705**	**0.835**	0.272	96,352.80
	Birch	0.709	0.572	0.636	0.370	0.564	0.284	**45,457.64**
3000	K-Means	0.744	0.735	0.736	0.545	0.731	**0.344**	255,790.18
	Hierarchical	0.699	0.690	0.694	0.453	0.668	0.286	220,325.90
	Spectral	**0.824**	**0.831**	**0.832**	**0.697**	**0.821**	0.277	**214,712.31**
	Birch	0.724	0.524	0.610	0.297	0.512	0.274	337,003.50
4000	K-Means	0.797	0.782	0.788	**0.680**	**0.790**	**0.349**	377,794.82
	Hierarchical	0.650	0.702	0.674	0.399	0.652	0.268	406,524.12
	Spectral	**0.778**	**0.785**	**0.780**	0.573	0.771	0.285	**338,876.22**
	Birch	0.730	0.536	0.615	0.368	0.532	0.271	391,132.86
5000	K-Means	0.776	0.758	0.762	0.609	0.749	**0.349**	680,925.54
	Hierarchical	0.653	0.685	0.665	0.392	0.646	0.251	636,428.78
	Spectral	**0.800**	**0.812**	**0.807**	**0.640**	**0.803**	0.290	706,719.43
	Birch	0.727	0.581	0.648	0.383	0.583	0.252	**628,563.14**

Figures 3–6 show the execution time of the algorithms during the encryption of plain data and the calculation of the distance matrix of the encrypted data for each algorithm. If the amount of data and key size increases, the time complexity of the algorithm increases.

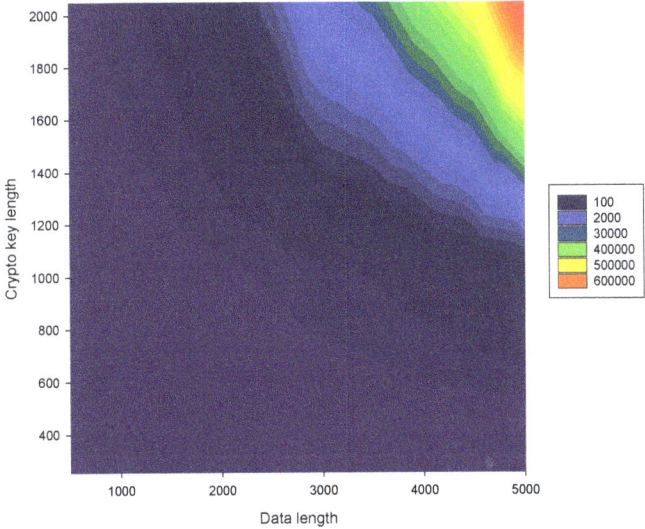

Figure 3. Calculation time graph for the KMeans algorithm.

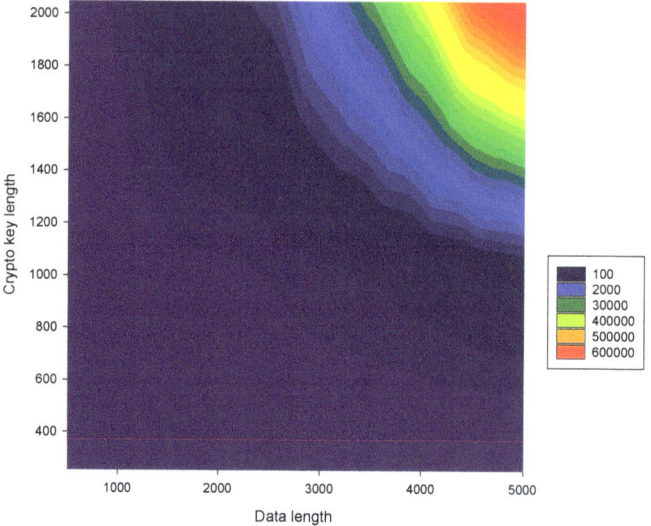

Figure 4. Calculation time graph for the Hierarchical algorithm.

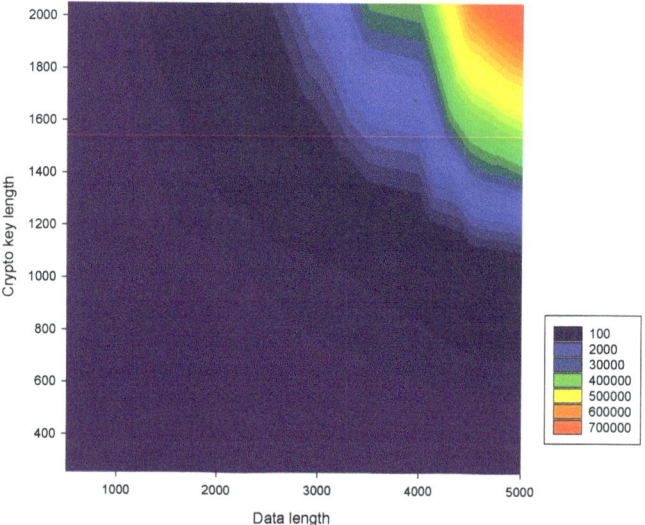

Figure 5. Calculation time graph for the Spectral algorithm.

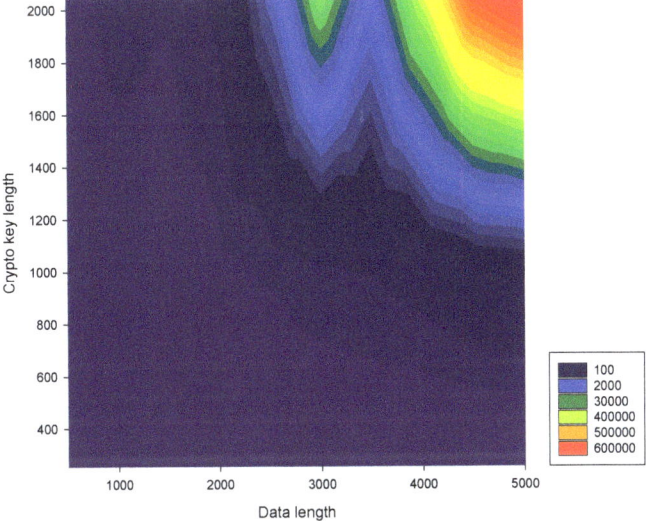

Figure 6. Calculation time graph for the Birch algorithm.

5.3. Results

As expected, based on the experimental results, the encrypted domain execution time is more than the plain data execution. As the Paillier crypto key size increases, the execution time also changes accordingly in the encrypted data execution. In the case of plain data, the execution time only changes by changing the data size.

Considering all six cluster evaluation metrics and time, we proposed a practical implementation of privacy preservation on the clustering training model using a partially homomorphic Paillier crypto system. Our proposed method is more efficient as we obtained better scores on the evaluation metrics and stronger key usage.

Based on the results presented in Section 5.2 and the computation results shown in Tables 2–5, we conclude that 2048 bit crypto key is the approach that provides the best results for all evaluation metrics. When it comes to the execution time of proposed training model it is clearly evident from the figures that this approach is not efficient enough. Even though we have achieved the best results in the clustering model performance with a crypto key with a 2048 bit length, it seems that the training time takes quite a long time. The training phase of the 2048 bit length crypto key is about eight times more of the training with the 1024 bit length crypto key. Security standards state that it is necessary to use crypto keys with a minimum length of 2048 bits in order to ensure privacy [27]. Although the 2048 bit key length is required by security standards, we also shared the results of other key lengths in this study. Figure 7 shows the duration of the clustering algorithm's training phase with a 5000 row dataset when using crypto keys from 256, 512, 1024, and 2048 bit lengths.

If we examine the resulting Tables 2–5 that show cluster evaluation metric results of plain data clustering, we can conclude that the plain data would get perfect evaluation metric scores except silhouette coefficient.

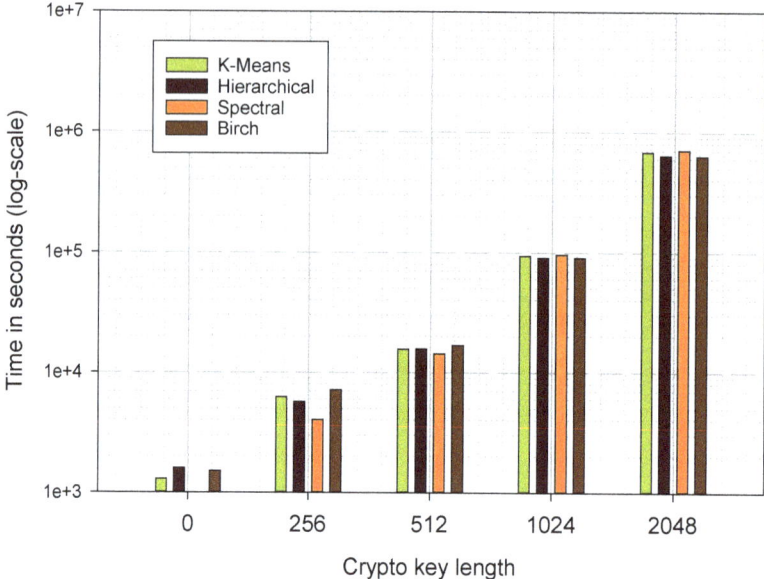

Figure 7. The training times on a logarithmic scale using the 5000 row dataset. The crypto key length with 0 refers to the plain domain execution time of the proposed method. This figure shows the nonlinear relationship between the increase in key size and the duration of algorithm training by using a logarithmic scale on y-axis.

6. Conclusions and Future Work

Currently, data privacy is one of the most crucial problems. Data come from many sources in online environments, including mobile phones, computers, and Internet of Things (IoT) tools and based on this data we can make various data analyses. As a result, we can improve the comfort of humanity. Due to the high volume of the data generated and collected by different sources, we have to conduct data analysis using cloud service providers, while taking into consideration that a mistrust of these providers as a well-known fact.

In our previous study, we represented the usage of the Paillier crypto system together with the classification algorithms. In this study, we have examined clustering algorithms, which are frequently used techniques in the field of machine learning. Privacy-preserving clustering algorithms are an important direction enabling knowledge discovery without requiring the disclosure of private data.

In this research, we applied four different clustering algorithms and their results are compared based on six different cluster evaluation metrics and their execution times. Each clustering algorithm obtained relatively similar results, but in the details they have differences.

The evaluation scores of our proposed privacy-preserving clustering models of the plain domain and encrypted domain are almost the same. Thus, the conversion from floating point numbers to integers creates only trivial metric differences for each clustering algorithm.

Our proposed cluster training model is a privacy-preserving method using the Paillier encryption system for outsourced sensitive datasets. The client builds a final clustering model with aggregation of each encrypted distance matrix calculated at each party. As a result, the final model is in the plain domain and some information such as cluster centroids are plain. If the client wants to share the model, then some information leakage may occur.

As a future work, in order to prevent data disclosure from the model, the model itself should also be encrypted using homomorphic encryption algorithms. In order to encrypt the model, the cluster centroids should also be encrypted. To allow the client to use the encrypted clustering model, a new model must be developed.

Author Contributions: Conceptualization was conceived by F.O.C., methodology, implemetation, validation was done by I.A. and F.O.C. The manusript's writing, review and editing by F.O.C., I.A., O.E. and S.Y.-Y. All authors have read and agreed to the published version of the manuscript.

Funding: This research received no external funding

Conflicts of Interest: The authors declare no conflict of interest.

References

1. Nasrabadi, N.M. Pattern recognition and machine learning. *J. Electron. Imaging* **2007**, *16*, 049901.
2. Tankard, C. What the GDPR means for businesses. *Netw. Secur.* **2016**, *2016*, 5–8. [CrossRef]
3. Edwards, S. Review of a medical illustration department's data processing system to confirm general data protection regulation (GDPR) compliance. *J. Vis. Commun. Med.* **2019**, *42*, 140–143. [CrossRef] [PubMed]
4. Simmons, G.J. Symmetric and asymmetric encryption. *ACM Comput. Surv. (CSUR)* **1979**, *11*, 305–330. [CrossRef]
5. Cuzzocrea, A. Privacy and Security of Big Data. In Proceedings of the First International Workshop on Privacy and Secuirty of Big Data—PSBD-14, Shanghai, China, 3–7 November 2014. [CrossRef]
6. Verykios, V.S.; Bertino, E.; Fovino, I.N.; Provenza, L.P.; Saygin, Y.; Theodoridis, Y. State-of-the-art in privacy preserving data mining. *ACM Sigmod Rec.* **2004**, *33*, 50–57. [CrossRef]
7. Catak, F.O.; Mustacoglu, A.F. CPP-ELM: Cryptographically Privacy-Preserving Extreme Learning Machine for Cloud Systems. *Int. J. Comput. Intell. Syst.* **2018**, *11*, 33–44. [CrossRef]
8. Lindell, Y.; Pinkas, B. Privacy Preserving Data Mining. In *Advances in Cryptology—CRYPTO 2000*; Bellare, M., Ed.; Springer: Berlin/Heidelberg, Germany, 2000; pp. 36–54.
9. Chaudhuri, K.; Monteleoni, C. Privacy-preserving logistic regression. In *Advances in Neural Information Processing Systems 21*; Koller, D., Schuurmans, D., Bengio, Y., Bottou, L., Eds.; Curran Associates, Inc.: Dutchess County, NY, USA, 2009; pp. 289–296.
10. Agrawal, R.; Srikant, R. Privacy-preserving Data Mining. *SIGMOD Rec.* **2000**, *29*, 439–450. [CrossRef]
11. Xu, K.; Yue, H.; Guo, L.; Guo, Y.; Fang, Y. Privacy-Preserving Machine Learning Algorithms for Big Data Systems. In Proceedings of the 2015 IEEE 35th International Conference on Distributed Computing Systems, Columbus, OH, USA, 29 June–2 July 2015; pp. 318–327. [CrossRef]
12. Merugu, S.; Ghosh, J. Privacy-preserving distributed clustering using generative models. In Proceedings of the Third IEEE International Conference on Data Mining, Melbourne, FL, USA, 22 November 2003; pp. 211–218. [CrossRef]
13. Shokri, R.; Shmatikov, V. Privacy-Preserving Deep Learning. In Proceedings of the 22nd ACM SIGSAC Conference on Computer and Communications Security, CCS '15, Denver, CO, USA, 12–16 October 2015; pp. 1310–1321. [CrossRef]
14. Yang, Z.; Zhong, S.; Wright, R.N. Privacy-Preserving Classification of Customer Data without Loss of Accuracy. In Proceedings of the 2005 SIAM International Conference on Data Mining, Newport Beach, CA, USA, 21–23 April 2005; pp. 92–102. [CrossRef]
15. Emekci, F.; Sahin, O.; Agrawal, D.; Abbadi, A.E. Privacy preserving decision tree learning over multiple parties. *Data Knowl. Eng.* **2007**, *63*, 348–361. [CrossRef]
16. Li, P.; Li, J.; Huang, Z.; Li, T.; Gao, C.Z.; Yiu, S.M.; Chen, K. Multi-key privacy-preserving deep learning in cloud computing. *Future Gener. Comput. Syst.* **2017**, *74*, 76–85. [CrossRef]
17. Yi, X.; Zhang, Y. Privacy-preserving naive Bayes classification on distributed data via semi-trusted mixers. *Inf. Syst.* **2009**, *34*, 371–380. [CrossRef]
18. Secretan, J.; Georgiopoulos, M.; Koufakou, A.; Cardona, K. APHID: An architecture for private, high-performance integrated data mining. *Future Gener. Comput. Syst.* **2010**, *26*, 891–904. [CrossRef]
19. Wu, C.H.; Ouyang, C.S.; Chen, L.W.; Lu, L.W. A new fuzzy clustering validity index with a median factor for centroid-based clustering. *IEEE Trans. Fuzzy Syst.* **2015**, *23*, 701–718. [CrossRef]

20. Banfield, J.D.; Raftery, A.E. Model-Based Gaussian and Non-Gaussian Clustering. *Biometrics* **1993**, *49*, 803–821. [CrossRef]
21. Duan, L.; Xu, L.; Guo, F.; Lee, J.; Yan, B. A local-density based spatial clustering algorithm with noise. *Inf. Syst.* **2007**, *32*, 978–986. [CrossRef]
22. White, S.; Smyth, P. A spectral clustering approach to finding communities in graphs. In Proceedings of the 2005 SIAM International Conference on Data Mining, Newport Beach, CA, USA, 21–23 April 2005; pp. 274–285.
23. Rosenberg, A.; Hirschberg, J. V-measure: A conditional entropy-based external cluster evaluation measure. In Proceedings of the 2007 Joint Conference on Empirical Methods in Natural Language Processing and Computational Natural Language Learning (EMNLP-CoNLL), Prague, Czech Republic, 28–30 June 2007.
24. Samet, S.; Miri, A. Privacy-preserving back-propagation and extreme learning machine algorithms. *Data Knowl. Eng.* **2012**, *79-80*, 40–61. [CrossRef]
25. Paillier, P. Public-Key Cryptosystems Based on Composite Degree Residuosity Classes. In *Advances in Cryptology—EUROCRYPT '99*; Stern, J., Ed.; Springer: Berlin/Heidelberg, Germany, 1999; pp. 223–238.
26. Min, Z.; Yang, G.; Shi, J. A privacy-preserving parallel and homomorphic encryption scheme. *Open Phys.* **2017**, *15*, 135–142. [CrossRef]
27. Barker, E.; Barker, W.; Burr, W.; Polk, W.; Smid, M.; Gallagher, P.D.; For, U.S. *NIST Special Publication 800-57 Recommendation for Key Management—Part 1: General*; National Institute of Standards and Technology: Gaithersburg, MD, USA, 2012.

© 2020 by the authors. Licensee MDPI, Basel, Switzerland. This article is an open access article distributed under the terms and conditions of the Creative Commons Attribution (CC BY) license (http://creativecommons.org/licenses/by/4.0/).

Article
Intrusion Detection Based on Spatiotemporal Characterization of Cyberattacks

Jiyeon Kim [1],* and Hyong S. Kim [2]

1 Center for Software Educational Innovation, Seoul Women's University, Seoul 01797, Korea
2 Department of Electrical and Computer Engineering, Carnegie Mellon University, Pittsburgh, PA 15213, USA; kim@ece.cmu.edu
* Correspondence: jykim07@swu.ac.kr; Tel.: +82-2-970-2835

Received: 3 December 2019; Accepted: 29 February 2020; Published: 9 March 2020

Abstract: As attack techniques become more sophisticated, detecting new and advanced cyberattacks with traditional intrusion detection techniques based on signature and anomaly is becoming challenging. In signature-based detection, not only do attackers bypass known signatures, but they also exploit unknown vulnerabilities. As the number of new signatures is increasing daily, it is also challenging to scale the detection mechanisms without impacting performance. For anomaly detection, defining normal behaviors is challenging due to today's complex applications with dynamic features. These complex and dynamic characteristics cause much false positives with a simple outlier detection. In this work, we detect intrusion behaviors by looking at number of computing elements together in time and space, whereas most of existing intrusion detection systems focus on a single element. In order to define the spatiotemporal intrusion patterns, we look at fundamental behaviors of cyberattacks that should appear in any possible attacks. We define these individual behaviors as basic cyberattack action (BCA) and develop a stochastic graph model to represent combination of BCAs in time and space. In addition, we build an intrusion detection system to demonstrate the detection mechanism based on the graph model. We inject numerous known and possible unknown attacks comprising BCAs and show how the system detects these attacks and how to locate the root causes based on the spatiotemporal patterns. The characterization of attacks in spatiotemporal patterns with expected essential behaviors would present a new effective approach to the intrusion detection.

Keywords: intrusion detection; spatiotemporal pattern; cyberattacks; cybersecurity

1. Introduction

Cyberattacks are becoming increasingly more sophisticated. For example, zero-day attacks exploit undisclosed vulnerabilities and advanced persistent threats (APT) attacks consist of multiple phases of attacks for a long period of time. With traditional intrusion detection systems based on signature and anomaly, it is challenging to detect these sophisticated attacks.

Signature-based intrusion detection systems (S-IDS) depend on known signatures to detect cyberattacks. There are two issues with S-IDSs. First, new attacks cannot be detected because new signatures are only obtained through post-analysis of attack events [1,2]. Even variant attacks are hard to detect as attackers work around the known signatures. Second, as the number of signatures increases, it is challenging to scale the detection mechanisms without impacting performance [3].

Anomaly-based IDSs (A-IDS) detect cyberattacks by comparing the system behavior with pre-defined normal behavior [2,4]. A-IDS can be effective for unknown attacks as it does not rely on known signatures. The major issue with A-IDS is the large number of false positives generated [5]. In simple applications, it is easy to define a normal behavior of the system. However, it is challenging to define a normal behavior in today's complex applications running in an N-tier architecture with dynamic features [6]. These applications obfuscate normal behaviors and thus create much false positives with anomaly detection based on a simple outlier detection.

Machine learning (ML) techniques are being actively employed in anomaly detection as an alternative to these issues. ML-based anomaly detection trains historical datasets to define normal behaviors and detect outlier events as attacks [5,7]. Although processing of massive datasets would help to set a flexible threshold of detection, there are still issues with false positives due to overfitting and unoptimized hyperparameters [8,9].

Most of existing S-IDS and A-IDS, including ML-based A-IDS, focus on a single computing or network element, whereas we focus on multiple elements. We use the terms *element* or *host* interchangeably to denote the computing or network element. Focusing on multiple elements in time and space rather than that of a single element would provide further evidence of an attack. Furthermore, this approach contributes to locate root causes by tracking the spatiotemporal behaviors.

In order to define the spatiotemporal attack patterns, we develop fundamental and essential behaviors that should appear in any attacks. We carefully study intrusion datasets as well as attack classifications including CAPEC [10] and characterize system and network features caused by intrusions. We define these behaviors of a single element as Basic Cyberattack Actions (BCAs).

BCAs allow detection of novel and complex cyberattacks as long as the attacks show any combination of BCA patterns. Future attacks could also consist of many combinations of BCAs. We propose to look at number of computing and network elements together in space (i.e., networked groups of hosts) and time rather than relying on individual BCA of a single element. Combination of BCAs describe the spatiotemporal characterization of an attack and would provide further insight into the attack. We also develop a stochastic graph model to represent the combination of BCAs.

In order to demonstrate our detection idea based on the spatiotemporal patterns, we develop an IDS in our production datacenter. We inject known and possible unknown attacks comprising BCAs and illustrate how the system detects these attacks and locates the root causes by tracking BCAs in time and space. The performance evaluation with extensive attacks comprising complex BCAs is not the focus of this paper and will be addressed in the forthcoming paper.

The remainder of this paper is organized as follows: We review related works in Section 2. Section 3 defines BCAs based on existing attack classifications. Section 4 defines a stochastic graph model to describe the behavior of BCA in time and space. In Section 5, we describe our BCA detection system. In Section 6, we evaluate our system with numerous attacks. Finally, the conclusions are presented in Section 7.

2. Related Work

S-IDSs detect signatures of known attacks. Kumar and Spafford [11] propose a pattern matching model for S-IDS based on Colored Petri Nets. Honeycomb [12] automatically generates attack signatures using a honeypot system and detects these signatures using pattern matching techniques. Josue et al. [13] propose a pattern matching algorithm to filter out the audit trail. Koral et al. [14] define a set of state transition signatures and detects an attack sequence of the transition. Zhengbing et al. [15] employ data mining techniques to develop more accurate signatures. These systems use known signatures and they are focused on improving the search and pattern matching speed. They do not consider unknown attacks without matching signatures.

A-IDSs define normal behaviors and detect outlier events as attacks. Although A-IDS is able to detect unknown attacks, it suffers from large numbers of false positives. Collaborative detection mechanisms are proposed to reduce false positives [16–20]. They aggregate and correlate a number of alerts generated by different IDSs. IDES [17] first proposes the IDS collaboration and EMERALD [18] refines IDES. Cuppens and Miege [16] used an expert system to develop an aggregation and correlation module. Valdes and Skinner [19] employed a probability-based approach for a similarity recognition. Yu et al. [20] develop a knowledge-based alert aggregation system. They collect a number of false alerts and process them based on correlation rules.

Numerous studies employ ML to identify legitimate behaviors. They define normal behavior patterns based on historical data from numerous system metrics. Bayesian network, decision Tree, and SVM (Support Vector Machine) are widely used in intrusion detection systems based on ML techniques. Kruegel et al. [21] propose an event classification scheme based on Bayesian network to mitigate false alarms. Bilge et al. [22,23] detect malicious domains by employing a passive DNS analysis based on a decision tree. Feng et al. [24], Kuang et al. [25], Thaseen and Kumar [26] apply SVM for better performance in intrusion detection. There are numerous studies that employ deep learning that belongs to ML. Khan et al. [27], Li et al. [28], Liu et al. [29] and Kim et al. [30] transform intrusion datasets to images and then detect attacks based on a convolutional neural network (CNN). Bontemps et al. [31], Staudenmeyer and Omlin [32] suggest an IDS model based on a long short-term memory recurrent neural network (LSTM) using the KDD dataset [33]. There are further IDS studies that perform binary and multiclass classifications based on a recurrent neural network [30,34–36].

In addition, Dokas et al. [37], Hu and Panda [38] employ data mining techniques. Stephenson [39] combines forensics with the intrusion detection and response. Ren and Jin [40] develop a framework for the real time intrusion forensic system.

Although numerous studies on S-IDS and A-IDS have been addressed, most of the studies focus on a single element. Our focus is on behaviors of multiple elements in time and space rather than that of a single element. As an existing study considering the concept of time and space, Chen et al. [41] identify spatiotemporal patterns of cyberattacks by analyzing victims' IP addresses collected by Honeypots. The biggest difference from our work is that they define every packet arriving at Honeypots as attacks and analyze characteristics of attack traffic in order to predict cyberattacks, whereas our focus is on defining a novel method of detecting cyberattacks based on fundamental attack behaviors in time and space. They focus on the macroscopic characteristics of attack traffic and identify deterministic and stochastic patterns among a wide range of consecutive IP addresses. In addition, they only use IP addresses observed from the victim side, whereas we monitor not only the states of both attackers and victims but their spatiotemporal relationships.

3. Basic Cyberattack Action (BCA)

In order to detect an attack by looking at number of computing and network elements together, we carefully study existing attack classifications as well as intrusion datasets. We focus on system and network characteristics by intrusions. We finally define BCAs, fundamental behaviors of attacks. BCAs observed from multiple elements naturally lend themselves to be described in space and time.

CAPEC [10] organizes more than 500 attack patterns employed to exploit vulnerabilities. CAPEC contains a comprehensive list with detailed information about each pattern. By analyzing CAPEC, we find that all attack patterns can be described with 10 essential methods of attack (MA) as shown in Table 1. Every attack pattern in CAPEC consists of some combination of MAs. We define five types of BCAs associated with relevant MA. In this work, we do not include MA10 as it depends on the human trust behavior during an attack. For example, CAPEC-98 (phishing attacks) trick people into offering access to their sensitive information. It deals with the human trust issue and it does not manifest in a particular system behavior that can be attributed to particular BCAs.

Table 1 also shows how each MA maps to BCAs. We analyze two types of intrusion dataset and as well as CAPEC to find out the mapping. Table 1 lists possible attacks corresponding to the mapping. The first intrusion dataset is KDD, the most widely-used dataset in intrusion detection. KDD classifies attacks into denial of service (DoS), remote-to-local (R2L), user to root (U2R) and Probing for IDS evaluation in the 1998 DARPA project. Numerous attacks belonging to the four classifications has been injected for dataset generation. The other one is CSE-CIC-IDS 2018 [42] that has been actively used in recent intrusion studies. CSE-CIC-IDS 2018 was generated by injecting 6 types of attack, such as brute force, DoS and botnet.

There are many proposed methods to detect MAs. In this work, we are mainly interested in BCAs and we could use any of these methods for MA detection. Focusing on common and fundamental features of cyberattacks rather than specific characteristics of each attack would become increasingly necessary to detect new and variant attacks.

Table 1. Mapping of MAs and BCAs through analyzing attacks.

ID	MA	CAPEC (ID)	KDD	CSE-CIC-IDS 2018	BCA
MA1	Flooding	XML Ping of the death (147)	DoS	DoS, DDoS, Botnet	BCA-1, BCA-4
MA2	Protocol manipulation	HTTP attacks (33, 34, 105)	-	DoS-Slowloris, SlowHTTPTest, Hulk	BCA-1, BCA-4
MA3	Time and state	Forced deadlock (25) Race condition (27, 29)	-	-	BCA-1
MA4	API abuse	Inducing account Lockout (2)	U2R	-	BCA-1
MA5	Injection	Buffer overflow (10, 24, 42, 46,47, 67) Command injection (136) SQL injection (108, 109)	-	SQL Injection	BCA-1
MA6	Analysis	Port scanning (300-308)	Probing	Infiltration attack	BCA-3
MA7	Spoofing	Resource location spoofing (38, 132)	DoS-smurf	-	BCA-5
MA8	Brute force	Password attack (16, 55, 70)	R2L	BruteForce-SSH, FTP, XSS	BCA-2
MA9	Modification of resources	Leverage alternate encoding (71) Block access to libraries (96)	U2R	-	BCA-1
MA10	Social engineering	Phishing (98) Clickjacking (103)	-	-	-

The five types of BCA are characterized as follows:

• BCA-1. Sudden performance degradation

Most attacks target to disrupt services offered by computing and network elements [43–45]. A sudden performance degradation would describe an essential behavior of affected hosts. This behavior can also occur in hardware and software failures. MA1 (*Flooding*), MA2 (*Protocol manipulation*), MA3 (*Time and state*), MA4 (*API abuse*), MA5 (*Injection*) and MA9 (*Modification of resources*) would result in sudden performance degradation. According to CAPEC, these MAs cause resource consumption, instability, crash, unexpected state, or execution logic change. Known attacks such as DoS [43,46–48], malware injection [43,46,49], buffer overflow [43,48], race condition [48], symbolic link [48] and a single point of failure [46] belong to BCA-1. Detecting attacks based on a single MA may lead to many false positives. However, BCA detection that combines several essential attack behaviors would decrease false positives significantly.

• BCA-2. Iterative behavior

Many cyberattacks begin by obtaining an access to a target element. Most common method to obtain an access is through the brute force method of login trials with different passwords [43,46,48]. The resulting behavior is iterative access requests and corresponding responses. MA8 (*Brute force*) is based on a repetitive trial-and-error method. Known attacks such as login attempts [43] and authentication attacks [46,48] belong to BCA-2. This pattern manifests distinctively from common application requests and responses. Normal transactions in client-server systems do not exhibit this iterative behavior. Therefore, these iterative actions result in an essential attack behavior of a computing element.

• BCA-3. Propagating behavior

Many attacks do not remain in a single target element. They tend to propagate to increase the number of infected hosts [43,50]. Attackers initially search for a vulnerable target. Once the target is infected by an attacker, the target becomes an attacker and starts propagating its search and infect tasks. This behavior is quite distinct from common application behavior. The resulting behavior is the increasing number of infected hosts as the time increases and such behavior translates to a spatiotemporal behavior of increasing infected elements. MA6 (*Analysis*) corresponds to the initial search such as probing [47] and scanning [43]. Known attacks such as worm [43,48] and port scanning [10] belong to BCA-3.

- BCA-4. Sudden increase or decrease in ingress and egress traffic

In additional to the performance degradation, the resulting behavior of attacks can be observed in either sudden increase or decrease of both ingress and egress traffic at the same time [51–54]. Usually the performance degradation would decrease the egress traffic corresponding to the responses of a server but the ingress traffic corresponding to the requests would remain the same. Decrease or increase in ingress and egress traffic usually result from malicious operation in computing or network elements. DDoS attacks [43,48] and flooding attacks [49] belong to BCA-4. In addition, BCA-4 could occur in combination with BCA-1 because this type of attacks could decrease the server performance.

- BCA-5. Uncorrelated ingress and egress traffic

We observe in any servers that the ingress traffic is highly correlated to the egress traffic. As the number of requests increases, we expect the number of responses to increase. This behavior is true when the server is working in desired operational range. As long as the server is capable of responding all requests immediately, we expect the number of responses to closely track the number of requests. When the server is congested or malfunctioning, the ingress and egress traffic are not correlated. Many attacks manifest in this uncorrelated ingress and egress traffic behavior. For example, when attackers spoof their identities during an attack, they do not receive any responses while it sends large numbers of requests [55]. Uncorrelated ingress and egress traffic would describe an essential behavior of a computing element with forged identity. In the existing works, masquerade [48] belongs to BCA-5. Figure 1 illustrates each BCA with the key behavioral features described above.

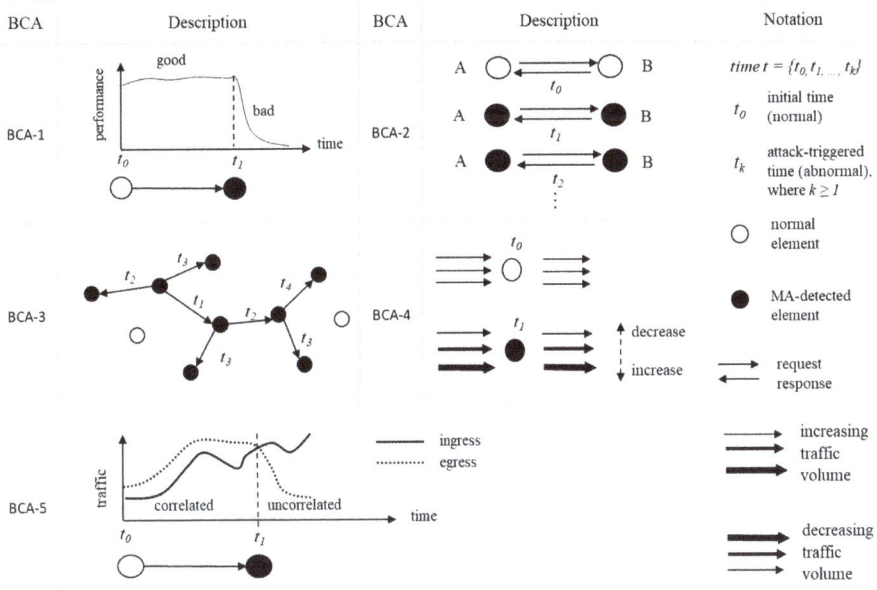

Figure 1. Key system behavior features of BCAs.

4. BCA Description and Composition

We now describe each BCA with associated MAs and spatiotemporal patterns. We use a stochastic graph model to describe the behavior of BCA in time and space. We define the stochastic graph model as follows. Table 2 shows the notation of the graph model.

Table 2. Notation of the graph model G.

Notation	Description
$G(t)$	overall stochastic graph comprising of multiple elements at time t
$G_i(t)$	stochastic graph related to a single element i at time t
$V_i(t)$	a set of states of a single element i at time t
$E_i(t)$	a set of edges between a single element i and other elements at time t
$\Lambda_i(t)$	a set of traffic volume from a single element i to other elements at time t
$\lambda_{i,j}(t)$	a stochastic random variable associated with an edge between a single element i and j

Definition 1. *A stochastic graph G(t) represent the overall stochastic graph comprising of all elements i at time t. $G_i(t)$ represents the stochastic graph only related to element i, where $i \in I$, thus a subset graph of $G(t) = \{G_i(t)\}$. $G_i(t)$ is modelled by 3-tuple of graph $G_i(t) = (V_i(t), E_i(t), \Lambda_i(t))$. $V_i(t) = \{null, MA1, MA2, \ldots, MA9\}$ represent the state of the element i. If the element detects MA3, its state is MA3, for example. Null represent that no MA is detected and is operating normally. $E_i(t) = \{e_{i,j}\}$ where $i, j \in I$, is the set of edges that represent the communication between the node i and j. $\Lambda_i(t) = \{\lambda_{i,j}(t)\}$ is the set of traffic volume from node i to all j, $\lambda_{i,j}(t)$, where $i, j \in I$. $\lambda_{i,j}(t)$ represents the stochastic random variable associated with $e_{i,j}$. $G_i(t)$ does not contain any vertices not connected to element i. BCAs can now be modelled using the stochastic graph G as follows.*

- BCA-1. Sudden performance degradation

$$V_i(t) = \{MA1, MA2, MA3, MA4, MA5, MA9\}, |E_i(t)| \geq 0 \text{ and } \sum_{\forall j} \lambda_{i,j}(t) \geq 0 \quad (1)$$

MA1 (*Flooding*), MA3 (*Time and state*), MA4 (*API abuse*), MA5 (*Injection*) and MA9 (*Modification of resources*) degrade performance of computing elements. MA2 (*Protocol manipulation*) would disrupt services offered by an application server. Host i then has suddenly degraded performance. $E_i(t)$ and $\sum_{\forall j} \lambda_{i,j}(t)$ could be zero or non-zero. BCA-1 is detected when host i has sudden performance degradation.

- BCA-2. Iterative behavior

$$V_i(t) = \{MA8\}, |E_i(t)| \leq |E_i(t + \Delta)| \text{ and } \frac{\int_t^{t+w} \sum_{\forall j} \lambda_{i,j}(t)}{w} \approx \frac{\int_{(t+\Delta)}^{t(t+\Delta)+w} \sum_{\forall j} \lambda_{i,j}(t+\Delta)}{w} \quad (2)$$

MA8 (*Brute force*) manifests in an iterative behavior. When host i repeats the same behavior such as continuous login trials, the host would generate consistent traffic during the period of attack. Δ and w denote the time period for the successive iteration and the window size for the traffic analysis respectively. During the password attack, iterative access requests and responses between the attacker and the target server generate consistent traffic volume. The password attack could target one or multiple servers. Brute force attacks may target different victims and the number of neighbors the stochastic graph would increase in time.

- BCA-3. Propagating behavior

$$V_i(t) = \{MA6\}, |E_i(t)| < |E_i(t+\Delta)| \text{ and } \sum_{\forall j} \lambda_{i,j}(t) \leq \sum_{\forall j} \lambda_{i,j}(t+\Delta) \quad (3)$$

In propagating attacks, an infected host i becomes the attacker and starts infecting another host. Host i would infect more and more hosts as time increases. Host i keeps scanning other hosts j to find vulnerable hosts. MA6 corresponds to the scanning behavior. The total volume does not play significant role here. Traffic from i to all connected elements would usually increase but it is not necessary to show BCA-3 behavior.

- BCA-4. Sudden increase or decrease in ingress and egress traffic

$$V_i(t) = \{MA1, MA2\}, |E_i(t)| \leq |E_i(t+\Delta)| \text{ and } \frac{d^2 \sum_{\forall j} \lambda_{i,j}(t)}{dt^2} > \alpha \text{ or } \frac{d^2 \sum_{\forall j} \lambda_{i,j}(t)}{dt^2} < \beta \quad (4)$$

In DDoS attacks, the traffic volume of both attackers and targets would suddenly increase exponentially. Host i under the DDoS attack results in $\frac{d^2 \sum_{\forall j} \lambda_{i,j}(t)}{dt^2} > \alpha$. The traffic volume increases greater than the acceleration rate α. HTTP DoS attacks disrupt a web application server by depleting the web resources. Ingress and egress traffic of the server i would suddenly decrease exponentially. In this attack, $\frac{d^2 \sum_{\forall j} \lambda_{i,j}(t)}{dt^2} < \beta$. The traffic volume decreases faster than rate β. MA1 (Flooding) and MA2 (Protocol manipulation) are the essential methods for the DDoS and HTTP DoS attack, respectively. Usually multiple new hosts show up in the DDoS attack, but it is not necessarily required.

- BCA-5. Uncorrelated ingress and egress traffic

$$V_i(t) = \{MA7\}, |E_i(t)| \geq |E_i(t+\Delta)| \text{ and } \Lambda_i(t) = R_{i,j}(\sum_{\forall j} \lambda_{i,j}(t), \sum_{\forall j} \lambda_{j,i}(t)) < \gamma \quad (5)$$

MA7 (spoofing) belongs to BCA-5. In IP spoofing attacks, the attacker does not receive any responses while it sends requests. This behavior results in uncorrelated ingress and egress traffic. When Host i spoofs its identity, $\Lambda_i(t)$ satisfies $R_{i,j}(\sum_{\forall j} \lambda_{i,j}(t), \sum_{\forall j} \lambda_{j,i}(t)) < \gamma$. $R_{i,j}$ is the cross correlation of ingress and egress traffic of i. γ is a threshold coefficient of $R_{i,j}$ and $0 < \gamma < 1$. Host i is hidden to other elements due to its spoofed IP address. As Host i is unknown to other elements during the attack, the number of its neighbor deceases.

5. BCA Detection System

Our system detects BCAs by monitoring spatiotemporal patterns according to the stochastic graph model. The spatiotemporal pattern describes the change of interactions among elements in time and space. There are many existing detection methods for MAs. We deploy any one of existing effective MA detection mechanisms. Periodically we generate a graph G_i for element i. MAs are associated to G_i when they are detected for element i. We match G_i against the stochastic graph models of BCAs to detect intrusions. We demonstrate the effectiveness of spatiotemporal patterns in detecting existing attacks as well as unknown attacks.

5.1. MA Detection

We apply existing mechanisms for MA detection in the host. Many of these mechanisms monitor system metrics and correlate metrics to detect a particular MA in a single computing or network element. We apply common MA detection mechanisms in the literature as shown in Table 3. S_i denotes the system metrics for MA detection mechanisms.

Our focus is not on performance of particular MA detection mechanisms but to demonstrate the advantage of BCA and their spatiotemporal patterns. Improvement in existing MA detection would improve our overall system. Again MA detection is limited to a single element and tends to have many false positives and false negatives.

Table 3. Correlation between MAs and metrics.

Type		Metric	MA	Reference
System measurements	S_1	CPU usage	MA1, MA3, MA4, MA5, MA9	[44,45]
	S_2	Memory usage		
	S_3	Disk usage		
Network measurement	S_4	inbound traffic/sec	MA1	[51–54]
	S_5	outbound traffic/sec		
	S_6	ratio of inbound and outbound traffic	MA7	
	S_7	Number of neighbors	MA6	[50]
Application measurement	S_8	Requests/sec	MA2, MA8	[53,56]
	S_9	Responses/sec		
	S_{10}	ratio of response over requests	MA2	
	S_{11}	Response time	MA2	[44,45,57,58]

5.2. BCA Detection

Individual host i monitors any change in MA, traffic volume, or temporal and spatial relationship among elements. Periodically its stochastic graph G_i is generated. The spatiotemporal pattern of G_i is then compared to BCA models. When there is a match between G_i and any of BCA models, we determine there is an intrusion and cyberattack to Host i and its associated elements.

Here is an example of BCA detection mechanism. Assume that host A generates $G_A(t)$ as shown in Figure 2.

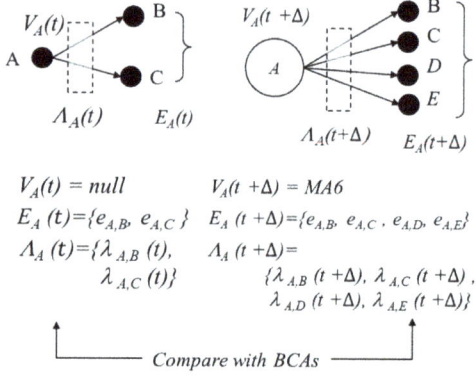

Figure 2. Generation of G_A.

We then proceed to match with BCA graphs. Assume that $t_0 = t$ and $t_1 = t + \Delta$. BCA-3 matches $G_A(t_1)$ in the example as shown in Figure 3.

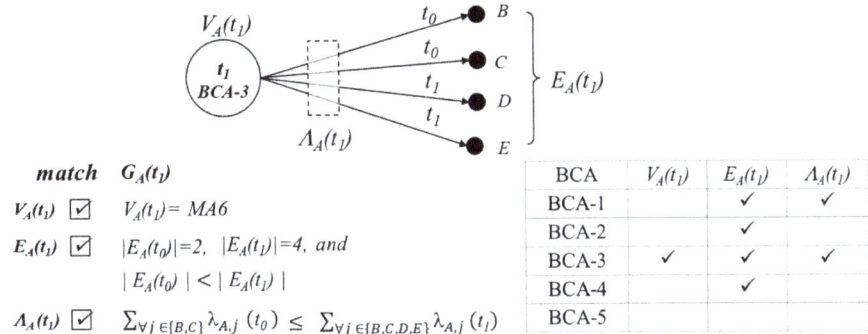

Figure 3. BCA-3 detection of G_A.

5.3. Combination of BCA Detections

The stochastic graph G contains all elements with detected MAs. Each element carries out BCA detection through matching its own stochastic graph with BCA graphs. We then see all BCA detected elements collectively. If any of these graphs are connected, meaning that there is a connecting edge between these graphs, we consider the validity of given BCA detections by those elements. By considering multiple elements together, we reduce additional false positives by finding contradicting combination of BCAs. We also further reassure the accuracy of the detection by examining multiple elements. Here are examples to illustrate further reduction in false positives as well as improving detection accuracy. Figure 4 shows a worm attack.

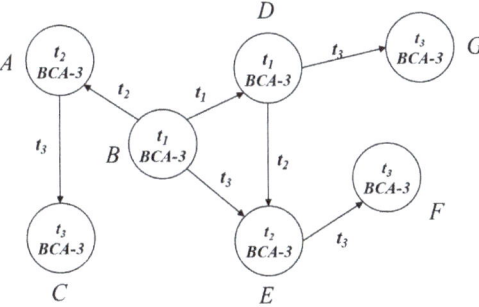

Figure 4. Combination of BCA-3.

Assume that hosts A-G detect BCA-3 at different times, t_1, t_2 and t_3. Each host generates $G_i(t)$, where $i \in \{A, B, \ldots, G\}$, according to BCA matching as shown in Figure 4. Once a host is infected by the worm attack, the host starts propagating the worm to other hosts continuously. Hosts A-G detect BCA-3 as the worm propagates in space and time. Assume that only one host detects BCA-3 and others do not detect any BCA patterns. There is no evidence of propagation and we determine that particular single BCA-3 detection has to be false positive. Combination of multiple $G_i(t)$ help us to reduce false positive. On the other hand, if there are multiple connected elements detecting BCA-3, then it confirms the propagating attack. Thus, $G(t)$ comprising of all $G_i(t)$ gives overall view of elements and helps to reduce false positives in many attack scenarios.

Figure 5 shows another example of advantage of having more comprehensive $G(t)$. Host A guesses a B's password using the brute force password attack. Host A sends login requests continuously until it finds out the correct password. During the attack, host B repeats the same behavior to authenticate the passwords. In the password attack, the iterative behavior of either side of the host requires similar behavior from the other host. If only A or B detects BCA-2, we cannot definitely determine it as the brute force attack or a false positive. The combination of BCA-2 detected by A and B increases the confidence in detecting the attack.

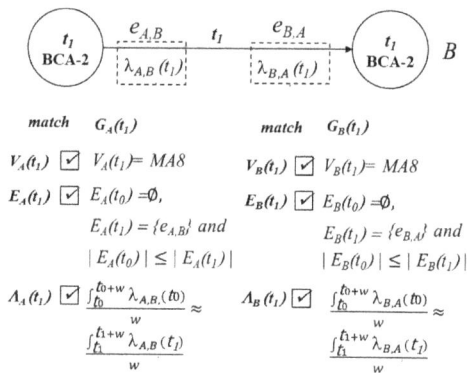

Figure 5. Combination of BCA-2.

5.4. Root Cause Analysis

Another advantage of using BCA graph is its ability to find possible root cause and location of the attack. The BCA graphs contain temporal and spatial relationship among elements. It is possible to trace the attack pattern to the originator using BCA graphs. Figure 6 shows an example of locating the root cause.

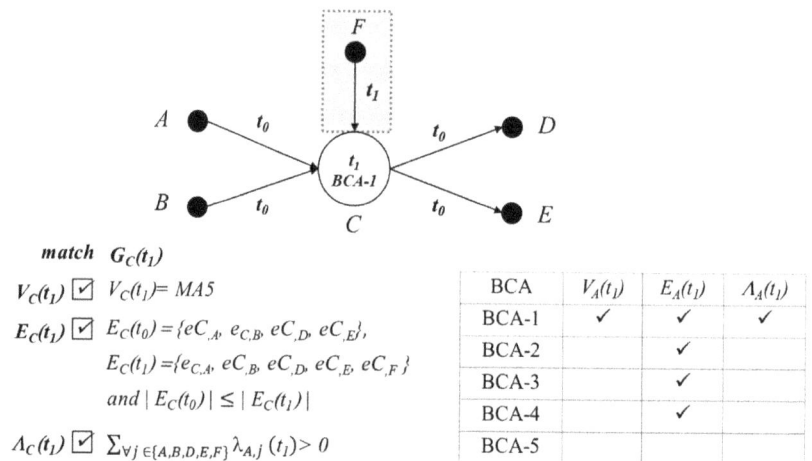

Figure 6. Example of root cause analysis.

At t_0, host A, B, C, D and E are running normally in a multi-tier application. When C detects BCA-1 due to performance degradation, $E_C(t_1) - E_C(t_0) = \{e_{F,C}\}$. Only $e_{F,C}$ shows up at t_1 while other edges appeared at t_0. Host F would be the attacker who disrupts host C by injection.

6. Experimental Evaluation

We deploy several experiments in our datacenter with a controlled VM cluster. We evaluate our system's performance in known attack and unknown attack detection. We also compare our system with those only relying on existing MA detection. We demonstrate how BCAs reduce false positives in several scenarios. We demonstrate that the spatiotemporal characterization of attack patterns helps in accuracy and reliability of intrusion and cyberattack detection. More extensive performance evaluation is not the focus of this paper and will be addressed in the forthcoming paper.

6.1. Experimental Setup

We implement our system and deploy in our production datacenter with a controlled VM cluster. We run a small agent in virtual machines (VMs). Each agent runs MA detection and BCA detection using its own stochastic graph. The agent creates its stochastic graph periodically. The agent then match its graph to BCA graphs. When it finds the matching BCA, the agent sends an alarm along with its graph to the management server. The management server compiles graphs from all elements to generate and update overall stochastic graph, G. The management server then examines all connected graphs G_i to determine the attacks and possible root causes. The infrastructure for the experiments consists of the following components:

- Physical servers: Fedora 21, QEMU 1.6.2 hypervisor
- VM: Ubuntu 14.04 and Fedora 22, 1v CPU, 1024MB RAM
- Cloud web application: Rubbos application [59]

For the high reliability of the experimental evaluation, we deploy the Rubbos web application running in an N-Tier architecture. During attacks, web servers and database servers keep processing service requests from 100 clients on average per second.

6.2. Known Attacks

As shown in Table 4, we inject four known attacks selected by analyzing intrusion datasets as well as CAPEC as described in Section 3. We use released attack scripts as well as a penetration software for attack injection. Both Scenarios 1 and 4 detect multiple BCAs including BCA-4. Scenario 1 detects sudden decrease in traffic while scenario 4 detects sudden increase in traffic in BCA-4 detection.

Table 4. Scenarios of known attacks.

	Scenario	BCA				
		1	2	3	4	5
1	Slowloris attack	✓			✓	
2	Password attack		✓			
3	SSH worm attack			✓		
4	Spoofed DDoS attack	✓			✓	✓

6.2.1. Scenario 1

Slowloris attack is a DoS attack targeting an application layer. The attacker modifies HTTP headers with wrong termination characters. The attacker then sends the packets to a web application server. This attack disrupts the web server due to a large number of incomplete open HTTP connections. The attacker consumes all connections on the server.

Existing HTTP DoS detection systems manually configure the web application parameters or set appropriate firewall rules to drop the suspicious packets [60]. Our system monitors the application metric S_8 (requests/s), S_9 (responses/s) and S_{10} (ratio of requests and response). Here we have four hosts A, B, C and D as shown in Figure 7a. We deploy A (client), B (web server), and C (DB server) running a Rubbos application at t_0. We inject the Slowloris attack into D using a released script [61]. Host D sends the modified HTTP requests (200 packets/s) to B.

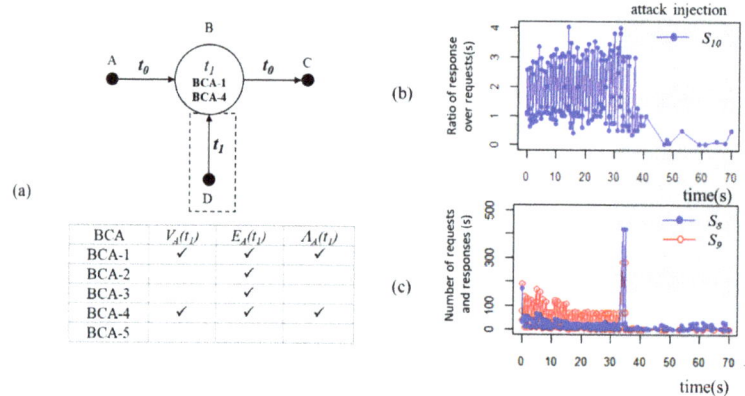

Figure 7. (**a**) Network structure of Scenario 1 and the BCA status of host B. (**b**) Experimental results of S_{10}. (**c**) Experimental results of S_8 and S_9.

- BCA detection

S_{10} (ratio of response over requests) indicates the performance of B for processing HTTP requests. When B is operating normally, the value of S_{10} fluctuates from 1 to 4, as shown in Figure 7b. S_{10} suddenly decreases when the attack is injected. S_{10} decreases as S_9 (responses/s) suddenly decreases due to the performance degradation, as shown in Figure 7c. Host B detects MA2 (*Protocol manipulation*) based on S_9 and S_{10}.

Host B detects BCA-1 and BCA-4 based on $G_B(t_1)$ as shown in Table 5. B satisfies $E_B(t_1)$ and $\Lambda_B(t_1)$ as well as $V_B(t_1)$ for BCA-1 and BCA-4. Existing systems would also detect this attack by monitoring only MA2 using S_8, S_9 and S_{10}.

Table 5. Detection of BCA-1 and BCA-4 on host B at t_1 in Scenario 1.

$G_B(t_1)$	Detection	BCA
$V_B(t_1)$	$V_B(t_1) = MA2$	**BCA-1**, **BCA-4**
$E_B(t_1)$	$E_B(t_0) = \{e_{B,A}, e_{B,C}\}$, $E_B(t_1) = \{e_{B,A}, e_{B,C}, e_{B,D}\}$, $|E_B(t_0)| = 2$, $|E_B(t_1)| = 3$ and $|E_B(t_0)| \leq |E_B(t_1)|$	**BCA-1**, BCA-2 BCA-3, **BCA-4**
$\Lambda_B(t_1)$	$\sum_{\forall j \in \{A,C,D\}} \lambda_{B,j}(t_1) \geq 0$	**BCA-1**
	$\frac{d^2 \sum_{\forall j \in \{A,C,D\}} \lambda_{B,j}(t_1)}{dt^2} > 0$	**BCA-4**
$G_B(t_1)$	**BCA-1 and BCA-4**	

Although our focus is not on detection methods of MAs, we analyze false positives in detecting MA2 for the validation. Because our attack scenario has 100 clients in the cloud application, we deploy 50 clients, 100 clients, and 200 clients without attacks. Table 6 shows the false positive rate (FPR) for S_8, S_9 and S_{10}, respectively.

Table 6. False positive rate of MA detection in Scenario 1.

Metric	50 Clients	100 Clients	150 Clients
S8	8.13%	0.81%	0%
S9	7.31%	0.81%	0%
S10	0%	0%	0%

- Root cause

According to $G_B(t_1)$, only $e_{B,D}$ shows up at t_1 as $E_B(t_1) - E_B(t_0) = \{e_{B,D}\}$. The new element D would be the attacker disrupting B's service by protocol manipulation (MA2).

6.2.2. Scenario 2

We inject a password attack that tries guessing a victim's password. In our experiment, we have two hosts, A and B as shown in Figure 8a. We inject the attack into A using Metasploit, a penetration software [62]. Metasploit is open-source software and allows us to inject a variety of attacks with our custom modules. Host A sends login requests more than 4000 times for 20 s guessing B's password. Host B is a MySQL server.

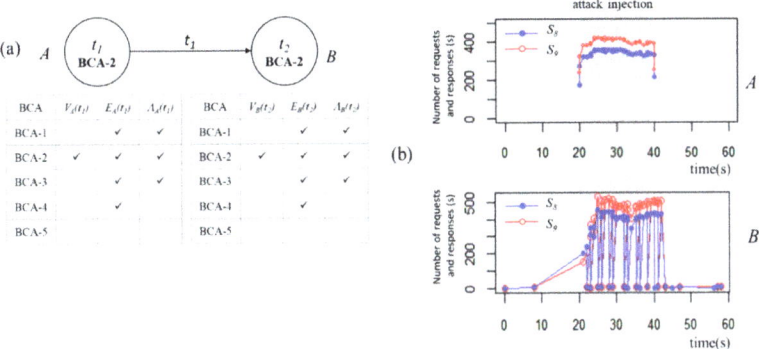

Figure 8. (a) Network structure of Scenario 2 and the BCA status of host A and B. (b) Experimental results of S_8 and S_9 on host A and B.

- BCA detection

Our system monitors S_8 (requests/s) and S_9 (responses/s). From A's perspective, S_8 shows the number of trials of guessing the password for one second. S_9 is the number of responses from the MySQL server, B. Host A and B detect MA8 (*Brute force*) according to large values of S_8 and S_9, as shown in Figure 8b.

Our system detects BCA-2 on both hosts from $G_A(t_1)$ and $G_B(t_2)$. Host A and B have a new neighbor ($E_A(t_1)$ and $E_B(t_2)$) and they generate very consistent traffic ($\Lambda_A(t_1)$ and $\Lambda_B(t_2)$) during the attack. $G_A(t_1)$ and $G_B(t_2)$ show that A and B satisfy all conditions for BCA-2 as shown in Table 7. Without MA detection ($V_A(t_1)$ and $V_B(t_2)$), it could be either BCA-1 or BCA-2.

Table 7. BCA-2 detection on host A and B at t_1 and t_2 in Scenario 2.

$G_i(t)$	Detection	BCA
$V_A(t_1)$	$V_A(t_1) = MA8$	BCA-2
$E_A(t_1)$	$E_A(t_0) = \varnothing, E_A(t_1) = \{e_{A,B}\}, |E_A(t_0)| = 0, |E_A(t_1)| = 1, and\ |E_A(t_0)| \leq |E_A(t_1)|$	BCA-1, **BCA-2**, BCA-3, BCA-4
$\Lambda_A(t_1)$	$\sum_{\forall j \in \{B\}} \lambda_{A,j}(t_1) \geq 0$	BCA-1, BCA-3
	$\dfrac{\int_{t_0}^{t_0+w} \lambda_{A,B}(t_0)}{w} \approx \dfrac{\int_{t_1}^{t_1+w} \lambda_{A,B}(t_1)}{w}$	BCA-2
$G_A(t_1)$		BCA-2
$V_B(t_2)$	$V_B(t_2) = MA8$	BCA-2
$E_B(t_2)$	$E_B(t_0) = \varnothing, E_B(t_2) = \{e_{B,A}\}, |E_B(t_0)| = 0, |E_B(t_2)| = 1, and\ |E_B(t_0)| \leq |E_B(t_2)|$	BCA-1, BCA-2, BCA-3, BCA-4
$\Lambda_B(t_2)$	$\sum_{\forall j \in \{A\}} \lambda_{B,j}(t_2) \geq 0$	BCA-1, BCA-3
	$\dfrac{\int_{t_0}^{t_0+w} \lambda_{B,A}(t_0)}{w} \approx \dfrac{\int_{t_2}^{t_2+w} \lambda_{B,A}(t_2)}{w}$	BCA-2
$G_B(t_2)$		BCA-2

Associating MA improves detection capability of our system. BCA-2 requires similar BCA-2 behavior from connected elements. The combination of BCA-2 detected by A and B in our system increases the confidence of correct detection. Existing systems that analyze elements independently could introduce many false positives.

For MA detection using S_8 and S_9, we have no false positive found. This is because number of requests to the database server is less than 70 per second for all 50 clients, 100 clients, 150 clients in normal state. However, FPR could increase if the application has much more clients than 150 clients.

- Root cause

According to $G(t)$ comprising of $G_A(t_1)$ and $G_B(t_2)$, the new edge between A and B appears at t_1 when A detects BCA-2. $E_A(t_1) - E_A(t_0) = \{e_{A,B}\}$. A is more likely to be the attacker sending the login requests to B, because A detects BCA-2 earlier than B.

6.2.3. Scenario 3

We inject a worm spreading over a local network. An attacker infects a target via an SSH. The attacker usually uses *known_hosts* file to collect target addresses and to bypass the authentication process. In our experiment, we deploy 10 hosts A to J which have all of other hosts' credentials. We first inject the worm into A using Metasploit. Host A repeats infecting other hosts. Once a target host is infected by the worm, it becomes the attacker and starts infecting another host.

- BCA detection

Our system monitors S_7 (number of neighbors) to detect the worm. S_7 refers to the number of trials to infect the worm via the SSH connection. Every host detects MA6 (*Analysis*) as S_7 increases as time increases as shown in Figure 9b. Our system detects MA6 when the number of neighbors (S_7) is greater than 5 (more than half of the entire hosts). Our system detects BCA-3 on every host based on $G_i(t)$ where $i \in \{A, B, C, \ldots, J\}$. Table 8 shows an example of the BCA detection of $G_A(t_1)$. Host A has new neighbors ($E_A(t_1)$). Traffic volume, ($\Lambda_A(t_1)$), increases as the infection propagates through elements. $G_A(t_1)$ matches all conditions of BCA-3. All hosts detect BCA-3 as the time increases. Overall $G(t)$ consisting of multiple BCA-3 elements is consistent with the expected behavior of BCA-3 with propagating attacks. Again the overall view of all related hosts increases the confidence of correct detection in this scenario.

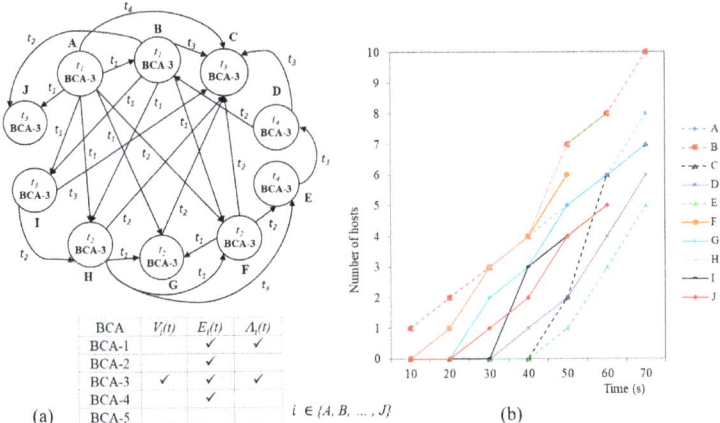

Figure 9. (a) Network structure of Scenario 3 and the status of BCA from host A to host J. (b) Experimental results of S_7 on host A to host J.

Table 8. BCA-3 detection on host A at t_1 in Scenario 3.

$G_A(t_1)$	Detection	BCA
$V_A(t_1)$	$V_A(t_1) = MA6$	**BCA-3**
$E_A(t_1)$	$E_A(t_0) = \varnothing, E_A(t_1) = \{e_{A,B}, e_{A,G}, e_{A,H}, e_{A,I}, e_{A,J}\}$ $\lvert E_A(t_0)\rvert = 0, \lvert E_A(t_1)\rvert = 5, \text{ and } \lvert E_A(t_0)\rvert < \lvert E_A(t_1)\rvert$	BCA-1, BCA-2, **BCA-3**, BCA-4
$\Lambda_A(t_1)$	$\sum_{\forall j \in \{B,G,H,I,J\}} \lambda_{B,j}(t_1) \geq 0$	BCA-1, **BCA-3**
$G_A(t_1)$		**BCA-3**

In order to analyze false positives in MA detection based on S_7, we monitor clients in our datacenter. Because the clients usually communicate with a web server, the number of neighbors is not proportional to the number of clients. In our experiment without attacks, the number of neighbors is less than 3 with a normal application running.

- Root cause

According to $G(t)$, A and B first detect BCA-3 at t_1, while other hosts detect BCA-3 at between t_2 and t_4. Either A or B could be the attacker that initiated the worm among the hosts.

6.2.4. Scenario 4

We inject a distributed SYN flooding attack with a spoofed IP address using Hping3 [63]. The attacker sends massive SYN packets to zombies with the victim's IP address. The zombies then send SYN-ACK packets to the victim. The massive SYN-ACK packets deplete bandwidth of the victim. In this experiment, we deploy 6 hosts (A–F) as shown in Figure 10a. A is the attacker. Host A keeps sending SYN packets to B-E with F's IP address.

- BCA detection

Our system monitors S_4 (inbound traffic/s), S_5 (outbound traffic/s) and S_6 (ratio of inbound and outbound traffic). In this experiment, S_4 and S_5 are used for MA1 (*Flooding*) detection. S_6 is used to detect MA7 (*Spoofing*). Host A sends massive SYN packets but does not receive any responses during the attack. These SYN packets increase S_5 as shown in Figure 10b and S_6 increases accordingly. Host A detects MA1 and MA7 due to the high value of S_5 and S_6 respectively. Host F receives massive SYN-ACK packets from four hosts (B-E). F detects MA1 due to the high value of S_4, as shown in Figure 10c. In our experiment, the four hosts (B–E) do not detect MA1 because each host does not meet the detection threshold (500 kb/s). The range of S_4 and S_5 are from 180 kb/s to 450 kb/s. The total amount of traffic going to F exceeds the threshold, thus host F detects MA1.

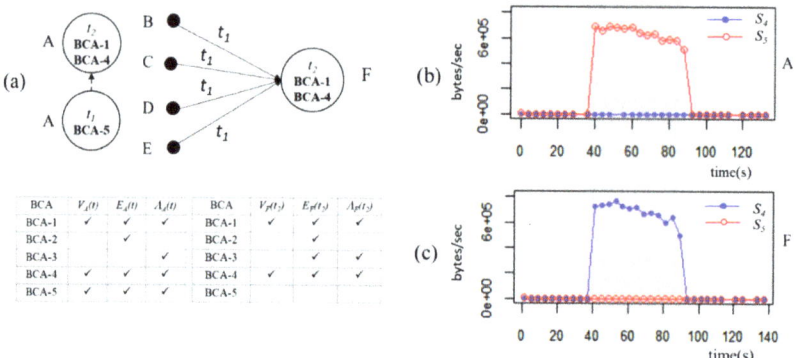

Figure 10. (a) Network structure of Scenario 4 and the BCA status of host A and F. (**b**,**c**) Experimental results of S_4 and S_5 on host A and F.

Our system detects BCA-4 and BCA-5 as shown in Table 9. Host A detects BCA-5 as it has a low correlation between inbound and outbound traffic. Host A and F detect BCA-1 as their outbound and inbound traffic suddenly increase respectively. Both A and F match all conditions for BCA-4 and BCA-5. By combining BCA graphs, our system correctly detects not only the DDoS attack to F but also the spoofing attack from A.

Table 9. BCA detection in Scenario 4 (BCA-5 detection on host A at t_1; detection of BCA-1 and BCA-4 on host A and F at t_2).

$G_i(t)$	Detection	BCA
$V_A(t_1)$	$V_A(t_1) = MA7$	BCA-5
$E_A(t_1)$	$E_A(t_0) = \varnothing, E_A(t_0) = \varnothing, |E_A(t_0)| = 0, |E_A(t_1)| = 0,$ and $|E_A(t_0)| \geq |E_A(t_1)|$	BCA-1, BCA-2 BCA-4, **BCA-5**
$\Lambda_A(t_1)$	$R_{A,j} \left(\sum_{\forall j = \{\text{"spoofed" ip}\}} \lambda_{A,j}(t_1), \sum_{\forall j = \{\text{"spoofed" ip}\}} \lambda_{j,A}(t_1) \right) < 0.1$	BCA-5
$G_A(t_1)$	BCA-5	
$V_A(t_2)$	$V_A(t_2) = MA1$	BCA-1, BCA-4
$E_A(t_2)$	$E_A(t_1) = \varnothing, E_A(t_2) = \varnothing, |E_A(t_1)| = 0, |E_A(t_2)| = 0,$ and $|E_A(t_1)| \geq |E_A(t_2)|$	BCA-1, BCA-2 BCA-4, BCA-5
$\Lambda_A(t_2)$	$\sum_{\forall j = \{\text{"spoofed" ip}\}} \lambda_{A,j}(t_2) \geq 0$	BCA-1
	$\frac{d^2 \sum_{\forall j = \{\text{"spoofed" ip}\}} \lambda_{A,j}(t_2)}{dt^2} > 0$	BCA-4
$G_A(t_2)$	BCA-1, BCA-4	
$V_F(t_2)$	$V_F(t_2) = MA1$	BCA-1, BCA-4
$E_F(t_2)$	$E_F(t_0) = \varnothing, E_F(t_2) = \{e_{F,B}, e_{F,C}, e_{F,D}, e_{F,E}\}, |E_F(t_0)| = 0, |E_F(t_2)| = 4,$ and $|E_F(t_0)| \leq |E_F(t_2)|$	BCA-1, BCA-2 BCA-3, **BCA-4**
$\Lambda_F(t_2)$	$\sum_{\forall j \in \{B,C,D,E\}} \lambda_{F,j}(t_2) \geq 0$	BCA-1, BCA-3
	$\frac{d^2 \sum_{\forall j \in \{B,C,D,E\}} \lambda_{F,j}(t_2)}{dt^2} > 0$	BCA-4
$G_F(t_2)$	BCA-1, BCA-4	

For detection of MA1 and MA7, we have no false positive found until we deploy 150 clients with normal behaviors. In the application, the values of S_4 and S_5 are less than 100 kb/s and 200 kb/s with 100 clients and 150 clients, respectively. In addition, S_6 has a value of at least 0.8 or higher with normal clients.

• Root cause

After host A detects BCA-5 at t_1, both A and F detect BCA-4 at t_2. Based on $G_A(t_1)$ and $G_A(t_2)$, we find host A spoofs its identity and sends massive traffic. According to $G_F(t_2)$, F has new edges between F and the 4 hosts (B–E). We can infer that A initiated DDoS attack to F using B–E's IP addresses.

6.3. Unknown Attack

We create an unknown attack based on the bait and switch method. It consists of a bait attack and the intended attack. The bait attack is designed to distract security managers' attention away from the intended attack. The ultimate goal of this attack is to distribute malicious codes. We deploy 3 malicious hosts (A, B, C), 4 clients (D, E, F, G), two web servers (H, I), and one DB server (J). Figure 11 shows seven hosts (D–J) running normally in the multi-tier application at t_0.

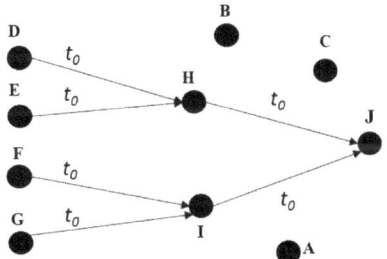

Figure 11. Initial state of unknown attack.

The unknown attack consists of three attacks as follows:

Password attack (intended attack) at t_1: This attack requires gaining access to the target server H. The attacker employs a slow password attack to find host H's password. The slow brute force attack is harder to detect using the existing brute force detection mechanisms. We inject the slow password attack into host B which is one of the malicious hosts. Host B repeatedly sends HTTP login requests to host H (web server) until it finds the correct password.

Flooding attack (bait attack) at t_2: The attacker employs a flooding attack to distract the security manager's attention from the intended attack. We inject the flooding attack to malicious host A. Host A starts sending large number of SYN packets to I in order to disrupt the server I.

Redirection attack (intended attack) at t_3: After host B gains access to host H through slow password attack, host B controls host H. Host B changes server H's configuration to redirect all incoming requests to host C (malicious host) instead of intended DB server J. When C receives requests from H, C sends malicious codes as a response to all clients.

• BCA detection

Figure 12 shows overall $G(t)$ from our system when the unknown attack is injected.

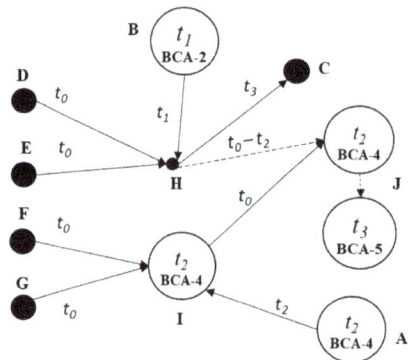

Figure 12. Detection of unknown attack.

Password attack (intended attack): According to $V_B(t_1)$, host B does not detect MA8 (*Brute force*) as shown in Table 10. Existing systems using metrics S_8 (requests/s) and S_9 (responses/s) would not detect the slow brute force attack. Our system detects the pattern of BCA-2 based on (E_B (t_1) and (Λ_B (t_1) at t_1. Host B finds a new edge to the target H and generates consistent requests during the attack as shown in Figure 13. BCA-2 requires similar BCA-2 behavior in the related host in the traditional password attack detection. In Figure 12, host H does not detect BCA-2 unlike host B. Host H fails to detect a slow rate of login request attack embedded among normal application requests. Our system detects host B's brute force behavior while existing systems fail to detect attacks on both B and H.

Table 10. BCA-2 detection on host B at t_1 during a password attack.

$G_i(t)$	Detection	BCA
$V_B(t_1)$	$V_B(t_1) = null$	-
$E_B(t_1)$	$E_B(t_0) = \emptyset$, $E_B(t_1) = \{e_{B,H}\}$, $|E_B(t_0)| = 0$, $|E_B(t_1)| = 1$ and $|E_B(t_0)| \leq |E_B(t_1)|$	BCA-2
$\Lambda_B(t_1)$	$\frac{\int_{t_0}^{t_0+w} \lambda_{H,B}(t_0)}{w} \approx \frac{\int_{t_1}^{t_1+w} \lambda_{H,B}(t_1)}{w}$	BCA-2
$G_B(t_1)$	BCA-2	

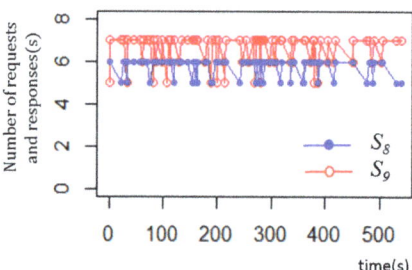

Figure 13. Iterative behavior of host B.

Flooding attack (bait attack): In the bait attack, host A and I detect MA1 (*Flooding*) due to high inbound (S_4) and outbound (S_5) traffic at t_2. According to $G_A(t_2)$ and $G_I(t_2)$ in Table 11, these hosts have a new edge between them and have a sudden increase in traffic as shown in Figure 14. This flooding attack also results in the sudden decrease of traffic in host J as host I is disrupted by flooding attack. The security manager is distracted by host I being attacked by host A through the flooding.

Table 11. BCA-4 detection on host A, I and J at t_2 during a flooding attack.

$G_i(t)$	Detection	BCA
$V_A(t_2)$	$V_A(t_2) = MA1$	BCA-4
$E_A(t_2)$	$E_A(t_0) = \varnothing, E_A(t_2) = \{e_{A,I}\}, |E_A(t_0)| = 0, |E_A(t_2)| = 1$ and $|E_A(t_0)| \leq |E_A(t_2)|$	BCA-4
$\Lambda_A(t_2)$	$\frac{d^2 \sum_{\forall j \in [I]} \lambda_{A,j}(t_2)}{dt^2} > 0$	BCA-4
$G_A(t_2)$	BCA-4 (increase)	
$V_I(t_2)$	$V_I(t_2) = MA1$	BCA-4
$E_I(t_2)$	$E_I(t_0) = \{e_{I,F}, e_{I,G}, e_{I,J}\}, E_I(t_2) = \{e_{I,F}, e_{I,G}, e_{I,J}, e_{I,A}\}, |E_I(t_0)| = 3, |E_I(t_2)| = 4$ and $|E_I(t_0)| \leq |E_I(t_2)|$	BCA-4
$\Lambda_I(t_2)$	$\frac{d^2 \sum_{\forall j \in \{F, G, J, A\}} \lambda_{I,j}(t_2)}{dt^2} > 0$	BCA-4
$G_I(t_2)$	BCA-4 (increase)	
$V_J(t_2)$	$V_J(t_2) = null$	-
$E_J(t_2)$	$E_J(t_0) = \{e_{J,H}, e_{J,I}\}, E_J(t_2) = \{e_{J,H}, e_{J,I}\}, |E_J(t_0)| = 2, |E_J(t_2)| = 2$ and $|E_J(t_0)| \leq |E_J(t_2)|$	BCA-4
$\Lambda_J(t_2)$	$\frac{d^2 \sum_{\forall j \in \{H, I\}} \lambda_{J,j}(t_2)}{dt^2} < 0$	BCA-4
$G_J(t_2)$	BCA-4 (decrease)	

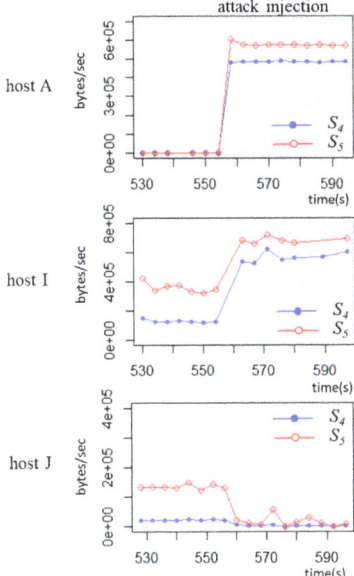

Figure 14. Sudden increase and decrease of traffic by the flooding attack.

Redirection attack (intended attack): According to $G_J(t_3)$ in Table 12, host J detects a removed edge between host H at t_3. The removed edge triggers the detection of BCA-5. The removed edge belongs to application elements. In normal operation, we do not expect any application element to be removed without prior notification. Thus it further confirms the attack behavior. Host J also detects a low correlation between inbound and outbound traffic due to the flooding and redirection attacks.

Table 12. BCA-5 detection on host J at t_3 during a redirection attack.

$G_i(t)$	Detection	BCA
$V_J(t_3)$	$V_J(t_3) = null$	-
$E_J(t_3)$	$E_J(t_2) = \{e_{J,H}, e_{J,I}\}, E_J(t_3) = \{e_{J,I}\}, |E_J(t_2)| = 2, |E_J(t_3)| = 1$ and $|E_J(t_2)| \geq |E_J(t_3)|$	BCA-5
$\Lambda_J(t_3)$	$R_{J,i} (\sum_{\forall i = \{I\}} \lambda_{J,i}(t_3), \sum_{\forall i = \{I\}} \lambda_{i,J}(t_3)) < 0.1$	BCA-5
$G_J(t_3)$	BCA-5	

Overall $G(t)$ graph indicates high possibility of the redirection attack based on other connected BCA detections. Our system correctly detects not only the bait attack but the intended attack where existing systems fail to detect the intended attack.

7. Conclusions

We have presented a different perspective on ways to detect cyberattacks. Rather than relying on traditional signatures and anomaly patterns, we proposed an approach based on fundamental and essential behaviors of cyberattacks. We defined these behaviors as Basic Cyberattack Action (BCA) and proposed five types of BCA such as a sudden performance degradation, iterative behavior, propagation behavior, sudden increase or decrease in ingress and egress traffic, and uncorrelated ingress and egress traffic. Individual BCA is detected by monitoring not only Methods of Attack (MAs) and traffic volume of a single element, but also the spatiotemporal relationship among elements. In order to represent combination of BCAs, we developed a stochastic graph model. The combination of BCAs describes the change of interactions among elements in time and space. By considering multiple elements together, we can reduce additional false positives by finding contradicting combination of BCAs. We also implemented and deployed our spatiotemporal-based intrusion detection system in our datacenter for preliminary validation of our idea. We demonstrated the effectiveness of BCAs in numerous known and unknown attack scenarios. For known attacks, we injected a Slowloris attack, password attack, SSH worm attack, and Smurf attack selected by analyzing intrusion datasets and CAPEC. Our experimental results showed that our system accurately detects all the known attacks comprising BCAs and locates possible root cause as well. Furthermore, we built an example of unknown attack based on a bait-and-switch method that combines three types of attacks such as a password attack, flooding attack, and redirection attack. The experimental results showed that such unknown attack is effectively detected by our system while existing detection mechanisms fail to detect the intended attack. Many existing systems may not be adequate for future unknown and advanced attacks. In addition, today's complex applications may trigger a significant number of false positives. We believe that the characterization of attacks in spatiotemporal patterns with expected essential behaviors of any attack presents a new effective approach to the intrusion detection. The performance evaluation with not only extensive attacks comprising complex BCAs but a variety of applications will be addressed in the future.

Author Contributions: Conceptualization, J.K. and H.S.K.; methodology, J.K. and H.S.K.; implementation and experiments, J.K.; writing—J.K. and H.S.K. All authors have read and agreed to the published version of the manuscript.

Funding: This research was supported by Basic Science Research Program through the National Research Foundation of Korea (NRF) funded by the Ministry of Education (NRF-2018R1D1A1B07050543).

Conflicts of Interest: The authors declare no conflict of interest.

References

1. Mell, P.M.; Lippmann, R.; Hu, C.T.; Haines, J.; Zissman, M. *An Overview of Issues in Testing Intrusion Detection Systems*; NIST Interagency/Internal Report (NISTIR)-7007; National Institute of Standards and Technology (NIST): Gaithersburg, MD, USA, 2003.
2. McCarthy, J.; Powell, M.; Stouffer, K.; Tang, C.; Zimmerman, T.; Barker, W.; Ogunyale, T.; Wynne, D.; Wiltberger, J. *Securing Manufacturing Industrial Control Systems: Behavioral Anomaly Detection*; National Institute of Standards and Technology (NIST): Gaithersburg, MD, USA, 2018.
3. Carter, E.; Hogue, J. *Intrusion Prevention Fundamentals*; Cisco Press: Indianapolis, IN, USA, 2006.
4. Chandola, V.; Banerjee, A.; Kumar, V. Anomaly detection: A survey. *ACM Comput. Surv. (CSUR)* **2009**, *41*, 1–58. [CrossRef]
5. Garcia-Teodoro, P.; Díaz-Verdejo, J.; Maciá-Fernández, G.; Vázquez, E. Anomaly-based network intrusion detection: Techniques, systems and challenges. *Comput. Secur.* **2009**, *28*, 18–28. [CrossRef]
6. Diffily, S. *The Website Manager's Handbook*; Lulu: Morrisville, NC, USA, 2006.
7. Sommer, R.; Paxson, V. Outside the closed world: On using machine learning for network intrusion detection. In Proceedings of the 2010 IEEE Symposium on Security and Privacy, Berkeley/Oakland, CA, USA, 16–19 May 2010; pp. 305–316.
8. Trinh, V.-V.; Tran, K.P.; Huong, T.T. Data driven hyperparameter optimization of one-class support vector machines for anomaly detection in wireless sensor networks. In Proceedings of the 2017 International Conference on Advanced Technologies for Communications (ATC), Quy Nhon, Vietnam, 18–20 October 2017; pp. 6–10.
9. Ikram, S.T.; Cherukuri, A.K. Intrusion detection model using fusion of chi-square feature selection and multi class SVM. *J. King Saud Univ. -Comput. Inf. Sci.* **2017**, *29*, 462–472.
10. MITRE. Common Attack Pattern Enumeration and Classification. Available online: http://capec.mitre.org (accessed on 5 March 2020).
11. Kumar, S.; Spafford, E.H. *A Pattern Matching Model for Misuse Intrusion Detection*; Technical Report CSD-TR-94-071; Purdue University: West Lafayette, IN, USA, 1994.
12. Kreibich, C.; Crowcroft, J. Honeycomb: Creating intrusion detection signatures using honeypots. *Acm Sigcomm Comput. Commun. Rev.* **2004**, *34*, 51–56. [CrossRef]
13. Kuri, J.; Navarro, G.; Mé, L.; Heye, L. A pattern matching based filter for audit reduction and fast detection of potential intrusions. In *International Workshop on Recent Advances in Intrusion Detection*; Springer: Berlin/Heidelberg, Germany, 2000; pp. 17–27.
14. Ilgun, K.; Kemmerer, R.A.; Porras, P.A. State transition analysis: A rule-based intrusion detection approach. *IEEE Trans. Softw. Eng.* **1995**, *3*, 181–199. [CrossRef]
15. Hu, Z.; Li, Z.; Wu, J. A novel Network Intrusion Detection System (NIDS) based on signatures search of data mining. In Proceedings of the 1st international Conference on Forensic Applications and Techniques in Telecommunications, information, and Multimedia and Workshop, Adelaide, SA, Australia, 23–24 January 2008; p. 45.
16. Cuppens, F.; Miege, A. Alert correlation in a cooperative intrusion detection framework. In Proceedings of the 2002 IEEE Symposium on Security and Privacy, Berkeley, CA, USA, 12–15 May 2002; pp. 202–215.
17. Lunt, T.F. IDES: An intelligent system for detecting intruders. In *Proceedings of the Symposium: Computer Security, Threat and Countermeasures*; Purdue University: West Lafayette, IN, USA, 1990; pp. 30–45.
18. Porras, P.A.; Neumann, P.G. EMERALD: Event monitoring enabling response to anomalous live disturbances. In Proceedings of the 20th National Information Systems Security Conference, Baltimore, MD, USA, 7–10 October 1997; pp. 353–365.
19. Valdes, A.; Skinner, K. Probabilistic alert correlation. In *International Workshop on Recent Advances in Intrusion Detection*; Springer: Berlin/Heidelberg, Germany, 2001; pp. 54–68.
20. Yu, J.; Reddy, Y.V.R.; Selliah, S.; Reddy, S.; Bharadwaj, V.; Kankanahalli, S. TRINETR: An architecture for collaborative intrusion detection and knowledge-based alert evaluation. *Adv. Eng. Inform.* **2005**, *19*, 93–101. [CrossRef]
21. Kruegel, C.; Mutz, D.; Robertson, W.; Valeur, F. Bayesian event classification for intrusion detection. In Proceedings of the 19th Annual Computer Security Applications Conference, 2003. Proceedings, Las Vegas, NV, USA, 8–12 December 2003; pp. 14–23.

22. Leyla, B.; Engin, K.; Christopher, K.; Marco, B. EXPOSURE: Finding Malicious Domains Using Passive DNS Analysis. In Proceedings of the NDSS 2011, 18th Annual Network and Distributed System Security Symposium, San Diego, CA, USA, 6–9 February 2011; pp. 1–17.
23. Bilge, L.; Sen, S.; Balzarotti, D.; Kirda, E.; Kruegel, C. Exposure: A passive dns analysis service to detect and report malicious domains. *ACM Trans. Inf. Syst. Secur. (TISSEC)* **2014**, *16*, 14. [CrossRef]
24. Feng, W.; Zhang, Q.; Hu, G.; Huang, J.X. Mining network data for intrusion detection through combining SVMs with ant colony networks. *Future Gener. Comput. Syst.* **2014**, *37*, 127–140. [CrossRef]
25. Kuang, F.; Xu, W.; Zhang, S. A novel hybrid KPCA and SVM with GA model for intrusion detection. *Appl. Soft Comput.* **2014**, *18*, 178–184. [CrossRef]
26. Thaseen, I.S.; Kumar, C.A. Intrusion detection model using fusion of PCA and optimized SVM. In Proceedings of the 2014 International Conference on Contemporary Computing and Informatics (IC3I), Mysore, India, 27–29 November 2014; pp. 879–884.
27. Khan, R.U.; Zhang, X.; Alazab, M.; Kumar, R. An Improved Convolutional Neural Network Model for Intrusion Detection in Networks. In Proceedings of the 2019 Cybersecurity and Cyberforensics Conference (CCC), Melbourne, Australia, 8–9 May 2019; pp. 74–77.
28. Li, Z.; Qin, Z.; Huang, K.; Yang, X.; Ye, S. Intrusion detection using convolutional neural networks for representation learning. In Proceedings of the International Conference on Neural Information Processing, Guangzhou, China, 14–18 November 2017; pp. 858–866.
29. Liu, Y.; Liu, S.; Zhao, X. Intrusion Detection Algorithm Based on Convolutional Neural Network. In Proceedings of the 4th International Conference on Engineering Technology and Application, Nagoya, Japan, 29–30 June 2017; pp. 9–13.
30. Kim, J.; Shin, Y.; Choi, E. An Intrusion Detection Model based on a Convolutional Neural Network. *J. Multimed. Inf. Syst.* **2019**, *6*, 165–172. [CrossRef]
31. Bontemps, L.; Cao, V.L.; Mcdermott, J.; Le-Khac, N.-A. Collective anomaly detection based on long short-term memory recurrent neural networks. In Proceedings of the International Conference on Future Data and Security Engineering, Can Tho City, Vietnam, 23–25 November 2016; pp. 141–152.
32. Staudemeyer, R.C.; Omlin, C.W. Evaluating performance of long short-term memory recurrent neural networks on intrusion detection data. In Proceedings of the South African Institute for Computer Scientists and Information Technologists Conference, East London, South Africa, 7–9 October 2013; pp. 218–224.
33. KDD CUP 1999 Data. Available online: http://kdd.ics.uci.edu/databases/kddcup99/kddcup99.html (accessed on 7 February 2020).
34. Yin, C.; Zhu, Y.; Fei, J.; He, X. A Deep Learning Approach for Intrusion Detection Using Recurrent Neural Networks. *IEEE Access* **2017**, *5*, 21954–21961. [CrossRef]
35. Kim, J.; Kim, H. An Effective Intrusion Detection Classifier Using Long Short-Term Memory with Gradient Descent Optimization. In Proceeding of the 2017 IEEE International Conference on Platform Technology and Service (PlatCon), Busan, South Korea, 13–15 February 2017; pp. 1–6.
36. Almiani, M.; AbuGhazleh, A.; Al-Rahayfeh, A.; Atiewi, S.; Razaque, A. Deep recurrent neural network for IoT intrusion detection system. *Simul. Model. Pract. Theory* **2019**, 102031. [CrossRef]
37. Dokas, P.; Ertoz, L.; Kumar, V.; Lazarevic, A. Data mining for network intrusion detection. In Proceedings of the NSF Workshop on Next Generation Data Mining, Marriott, Inner Harbor, Baltimore, 1–3 November 2002; pp. 21–30.
38. Hu, Y.; Panda, B. A data mining approach for database intrusion detection. In Proceedings of the 2004 ACM Symposium on Applied Computing, Nicosia, Cyprus, 14–17 March 2004; pp. 711–716.
39. Stephenson, P. The application of intrusion detection systems in a forensic environment. In Proceedings of the Third International Workshop on Recent Advances in Intrusion Detection (RAID), Toulouse, France, 2–4 October 2000.
40. Ren, W.; Jin, H. Distributed agent-based real time network intrusion forensics system architecture design. In Proceedings of the 19th International Conference on Advanced Information Networking and Applications (AINA'05), Taipei, Taiwan, 28–30 March 2005; pp. 177–182.
41. Chen, Y.-Z.; Huang, Z.-G.; Xu, S.; Lai, Y.-C. Spatiotemporal patterns and predictability of cyberattacks. *PLoS ONE* **2015**, *10*, e0131501. [CrossRef] [PubMed]
42. CSE-CIC-IDS2018 on AWS. Available online: https://www.unb.ca/cic/datasets/ids-2018.html (accessed on 7 February 2020).

43. eCSIRT. WP4 Clearinghouse Policy—Release 1.2. Available online: http://www.ecsirt.net/cec/service/documents/wp4-clearinghouse-policy-v12.html (accessed on 5 March 2020).
44. Falkenberg, A.; Mainka, C.; Somorovsky, J.; Schwenk, J. A new approach towards DoS penetration testing on web services. In Proceedings of the 2013 IEEE 20th International Conference on Web Services, Santa Clara, CA, USA, 28 June–3 July 2013; pp. 491–498.
45. Mirkovic, J.; Hussain, A.; Fahmy, S.; Reiher, P.; Thomas, R.K. Accurately measuring denial of service in simulation and testbed experiments. *IEEE Trans. Dependable Secur. Comput.* **2008**, *6*, 81–95. [CrossRef]
46. Iqbal, S.; Kiah, M.L.M.; Daghighi, B.; Hussain, M.; Khan, S.; Khan, K.; Choo, K.-K.R. On cloud security attacks: A taxonomy and intrusion detection and prevention as a service. *J. Netw. Comput. Appl.* **2016**, *74*, 98–120. [CrossRef]
47. Lippmann, R.; Haines, J.W.; Fried, D.J.; Korba, J.; Das, K. The 1999 DARPA off-line intrusion detection evaluation. *Comput. Netw.* **2000**, *34*, 579–595. [CrossRef]
48. Simmons, C. AVOIDIT: A cyber attack taxonomy. In Proceedings of the 9th Annual Symposium on Information Assurance (ASIA'14), Albany, NY, USA, 3–4 June 2014; pp. 2–12.
49. Islam, T.; Manivannan, D.; Zeadally, S. A classification and characterization of security threats in cloud computing. *Int. J. Next-Gener. Comput.* **2016**, *7*, 1–17.
50. Stafford, S.; LI, J. Behavior-based worm detectors compared. In *International Workshop on Recent Advances in Intrusion Detection*; Springer: Berlin/Heidelberg, Germany, 2010; pp. 38–57.
51. Bogdanoski, M.; Suminoski, T.; Risteski, A. Analysis of the SYN flood DoS attack. *Int. J. Comput. Netw. Inf. Secur. (IJCNIS)* **2013**, *5*, 1–11. [CrossRef]
52. Rana, D.S.; Garg, N.; Chamoli, S.K. A Study and Detection of TCP SYN Flood Attacks with IP spoofing and its Mitigations. *Int. J. Comput. Technol. Appl.* **2012**, *3*, 1476–1480.
53. Shea, R.; Liu, J. Performance of virtual machines under networked denial of service attacks: Experiments and analysis. *IEEE Syst. J.* **2012**, *7*, 335–345. [CrossRef]
54. Siaterlis, C.; Maglaris, V. Detecting incoming and outgoing DDoS attacks at the edge using a single set of network characteristics. In Proceedings of the 10th IEEE Symposium on Computers and Communications (ISCC'05), Murcia, Spain, 27–30 June 2005; pp. 469–475.
55. Templeton, S.J.; Levitt, K.E. Detecting spoofed packets. In Proceedings of the DARPA Information Survivability Conference and Exposition, Washington, DC, USA, 22–24 April 2003; pp. 164–175.
56. Grill, M.; Nikolaev, I.; Valeros, V.; Rehak, M. Detecting DGA malware using NetFlow. In Proceedings of the 2015 IFIP/IEEE International Symposium on Integrated Network Management (IM), Ottawa, ON, Canada, 11–15 May 2015; pp. 1304–1309.
57. Sachdeva, M.; Singh, G.; Singh, K. Performance analysis of web service under DDoS attacks. In Proceedings of the 2009 IEEE International Advance Computing Conference, Patiala, India, 6–7 March 2009; pp. 1002–1007.
58. Kim, J.; Kim, H.S. PBAD: Perception-based anomaly detection system for cloud datacenters. In Proceedings of the 2015 IEEE 8th International Conference on Cloud Computing, New York, NY, USA, 27 June–2 July 2015; pp. 678–685.
59. RuBBoS. Available online: http://jmob.ow2.org/rubbos.html (accessed on 5 March 2020).
60. Moustis, D.; Kotzanikolaou, P. Evaluating security controls against HTTP-based DDoS attacks. In Proceedings of the IISA 2013, Piraeus, Greece, 10 June–2 July 2013; pp. 1–6.
61. Slowloris. Available online: https://github.com/Ogglas/Orignal-Slowloris-HTTP-DoS/blob/master/slowloris.pl (accessed on 5 March 2020).
62. Metasploit. Available online: https://www.metasploit.com/ (accessed on 5 March 2020).
63. Hping3. Available online: https://tools.kali.org/information-gathering/hping3 (accessed on 5 March 2020).

© 2020 by the authors. Licensee MDPI, Basel, Switzerland. This article is an open access article distributed under the terms and conditions of the Creative Commons Attribution (CC BY) license (http://creativecommons.org/licenses/by/4.0/).

Article

A Comparative Analysis of Cyber-Threat Intelligence Sources, Formats and Languages

Andrew Ramsdale [1], Stavros Shiaeles [2,*] and Nicholas Kolokotronis [3]

[1] School of Computing, Electronics and Mathematics, Faculty of Science and Engineering, Plymouth University, Plymouth PL4 8AA, UK; andrew.ramsdale@postgrad.plymouth.ac.uk
[2] School of Computing, Faculty of Technology, University of Portsmouth, Portsmouth PO1 2UP, UK
[3] School of Economics and Technology, Faculty of Informatics and Telecommunications, University of Peloponnese, 22131 Tripolis, Greece; nkolok@uop.gr
* Correspondence: stavros.shiaeles@port.ac.uk

Received: 5 April 2020; Accepted: 13 May 2020; Published: 16 May 2020

Abstract: The sharing of cyber-threat intelligence is an essential part of multi-layered tools used to protect systems and organisations from various threats. Structured standards, such as STIX, TAXII and CybOX, were introduced to provide a common means of sharing cyber-threat intelligence and have been subsequently much-heralded as the de facto industry standards. In this paper, we investigate the landscape of the available formats and languages, along with the publicly available sources of threat feeds, how these are implemented and their suitability for providing rich cyber-threat intelligence. We also analyse at a sample of cyber-threat intelligence feeds, the type of data they provide and the issues found in aggregating and sharing the data. Moreover, the type of data supported by various formats and languages is correlated with the data needs for several use cases related to typical security operations. The main conclusions drawn by our analysis suggest that many of the standards have a poor level of adoption and implementation, with providers opting for custom or traditional simple formats.

Keywords: cyber-threat intelligence; threat exchange; vulnerability alerts; incident reporting; indicators of compromise; cyber-observables

1. Introduction

With the advent of the *Internet of things* (IoT), there has been an unprecedented increase of cyber-attacks, which have evolved and become more sophisticated. Adversaries now use a vast set of tools and tactics to attack their victims with their motivations ranging from intelligence collection to data destruction or financial gain. Understanding the attacker has become more complicated and even more important as this knowledge, if transformed into actionable information, can be used to adapt networks' defences in an automated manner to better protect the network against possible threats. *Cyber-threat intelligence* (CTI) focuses on the capabilities, motivations and goals of an adversary and how these could be achieved. Intelligence is the information and knowledge gained about an adversary through observation and analysis; intelligence is not just data, but the outcome of an analysis and must be actionable to meet the needs of current defensive systems that have to deal with and respond to cyber-attacks. Amongst others, examples of CTI include indicators (system artefacts or observables associated with an attack), security alerts, incident reports and threat intelligence, along with any other relevant information on recommended (or vulnerable) security tool configurations [1,2].

The efficient sharing of CTI is at the core of cyber-threat detection and prevention, as it allows building multi-layer automated tools with sophisticated and effective defensive capabilities that continuously analyse the vast amounts of the heterogeneous CTI related to attackers' *tactics, techniques and procedures* (TTPs), indicators of ongoing incidents, etc. [3,4]. Given the numerous architectures, products and systems being used as sources of data for information sharing mechanisms, standardised and structured representations of CTI are required to allow a satisfying interoperability level across the various stakeholders [2]. Therefore, considerable efforts have been put during the last decade to standardise the data formats and exchange protocols related to CTI, including recent efforts aiming at promoting the *CTI for "things"* [5]; the initiative *making security measurable* (MSM) constitutes the most prominent effort toward improving CTI sharing among the various stakeholders [6].

The analysis carried out in this paper considers prominent representatives of CTI formats and languages that have been proposed and further studied in the literature, such as the *structured threat information expression* (STIX) [7], *trusted automated exchange of indicator information* (TAXII) [8,9] and *cyber observable expression* (CybOX) [10]. Among the paper's goals are to explore the capabilities of the available formats and languages and their capacity to convey various CTI types, to correlate their features with the degree to which they are used from the vast number of CTI sources and to correlate their capabilities with the needs of typical security use cases to which they are to be used. The above (and other) standardised formats and languages were believed to be the answer to the problem of not having common mechanisms for sharing cyber-threat intelligence. According to [11], STIX is the de facto standard for describing threat intelligence. In a literature review of STIX, TAXII and CybOX, several issues were identified that should be addressed to allow their wide adoption; these include:

- The headline standards of STIX, TAXII and CybOX have been superseded.
- The apparent acceptance and utilisation of the standards appeared lower than expected.
- Much of the body of knowledge found in the literature is outdated mainly due to the rapid change and development of the CTI formats and use.

To address the above issues and provide a state-of-the-art view of the CTI formats, use cases and implementations, the publicly available sources of CTI that share such data were researched along with any related formats and languages.

The organisation of the paper is as follows. We first provide a quick overview of the literature and the current state-of-the-art in Section 2, to have a knowledge base and an informed perspective on the findings and issues encountered. This is followed by Sections 3–5 that investigate CTI sources and formats and present the main result of our analysis. We conclude in Section 6.

2. Related Work

Much work has been carried out into investigating the sources, methods and platforms for sharing CTI. The science and technology used in practice, moves at a rapid pace, which results in literature becoming rapidly out of date with regards to the formats and languages currently in use. Irrespective of this, it still provides a valuable and relevant background to the research, with many of the findings still being valid regardless of the actual CTI format or platform used.

An exploratory study of software vendors and sharing perspectives was carried out in [11,12], where [12] focused more on the relationships between CTI sharing vendors and how these affect the sharing practices, whilst Sauerwein, et al. [11] targeted more on analysing threat intelligence sharing platforms and protocols. The applicable key findings are that there is no common definition of threat intelligence sharing platforms and that STIX is the de facto industry standard for describing threat intelligence. The authors of [11] carried out a broad literature review that identified 22 threat intelligence sharing platforms, comparing protocols and methods used for sharing CTI. According to Brown, et al. [13], there is an ever-increasing need to obtain greater amounts of threat intelligence, with the challenge of dealing with the large volumes of data effectively. A target-centric approach was proposed, where CTI is filtered given an understanding of the threat landscape and what the targets in an organisation are likely to be. The intelligence can be enriched from many sources to provide data that are relevant and applicable, while sharing is performed in a controlled manner, ensuring data privacy and security. The paper discusses standard and open formats for the sharing of threat information and concludes that the adoption of STIX and TAXII by industry has led to many interoperable cyber information-sharing systems being developed. Given the vast quantity of CTI sources and feeds identified, the proposed target-centric approach merits further discussion. Another method to assess the relevancy of CTI sources according to the observables that they provide in allowing the early detection of cyber-attacks was proposed in [14]; the main idea relied on CTI content analysis and the "appearance-burst-disappearance" overall trend model. Likewise, content analysis techniques were also applied in [15], but with the different goal of introducing a new taxonomy of the CTI information conveyed by a data source: vulnerabilities, threats, countermeasures, attacks, risks and assets. In addition, this has been correlated with the type of the CTI source (i.e., blogs, forum, vendors, mailing lists, etc.) to gain some insight regarding the use of structured (or unstructured) CTI formats, the support of interfaces and APIs, the frequency of updating/sharing, the trustworthiness of the CTI and its originality. The latter is also considered in this paper, but for a much broader type of sources than those in [15], which are mostly limited (with few exceptions) to our class of external open-source intelligence sources that is next introduced.

The web-based research on cyber-threat intelligence that was carried out by Abu, et al. [16] concluded that the academic material available is limited due to the immaturity and instability in this relatively new field and therefore grey papers (as called therein) from various organisations and vendors must be the main information source. Along the same lines, Pala and Zhuang [17] reviewed research papers and approaches in cybersecurity information sharing and identified that techniques trying to optimally balance between cyber-investment/cyber-risk/privacy and CTI sharing (e.g., by using game theory) are gaining more attention. In contrast to the above approach, our research heavily relies on the direct inspection of the actual CTI obtained from various sources, with use of open-source tools whenever required and on the original documentation and articles by organisations and community sources. A survey focusing on technical aspects of threat intelligence was carried out in [18], where the types of intelligence, the benefits of sharing and the reasons for not sharing data were given. The authors also looked at the matter of quantity versus quality of CTI and the limitations in representing *indicators of compromise* (IoC), with a review of threat sharing formats and related platforms and their flexibility in sharing CTI. The paper adds to the data quantity issues found and highlights the need for quality and applicability of CTI. The analysis carried out in [18] assumes that CTI is classified into strategic, operational, tactical and technical, which differs from the one utilised in this paper and puts emphasis on CTI sharing platforms and their data enrichment, tools' integration and sharing capabilities.

On the other hand, Menges and Pernul [19] as well as Mavroeidis and Bromander [20] provided detailed analyses on the CTI sharing standards and incident reporting formats, along with certain associated threat taxonomies. More precisely, a different subset of the *malware attribute enumeration and characterisation* (MAEC), the *incident object description exchange format* (IODEF), the *vocabulary for event recording and incident sharing* (VERIS), the *extended abuse reporting format* (X-ARF), STIX and OpenIOC was considered in each paper with the analysis considering different features/criteria than those established herein. As an example, Menges and Pernul [19] was mostly concerned with general evaluation criteria (e.g., machine/human readability, interoperability, extensibility, aggregability, etc.), additional evaluation criteria (licensing, documentation and maintenance costs) and less with structural evaluation criteria (indicators, attacker, attack and defender), which are much more detailed in this paper and linked with typical security use cases. Although the latter type of criteria is rather the one that Mavroeidis and Bromander mostly considered [20], the particular criteria established (e.g., identity, motivation, goal, IoC, tool, target, strategy and TTP) allowed the comparative evaluation to be performed at a very high, non-technical level; the same criteria were used in [20] to evaluate threat taxonomies, such as CVE, CWE, CVSS, etc. Finally, Burger, et al. [21] as well as Asgarli and Burger [22] focused on segmented landscape of CTI standards and further investigated the use of CTI ontologies to allow for a better understanding of the security semantics and make inferences about ongoing cyber-security threats and incidents.

Although mainly concerned with STIX 1.x as a solution for sharing CTI, Serrano, et al. [23] highlighted several areas of importance in the context of CTI sharing. These include the legal and privacy implications in sharing CTI across borders and jurisdictions (also the focus in [24] and [25]), which have recently received great attention due to the *general data protection regulation* (GDPR), the requirement of a critical mass for CTI sharing sources that characterises its effectiveness, along with the belief that the main impediment to security data sharing is the lack of a suitable platform that addresses the issues of formats and legal boundaries for CTI data. Practices in sharing CTI were also studied in [26], where the results obtained from an online survey were used to classify potential barriers (and benefits) into areas such as *operational, organisational, economic* and *policy*; the quality and accuracy of CTI; the risk of privacy violation; the redundancy/relevancy of CTI; and the infrastructure costs were identified as the primary barriers. The lack of such a suitable platform was addressed in [27], where the *malware information sharing platform* (MISP) and the technical solutions used for sharing and synchronising threat information and taxonomies were described, as well as possible ways of extending the system's functionality. The MISP web interface and the use of the platform to present statistical information on the collected threats was discussed. Next, we further examine the MISP platform and the custom formats it uses for sharing CTI, along with the use of the *traffic light protocol* (TLP) that deals with the sharing of sensitive information.

In contrast to the aforementioned works, this paper's contributions are summarised as follows: (a) the research methodology relies on actual CTI obtained from a very large number of sources that are typically being used by today's security systems and products, instead of relying on previously published academic papers; (b) the types of sources considered are much broader, by considering internal, external and open sources to get representative results; (c) several tools/scripts were employed during the CTI collection process to allow for a comparison of the CTI against the original documentation and related technical/research papers; (d) the CTI formats and languages investigated herein are broader than those of the previous works, either by including recent ones gaining more attention (e.g., CVRF) or classical ones (e.g., DNSBL) that, although efficient in certain use cases, are usually not considered; and (e) the assessment criteria used are much more detailed and technical due to our goal in determining the extent at which typical security use cases can be supported by the existing CTI formats and languages.

3. CTI Sources

This section presents several CTI sources that have been examined, which are characterised as being *internal, externally sourced observables or feeds* and *externally open-source intelligence* [1,28,29]. It is important to highlight that the examination of CTIs was carried out by installing and using the tools provided from the manufactures, as well as by reading and analysing their documentation and various other online resources.

3.1. Internally Sourced

The CTI obtained from internal sources is comprised of observable events that have happened on an organisation's internal network and hosts (referred to as *threat indicators* in [30]). It can provide indicators about threats having breached the security perimeter, having broken the internal access control rules, having infected a system, or having attempted to get access to a restricted system. Statistical data provide a baseline of the normal behaviour so that any abnormality can be highlighted and investigated; possible sources are given in Table 1. More details about internal CTI sources are provided below.

System logs and events. Such information is widely available on devices and applications; it can be easily forwarded to a central facility using tools such as *Syslog* or *Windows event forwarding* (WEF). As only certain log messages and events apply to CTI, any central logging system, e.g., a *security incident and event management* (SIEM) system, should apply filters and rule-sets to extract CTI.

Table 1. Internal sources of cyber-threat intelligence.

CTI	Systems	Description
System logs and events	All systems	System activity, principally errors and security events
Network events	Network equipment, (switches, routers, firewalls)	devices connecting/disconnecting, ACL alert, login/failed login, etc.
Network utilisation and traffic profiles	Network equipment, (switches, routers, probes)	SNMP, NetFlow, RMON, etc. to Network management platform
Alerts from boundary devices	IDS/IPS, Firewall, WAF	Alerts/events collected and analysed by SIEM or vendor-specific management portal
AV, system alerts	Corporate AV software installed on host systems, (client and Server)	Corporate AV system alerts from host AV software
Human	All systems	Observed anomalies or events
Forensic	All systems	Artefacts and intelligence gathered after an event

Network events. Network devices such as routers, switches and firewalls, support *simple network management protocol* (SNMP), which can be used to send (in near real-time) event messages, known as *SNMP traps*, to a central server for processing. SNMP traps can be configured for a variety of CTI events in internal network (e.g., connections requested, login event occurring, etc.).

Network utilisation and traffic profiles. These may indicate abnormal behaviour, such as untrusted or excessive traffic from a client or between clients. Statistics are available in many forms, from simple counters in SNMP and *Remote MONitoring* (RMON) to detailed IP and protocol data from *NetFlow* and similar equipped switches and probes.

Boundary security devices. In addition to the above events, proprietary boundary security devices, such as *network intrusion prevention systems* (NIDS) and *web application firewalls* (WAF), may have their own application-specific management console that also feeds security events to a SIEM. An example of an alert generated by *Suricata* NIDS in JSON format is provided below in Listing 1.

Listing 1. Example of CTI (alert) obtained from Suricata.

```
{
"timestamp": "2009-11-24T21:27:09.534255",
"event_type": "alert",
"src_ip": "192.168.2.7",
"src_port": 1041,
"dest_ip": "X.X.250.50",
"dest_port": 80,
"proto": "TCP",
"alert": {
"action": "allowed",
"gid": 1,
"signature_id":2001999,
"rev": 9,
"signature": "ET MALWARE BTGrab.com Spyware Downloading Ads",
"category": "A Network Trojan was detected",
"severity": 1
}
}
```

Anti-virus systems. Corporate anti-virus systems report malware events back to a central console, allowing a comprehensive coverage for the hosts within an organisation; as with boundary devices, this may also feed security events to a SIEM.

Human. An organisation's staff is often the quickest to recognise that something is wrong; the ability to rapidly spot and report events is something that can be achieved through user awareness and continuous professional security training programs.

Forensic. This CTI includes artefacts gathered from the investigation following a security incident and can be used to bolster security defences. The analysis of infected systems and log files can provide details about the *tactics, techniques and procedures* (TTPs) used during the attack.

3.2. Externally Sourced Observables

Locating, identifying and analysing the externally sourced observables or *feeds* formed the bulk of the research that was conducted in this work [30]. A selection of open and free to use sources of CTI was identified along with the formats and languages used, with an emphasis on sources using the STIX/TAXII standard. These community, open-source IoCs and observables typically consist of the observed malicious sources or data, e.g., IP address, domain, URL, file names and hashes. The principal use case is to explore this information to create rule sets for firewalls, network-based and host-based *intrusion detection and prevention systems* (IDPS), SIEM systems, etc., to block (or alert on seeing) the observable or a matching indicator.

To obtain samples of CTI data, the STIX sources having been identified to use the TAXII 1.x transport protocol were accessed with the *Cabby TAXII client* [31], while a simple Python script was written using the *CTI TAXII client* [32] for TAXII 2.x sources. Other simpler formats, such as text, CSV, JSON, etc., were accessed using a standard web browser or the Linux wget command to review the fields included. The CTI feeds and their respective formats were analysed and compared. Wherever available, the format documentation was downloaded from the source or authoring organisation to allow for a deep understanding of the format used and to contribute to the research and analysis of the formats and languages. Over 275 feeds were identified from the CTI sources, where the first 125 of these (all based on the STIX standard) were selected for analysis; the remaining >150 feeds identified were stored for future analysis. Table 2 shows the quantity and format of the 125 selected feeds obtained from each CTI source, where in case that a feed supports multiple formats, the most complex one was chosen. The formats and languages listed in Table 2 are further examined below (with certain indicative examples) and also discussed later in the paper.

Table 2. CTI Sources' Formats Used.

Source	Format								
	Text	CSV/RSS	JSON/XML	STIX 1.x	STIX 2.x	MISP	IDS	DNS	Total
abuse.ch	4	10	0	0	1	0	7	1	23
AbuseIPDB	0	0	1	0	0	0	0	0	1
Bambenek Consulting	0	1	0	0	0	0	0	0	1
blocklist.de	11	0	0	0	0	0	0	0	11
botvrij.eu	0	9	0	0	0	1	0	0	10
C1fApp	0	0	1	0	0	0	0	0	1
Censys	0	0	1	0	0	0	0	0	1
CINS Army (Sentinel)	1	0	0	1	0	0	0	0	2
cybercrime-tracker	1	2	0	0	0	0	0	0	3
Dshield (SANS)	3	3	1	0	0	0	0	0	7
FreeTAXII	0	0	0	0	11	0	0	0	11
Green Snow	1	0	0	0	0	0	0	0	1
HAIL A TAXII	0	0	0	9	0	0	0	0	9
Limo (Anomali)	0	0	0	0	11	0	0	0	11
Malc0de database	1	1	0	0	0	0	0	1	3
Malware Domain List	5	4	0	0	0	0	0	0	9
MISP (CIRCL)	0	0	0	0	0	1	0	0	1
PickUpSTIX (NC4/Soltra)	0	0	0	4	0	0	0	0	4
Spamhaus	3	1	0	0	0	0	0	6	10
TAXIIstand	0	0	0	1	0	0	0	0	1
ÜberTAXII	0	0	0	0	4	0	0	0	4
xavier.mertens.consulting	0	0	0	0	0	1	0	0	1

Among the above sources, abuse.ch makes several CTI feeds available through projects, such as *MalwareBazaar* and *URLhaus*, for sharing information about malware samples along with URLs being used for malware distribution, or the *SSL Blacklist* that provides information to detect malicious SSL connections and digital certificates used by botnet *command and control* (C&C) servers. The feeds provided by abuse.ch are comprehensive and are used and re-transmitted by several other providers. A typical example of the CTI shared (with the SHA1 fingerprints of the aforementioned certificates) in a CSV format is shown below in Listing 2.

Listing 2. Example of CTI obtained from abuse.ch.

```
################################################################
# abuse.ch SSLBL SSL Certificate Blacklist (SHA1 Fingerprints)  #
# Last updated: 2020-05-03 06:46:48 UTC                         #
#                                                               #
# Terms Of Use: https://sslbl.abuse.ch/blacklist/               #
# For questions please contact sslbl [at] abuse.ch              #
################################################################
#
# Listingdate,SHA1,Listingreason
2020-05-03 06:46:48,081cf50a56f59be9b1f9504858a225b80f233cb2,IcedID C&C
2020-05-02 07:48:30,19cf21e6326b6125b023c53df23b74060f4e786e,IcedID C&C
2020-05-02 07:41:15,e5d49e0b12012e40498cc991ae586b3ce05bf2f6,IcedID C&C
2020-05-01 18:01:48,8644711545fc8d1ba02fd4e4424290a06815c320,Adwind C&C
2020-05-01 17:59:19,20373e4d4d11ba0e839378737ee9fc49cb164bbd,ServHelper C&C
...
```

Another CTI provider is the service blocklist.de that takes reports from numerous active servers that use *fail2ban* and similar abuse blocking applications. The lists may be obtained through a direct download or via an API and are single-column text files that contain IP addresses; moreover, such information can be obtained by the DNS *real-time blackhole list* (RBL), which provides a simple DNS query response mechanism to determine the state of an individual IP address, as in the example that is shown in Listing 3.

Listing 3. Example of CTI obtained via blocklist.de with DNSRBL.

query:
host -t any 112.220.10.1.bl.blocklist.de
response:
112.220.10.1.bl.blocklist.de has address 127.0.0.21
112.220.10.1.bl.blocklist.de descriptive text "Infected System (Service: bruteforcelogin, Last-Attack: 1588509427), see http://www.blocklist.de/en/view.html?ip=1.10.220.112"

The list of IP addresses available for download by blocklist.de can also be protocol-specific (e.g., for the SSH, FTP, IMAP and SIP), targeting at bots, or other attacks such as the above brute-force attack against a web login; no metadata or other enrichment is provided. Similar information is also provided by *Spamhaus*, which is a well-known CTI source providing lists of IP address ranges that are involved in sending spam emails (SBL advisory), are compromised by malware and other exploits (XBL advisory), or belong in domains having low reputation (DBL advisory) amongst others. Further to the above, a subset of the SBL list is provided via the *don't route or peer* (DROP) list that can be used by firewalls and routers to drop malicious traffic; an example is given below in Listing 4.

Listing 4. Example of CTI obtained from Spamhaus.

```
; Spamhaus DROP List 2020/04/30 - (c) 2020 The Spamhaus Project
; https://www.spamhaus.org/drop/drop.txt
; Last-Modified: Thu, 30 Apr 2020 14:23:20 GMT
; Expires: Thu, 30 Apr 2020 15:41:23 GMT
1.10.16.0/20 ; SBL256894
1.19.0.0/16 ; SBL434604
1.32.128.0/18 ; SBL286275
2.56.255.0/24 ; SBL444288
2.59.151.0/24 ; SBL444170
...
```

On the other hand, the CTI provided from *Anomali Limo* is following the STIX 2.x standard and is delivered by means of the STAXX open source platform and Limo TAXII feed. The compliance with the STIX 2.x format is somewhat lazy, since many of the indicators' metadata are presented in the description field. Several collections are available, providing details about ransomware, cyber-crime, emerging threats (compromised or C&C servers), malware domains, phishing URLs, etc., but some of the feeds are re-transmissions of other sources (e.g., from abuse.ch).

3.3. External Open-Source Intelligence

For this type of CTI, we concentrated on *open sources of threat intelligence* (OSINT) from publicly available sources that contributed to building and understanding the threat landscape; although these tend to be more human (and more strategic, as highlighted in [30]) than machine-readable, they are often unstructured. Typical examples are: an announcement of a large data leak compromising user data that could be used to access other systems, in phishing attacks or in geopolitical tensions that may increase the risk of cyber-attack. Table 3 provides a brief list and description of the CTI sources that were identified.

Table 3. Externally sourced intelligence.

Source	Description
News feeds	News articles covering ongoing threats
Vulnerability	Alerts and advisories
Search automation	Using search technologies to find vulnerable systems: Google dorks, Shodan, etc.
Anti-virus vendors	Information, alerts, news feeds on malware activity and threats
Communications	Monitoring communication channels for intelligence: Slack, IRC, Twitter, etc.
Dark web	Intelligence available directly from the criminal underworld

A wealth of CTI information was available in the plentiful supply from news feeds, alerts, *antivirus* (AV) vendors, etc. In most of the cases, it was also available in RSS format, which is machine-readable; however, the news or alerts content typically contains a link redirecting to a free format web page that does not easily lend itself to automated consumption and understanding despite the considerable advances in the areas of *natural language processing* (NLP) and *artificial intelligence* (AI). Typical examples of such sources include CERT-EU, Schneier on security, Krebs on security, and SANS institute, amongst others.

Advisories and vulnerability alerts are sources having a standardised CTI format, in many cases using the *common vulnerabilities and exposures* (CVE) and *common weaknesses enumeration* (CWE), as well as the *common vulnerability reporting framework* (CVRF), which is next reviewed. This information is typically associated with a severity measure in the format of the *common vulnerability scoring system* (CVSS) and is also linked with the systems affected by the vulnerability through the *common platform enumeration* (CPE), therefore greatly helping in the dissemination of threat intelligence but with some limitations. Typical examples of such sources include the *national vulnerability database* (NVD), Cisco security advisories, Microsoft security portal, Oracle security advisories, Red Hat security advisories, SecurityFocus, etc. In contrast to the previous type of external OSINT sources, these ones contain (or can readily generate) actionable security information. For example, NVD's data feeds, apart from the incorporation of the CVSS string (giving granular information about a vulnerability's preconditions and impact) also includes labels to any external references, such as *exploit*, *patch*, *mitigation*, *technical description* and *product*, which can direct tools automating the extraction of actionable information. An example from NVD's feed in JSON format is provided in Listing 5.

Listing 5. Example of CTI obtained from NVD (truncated/simplified for illustration purposes).

```
{
"cve" : {
"CVE_data_meta" : {
"ID" : "CVE-2020-0001"
},
"problemtype" : {
"value" : "CWE-269"
},
"references" : [ {
"url" : "https://source.android.com/security/bulletin/2020-01-01",
"tags" : [ "Vendor Advisory" ]
} ],
/* vulnerability description */
},
"configurations" : {
"cpe_match" : [ {
"vulnerable" : true,
"cpe23Uri" : "cpe:2.3:o:google:android:10.0:*:*:*:*:*:*:*"
} ]
},
"impact" : {
"cvssV3" : {
"version" : "3.1",
"vectorString" : "CVSS:3.1/AV:L/AC:L/PR:L/UI:N/S:U/C:H/I:H/A:H",
"attackVector" : "LOCAL",
"attackComplexity" : "LOW",
"privilegesRequired" : "LOW",
"userInteraction" : "NONE",
"scope" : "UNCHANGED",
"confidentialityImpact" : "HIGH",
"integrityImpact" : "HIGH",
"availabilityImpact" : "HIGH",
"baseScore" : 7.8,
"baseSeverity" : "HIGH"
},
"exploitabilityScore" : 1.8,
"impactScore" : 5.9
}
}
```

The dark web search focused on finding intelligence, tools and services that are not available on the surface web. Our analysis was conducted using a TOR browser running on a disposable virtual machine to provide some insulation from malicious content. The speed and reliability of connections to .onion sites hampered and frustrated progress. Access to several forums was granted by using anonymised email addresses but it was quite limited without first having gained trust in the community.

4. CTI Formats and Languages

Many CTI formats were identified from CTI sources and the literature; these were selected for further analysis based on their popularity in the literature or the source feeds. Where available, the original specifications, documents, schemas, etc., were examined by installing the right tools and applications. Samples of the formats were identified either from the CTI sources under investigation or the literature. The formats and languages have been classified into four main categories:

- Standards that have been specifically published for representing the CTI
- Custom application-specific or vendor-specific formats
- Commonly used standards that were not designed for representing the CTI
- Legacy formats, commonly referred to in the literature, but no longer being supported or used

A brief overview of the ones selected for further analysis is provided in the following subsections.

4.1. CTI Standards

STIX is one of our principal research subjects; it is a rich and extensive XML format that was first released in 2012 [33], with the minor revision 1.2 being released in 2015. The aim of STIX was to be a flexible and expressive language for representing cyber information. Where existing formats were used, e.g., MAEC [34], the objective was to *integrate rather than duplicate* them [7]. This provided a highly flexible format that ultimately led to its downfall, as the nested structures present in the XML documents became too complex and difficult to parse. STIX 1.2 was superseded by the 2.0 and in 2017 by 2.1 release. TAXII is the preferred, but not compulsory, transport mechanism for STIX [35]; there are different versions of TAXII for each release of STIX, which are not compatible with each other.

CybOX provides STIX 1.x the means to express cyber observables, events and other properties [10]. With the advent of STIX 2.1, CybOX has been integrated and is now part of the STIX standard. The principal differences between STIX 2.x and STIX 1.x are in the serialisation from XML to JSON that was designed to make the protocol more lightweight and much simpler for programmers [35]. The structure in STIX 2.x is flat rather than nested, with *STIX domain objects* (SDO) defined at the top level of the document to simplify parsing and storage; the relationship between the SDOs is accommodated by the introduction of a *STIX relationship object* (SRO) [36]. The CybOX objects have become *cyber observable* objects in STIX 2.x (under CybOX 3.0 release [37]) along with MAEC, therefore considerably decreasing complexity. Such changes were accompanied by a change in the management of the STIX project, which moved from MITRE to the OASIS CTI technical committee [38]. The MAEC 5.0 standard was designed for characterising malware using attributes such as behaviours, artefacts and relationships between malware samples [34,39]. This latest release was updated in line with STIX 2.x to maintain compatibility using the same cyber observable objects and JSON serialisation.

CVRF is another standard, whose format is machine-readable, aiming for the submission and distribution of vulnerability advisories and reports [40]. The utilisation of CVRF by MITRE's CVE repository, the principal registry of vulnerabilities and exposures, along with active support and feeds from vendors, such as Cisco, Oracle and Red Hat, are expected to help to establish CVRF as the de facto standard for the distribution of vulnerabilities and security advisories.

4.2. Application and Vendor Specific Formats

CESNET operates a large network infrastructure providing service to higher education and research establishments throughout Czech Republic; it created the *intrusion detection extensible alert* (IDEA) to overcome the complexities of other CTI formats [41]. IDEA aims at the sharing of CTI data that are varying in nature, thus it has to be flexible, extensible while staying simple. The MISP format is the native protocol for communication between the MISP platform instances [42]; this JSON format is highly extensible and widely used by the MISP platform. The *collective intelligence framework* (CIF) is another widely used CTI aggregation and sharing platform that provides its JSON format for sharing CTI [43]. Finally, IDS/IPS rules are a long-lived CTI format that can be directly consumed by IDS/IPS applications such as Snort [44] and Suricata [45].

4.3. Commonly Used Standards

These formats were never designed or intended for use as a CTI sharing medium; despite this, the *DNS block list* (DNSBL), *DNS real-time black hole list* (DNSRBL) and Text/CSV are the oldest and most widely used formats identified. More precisely, DNSBL and DNSRBL are not downloadable lists of CTI host IPs [46]. Instead, they provide a rapid and efficient DNS-based request/response protocol to determine if an IP or domain exists on a blacklist or whitelist. It is likely one of the oldest methods used to get useful CTI information and is typically used by e-mail spam and malware filters.

Really simple syndication (RSS) is a lightweight XML format that is designed for the distribution of news items [47]. This format has been adopted by several sources for the distribution of CTI with detailed data available from a central repository. On the other hand, Text/CSV is the simplest and most widely used format of all the CTI source feeds sampled, either a single column text list of IPs or URLs (e.g., in the case of black lists), or as a rich, multi-IoC comma or tab-separated variables; they provide all the data in the most efficient and compact manner of any format.

4.4. Legacy Formats

The analysis of the final three CTI formats that we noted from the literature was curtailed due to the absence of current development, no active support or not being identified in any CTI source feeds examined.

Originally created by Mandiant Inc., under openioc.org, the OpenIOC format was designed to provide a common methodology and format for describing host-based or network-based indicators of compromise [48]. The legacy Mandiant resources and/or tools are available on GitHub, but there is currently no apparent activity [49]. The IODEF format was introduced by the Internet Engineering Task Force in RFC 5070 [50]; its current version 2 is described in RFC 7970. It is an XML-based format for exchanging CTI that is reported in the literature, but no evidence was identified about its current support, despite the second version's activity in 2016. Finally, the *open threat partner exchange* (OpenTPX) is an open-source and well-documented JSON format designed for sharing CTI [51]; no feeds were identified and there is no apparent evidence of updates since 2015.

5. Analysis

This section is mainly focused on externally sourced CTI feeds found in Sections 3 and 4. These sources are discussed after a brief analysis of the other CTI sources from our research.

5.1. Internally Sourced CTI

The CTI from internal sources appears to have a quite comprehensive coverage from the HIDS, SIEM and antivirus software provisions available; the majority of these were commercial offerings. It appears that the use of CTI, obtained from network activity such as network traffic flows, DNS requests, DHCP, ARP etc. (excluding NIDS), is not widely utilised and no further analysis was carried out to determine the effectiveness of current solutions on this type of CTI.

5.2. External Open Source Intelligence

The CTI examined from external open-source intelligence (OSINT) showed a very different context comparing to the machine-readable sources and formats. The analysis and application of this CTI is predominantly a manual process, converting this human-readable CTI into machine actionable formats where some of these were available, with some limitation, in machine readable formats such as RSS and CVRF. Advances in natural language processing and AI offer significant opportunity in this area. The availability and structure of vulnerabilities and exposures through the CVE standard is well known and widely used [39] but the main drawback of this system is the limited applicability of the information available in a standard format. It should be noted that some vendors provided CVE feeds (e.g., [52–54]) that were quite comprehensive in what the applicable software versions were. The consistency and quality of the CTI that was identified from the dark web was found to be poor and mired in unsavoury content, mostly due to the lack of indexing and controller access to forums and credible resource. As much of the malicious activity originates from those who inhabit the dark web, it cannot be ignored as a potential source of intelligence.

5.3. CTI Source Feeds, Formats and Languages

The analysis carried out on the CTI source feeds revealed several different types of formats including single-column text feeds, multi-column, rich CSV feeds and more complex formats such as STIX and RSS. Many of these feeds, particularly those available in the more complex formats, were found to be retransmissions of simpler plain text feeds from other CTI sources. Examination of the feeds for evidence of originality (instead of being retransmissions) was not always possible. It is worth noting that some sources were found to be informative, giving details of how or where the CTI data were obtained and, in some cases, how agents could be downloaded, etc. A selection of sources, typically CSV or RSS feeds, provided web portal interfaces to search and examine the CTI data in greater depth. Figure 1 gives an overview of the originality for the threat feeds examined.

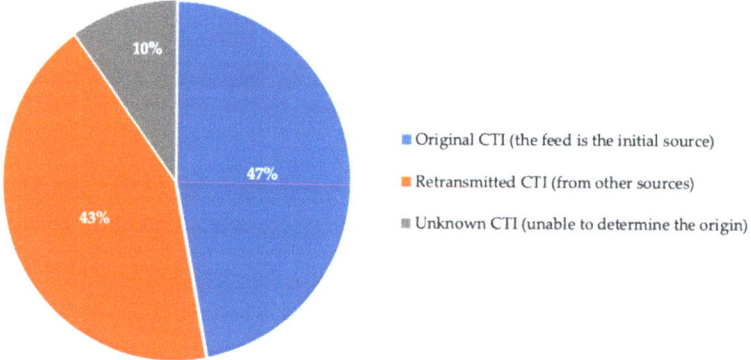

Figure 1. CTI source originality.

In the retransmission of CTI data, we found that some original source data can be lost or corrupted, which typically was attributed to the poor formatting, dates having been replaced so misrepresenting the freshness of the data, retransmitted or aggregated data appearing as a shadow sighting and giving false significance to the threat. We also observed a common practice of splitting the rich array of CTI types associated with a threat into separate, un-associated types, e.g., IP, domain, etc., diminishing the value of the original cohesive dataset.

In Figure 2, we illustrate the range of CTI types that were represented in the analysed CTI source feeds. IP addresses were the most common type, followed by the description of the threat or malware type and the URLs. From our analysis of the formats we knew that the rich intelligence source feeds could provide a more comprehensive dataset than that available from a simple block list. We compared how many of the sources using complex data formats provided rich CTI feeds. Here, we define *rich* as the CTI having more than two types represented in the feed, otherwise we consider it as being *sparse*. Our results are represented below.

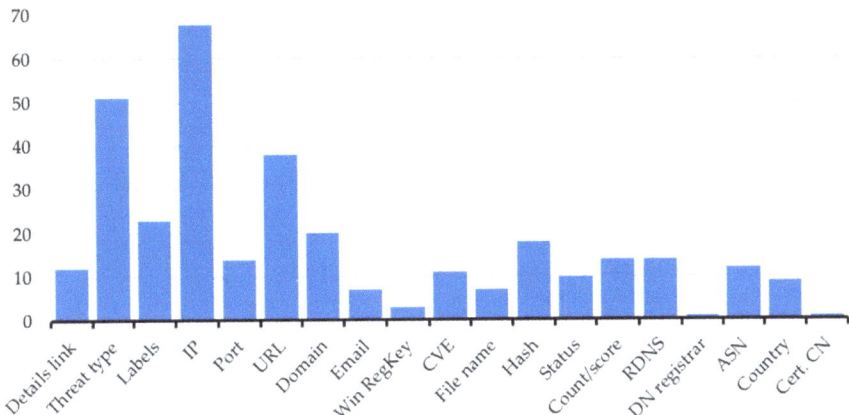

Figure 2. CTI types represented.

As highlighted in Figure 3, the capability of STIX to represent complex and rich CTI is somewhat underutilised, with most samples containing only sparse CTI. We carried out further analysis of the STIX 1.x format and compared the efficiency found in retransmitted CTI feeds. For example, a single entry <item> in the RSS *Malc0de database feed* [55] consumed 307 bytes. In contrast, the STIX 1.1 feed representing the indicators of same single entry from *PickUpSTIX* [56] consumed 18,153 bytes. Thus, it is clear that the used XML came with significant overhead and complexity.

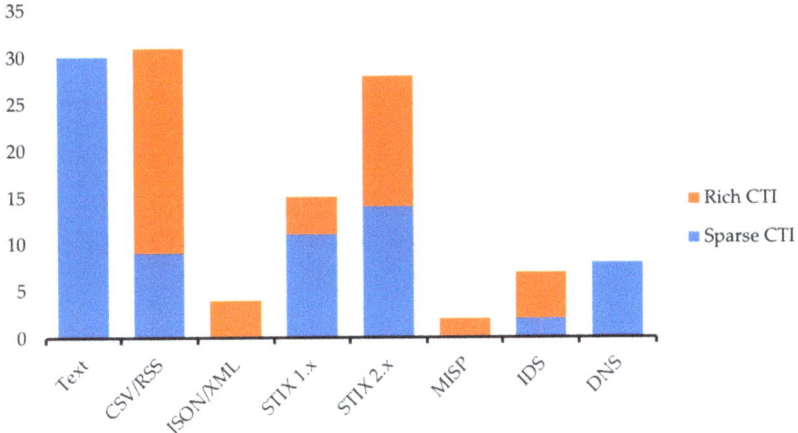

Figure 3. Rich vs. sparse CTI.

From the documentation of STIX 2.x, it is known that it can provide a more succinct representation than its 1.x predecessors. We still found that only half of the feeds analysed contained rich CTI data. A common approach taken was to put data in the description or title attributes rather than add additional observable objects or indicators to the feed. We refer to this as the *lazy implementation* of STIX format. We did note that the STIX feeds containing original content tend to be richer and much better implemented than those simply retransmitting data from other sources.

Complexity was one of the prime reasons for moving from STIX 1.x to 2.x, where the need for keeping things simple is also stated as a goal in MISP, CIF and IDEA formats. When analysing complex

CTI represented in MISP and STIX 2.x documentation, the strength of the formats to cross reference CTI comes to the fore. When we compare this to the implementations of simpler but still rich CTI, e.g., containing IPs, file names, file hashes and URLs, that are indicators for a strain of malware. However, without the need of TTPs, sequence of events, actor identities, etc., we see that the simpler formats can better express these.

To further examine how the use of the STIX versions varied between the providers, a common original source was chosen that was retransmitted by both STIX 1.x, 2.x sources. For our comparisons, the abuse.ch ransomware tracker feed was used [57]. The STIX 1.1 feed was sourced from PickUpSTIX [58], which contains better source metadata compared to with the *Anomali Limo* feed [59].

STIX 1.x and 2.x have similar capabilities to represent the data complexity as can be easily seen from Table 4. It was concluded that the Limo source appears to have a somewhat *lazy implementation* and further analysis was conducted on the STIX 2.x sources to reveal if this is a common practice or not. For this, sixteen samples of TAXII collections were examined from three STIX 2.x source providers to compare how well they utilised the capabilities of the format and structure. The observed data or indicator objects were analysed for containing multiple IoC types in the file (e.g., IP, URL, MD5, etc.); multiple IoC in an either observed-data.objects or indicator.pattern objects; and examples of rich content, e.g., multiple IoC, related objects, etc.

Table 4. STIX, PickUpSTIX and Limo metadata comparison.

Data	PickUpSTIX (STIX 1.1)	Limo (STIX 2.x)
Terms of use	Included per STIX package	-
TLP	TLP White per STIX package	Common Marking definition TLP: Green
Producer Description	Aggregator of Malware Sites	-
Producer Role	Aggregator	-
Producer Timestamp	Timestamp	-
Producer feed URL	Ransomware feed URL	-
Indicator Title	Description and IoC URL	Threat Stream ID, type, state, org, source
Observable Title	IoC	IoC
Observable condition/pattern	IoC	IoC
Observables per Indicator or related group of indicators	Multiple: IP, ASN, file, hash, URL, etc.	Single IoC type per feed (IP, Domain)
Labels	Unclassified (Public) marking	Malicious activity Threatstream severity Threatstream confidence

Our results in Table 5 indicate that the analysed STIX 2.x samples gained only a little advantage from using the STIX format.

Table 5. STIX 2.x Feature Use.

Source	Multiple IoC Types Per File	Multiple IoC Types Per Indicator	Rich CTI/Indicators
Limo (Anomali)	1 of 9 collections	None	1 of 9 collections
xavier.mertens.consulting	None	None	None
ÜberTAXII	5 of 6 collections	4 of 6 collections	3 of 6 collections

From the CTI samples identified in our research, many simpler formats such as CSV and RSS had grouped indicators for a given threat with a common label or tag. STIX uses a combination of observed data structures, indicator patterns and relationships. The STIX *bundle* object is only a container and does not imply any relationship between the objects contained therein; a *relationship* object is required to represent this, using the UUIDs of the related objects, along with its own UUID, markings, originator, etc. This can result in a complex document to represent a collection of CTI related to a single threat. This is an area in which the MISP format excels; the sharing of data between MISP instances is threat-centric. Here, a single *event* file contains all the CTI for a threat; UUIDs are used to cross-reference and form relationships the same as STIX; and the attribute array structures are similar to STIX observables. However, the relationships are embedded with no additional objects or complexity required.

We find a similar situation with STIX markings when compared to MISP tags. In STIX, a marking definition is typically a global object and the indicator objects reference these directly. MISP, which has a rich tag and taxonomy implementation, embeds the tag objects directly. This is very simple but creates the potential for inconsistency between versions of the same tag. As the name suggests, *universally unique identifiers* (UUID) RFC4122 provide unique IDs [60]. Several of the CTI formats examined use these to identify and reference CTI data, markings, relationships and more.

CVRF was found to be a rich format that can meet the need to share vulnerabilities; the addition of a revision history within the vulnerability structure would provide a clearer versioning of individual vulnerabilities. The biggest weaknesses observed was the limited compliance from major influencers and the dilution of the format with multiple, equally suitable alternatives and insufficient target data and remediations in a consistent and standardised manner. As noted above, there is good vendor support for identifying the applicability of a vulnerability and remediations.

MAEC has good support from Sandbox providers, although there is a dilution from the use of older versions and the widespread availability of platform-specific API formats. MAEC 5.0 leverages STIX 2.x cyber observables, types and languages, but there is no evidence of reciprocal support with no facility to reference or include MAEC content in STIX 2.x, as was available in STIX 1.x.

The platform or API custom formats such as MISP, IDEA, CIF, etc., had an enthusiastic use of the formats, and they were found to be better suited to their given use case and able to represent the CTI observables and indicators in a succinct yet comprehensive manner. The MISP format has grown from real-world use; the MISP project sites over 6K installations of the MISP platform, illustrating the wide support in both community and government organisations.

In Tables 6–8, the various CTI formats and languages that were researched and analysed are compared to determine how well they are able to convey CTI for different use cases. The criteria are applied to the representation of a single, complete cyber observable, where a single observable can be an event, indicator or similar such single entry, line or item in a list or structure. For example, the CTI indicating the presence of a malware compromise, source of the infection (IP, domain, URL, file, hash, etc.), the destination or target (IP, hostname, domain, vulnerability, etc.) and threat details (malware name, family, type, etc.). The test applies to dedicated fields or columns that are machine readable and unambiguous, inclusion of CTI data fields in general purpose descriptions is ignored.

Table 6. Format and Languages, Assessment Criteria.

Criteria/Feature	Assessment Criteria	Notes
Blocklist	Provides effective and simple representation of a blocklist. This can be an IP/domain list, or a go/no–go request/response mechanism	-
IP v4 Address	An IP v4 address or network and mask e.g., CDIR format or with netmask	▲ for supporting IP ranges/multiple IP's
IP v6 Address	An IP v6 address or network and mask e.g., CDIR format or with netmask	▲ for supporting IP ranges/multiple IP's
Hardware/product	Hardware or product information, system make, model, MAC address, etc. Expect 2	▲ for more, and ?▽ for less
Email address	Represent an email address, typically a known malware 'from' address or C&C address	▲ for multiple addresses
Hostname	The hostname	-
URL/URI	URL	-
Domain	Domain (FQDN)	▲ for details, RDNS
Attacker/Target	Specify the data refers to the attacker or network source and/or the Target or destination	▲ for source and destination details
Vulnerability	Details of a vulnerability, e.g., CVE or reference to similar source, OS/SW vendor etc.	-
Malware or Threat Type	Provide the name of the malware or threat	▲ for details of role, family, type
Ransomware	In addition to malware, specific ID as ransomware	▲ for details on virus total, etc.)
File	Details of a malicious file, e.g., file name, source path, destination path, file hash, alternate names, virus total, etc. Expect 2 or 3	▲ for more, and ▽ for less
Detailed system IoCs	Details of observable artefacts or indicators of system compromise, e.g., Windows registry values, files, executables, libraries infected, hashes. Expect 2 or 3	▲ for more, and ▽ for less
DDoS	Identify the CTI as belonging to DDoS, or indicating DDoS. May include: C&C server, botnet description, DDoS type, IP lists, ASN, IP/Port and rate or counts, Expect 2	▲ for more, and ▽ for less
Compromised host, RAT	Identify CTI as indicating or observed compromised host, Remote Access Trojan, or similar 'owned' host, network, website, etc. Not a bot net. Expect a host identifier (IP, URL) and threat/compromise.	-
Botnet	Identify the CTI as belonging to a botnet, should include botnet name along with indicators/observables, C&C servers, bots, target device/OS, etc. Expect 2	▲ for more, and ▽ for less
Spam	Identifies CTI as being concerned with Unsolicited Commercial Email, may include domains, IP, email addresses, subject lines, etc.	-
Phishing	Identifies CTI as being concerned with Unsolicited malicious email aimed at compromising, or some malicious act. May include domains, IP, email addresses, subject lines, file detail	-
Software	Details of software, operating system, version, etc. Expect 2 (e.g., OS and version)	▲ for more, and ▽ for less
Time Stamps	Timestamps such as: data produced, first seen, last seen, window, etc. Expect 2 or 3	▲ for more, and ▽ for less
CTI Source	Accreditation of the CTI source	▲ for references, or the collector/agent
Complexity	A measure of not being over complex, effectively doing what it says on the tin without being over packaged	▲ if succinct, and ▽ if over complex
Rich CTI data	The Format or language can represent 10 CTI attributes	▲ for more, and ▽ for 8–9
Patterns	Patterns to match observed data, e.g., LIKE text, Regular Expressions, Hex bytes, etc. Expect 2	▲ for more, and ▽ for less
Identity	Identify a person, user, threat actor or organization. Can include name, location, function, etc. Expect name and function/type	▲ for more, and ▽ for less
Course of Action	What to do, remediation, etc. to protect from a threat or fix a vulnerability, expect text and references	▲ for more, and ▽ for less
Versioning	The means to know that the CTI has been updated	-
Author	Organization, group or person who created this CTI, ref to ID is acceptable.	-
Confidence, count	Confidence, rating or simple count of observations	-
Markings	TLP or similar security of distribution marking, Tags, etc.	-
Artefact	Contain encoded CTI artefact data or link to data.	-

Table 7. Formats and languages, use case and features.

Typical Use Case							Criteria/Feature	Formats and Languages											
Email Blocklist	Spam/Email Filter	Firewall/Router ACL	NIDS	HIDS/SIEM	Malware Analysis	Human, SOC, DB		STIX 1.x	STIX 2.x	MAEC	CVRF	IDEA	CIF (Platform API)	MISP (Platform API)	Snort/Suricata Rules	DNSBL	RSS	Text CSV	Text List
▲	▲	▲	▽	▽		▽	Blocklist					▽			▽	▲	▽	✗	▲
✗	✗	✗	✗	✗	✗	✗	IP v4 Address	✗	✗	✗		▲	▲	▲	▲	✗	✗	✗	✗
✗	✗	✗	✗	✗	✗	✗	IP v6 Address	✗	✗	✗		▲	▲	▲	▲	✗	✗	✗	✗
			▽	✗	✗	✗	Hardware	▽	▽	▽	▲								
✗	✗		✗	✗	✗	✗	Email address	✗	✗	✗		✗		✗	▲		✗	✗	✗
			✗	✗	✗	✗	Hostname	✗	✗	✗		▲		✗	✗				
	✗		✗	✗	✗	✗	URL/URI	✗	✗	✗		▲		✗	✗	✗	✗	✗	✗
✗	✗		✗	✗	✗	✗	Domain	✗	✗	✗		▲	▲	✗	✗	✗	✗	✗	✗
			✗	✗		✗	Attacker/Target	✗	✗	✗		▲		✗	✗				
			✗	✗	▲	✗	Vulnerability	✗	▽	▲	▲	✗		▲					
	▽		✗	✗	▲	✗	Malware/Threat Type	✗	▽	▲	▲	▲		✗				✗	✗
	▽		✗	✗	▲	✗	Ransomware	✗		▲		▲		✗				✗	✗
	▽		✗	✗	▲	✗	File	▲	▲	▲				▲	✗			✗	✗
				▲	▲	✗	Detailed system IoCs	▲	▲	▲		▽		▲	✗				
		✗	✗			✗	DDoS					✗		✗			▽	▽	▽
			✗	✗	✗	✗	Compromised host		▽			▲		▽			✗	✗	✗
	▽		✗	✗	✗	✗	Botnet		▲			✗					▽	▽	▽
	▲					✗	Spam		✗			✗				✗	✗	✗	✗
	▲		✗	✗	✗	✗	Phishing		✗			✗				▽	▽	▽	▽
				▲	✗	✗	Software	✗	✗	✗	▲								
			✗	✗	✗		Time Stamps	✗	✗	✗	▲	▲		✗			✗	▽	▽
						✗	CTI Source	✗	✗	✗	▲	▲		✗				▽	▽
▲	▲	▲	▲	✗	✗	✗	Complexity	▽	✗	✗	✗	▲		✗	▲	▲	▲	▲	▲
	▽		✗	✗	▲	▲	Rich CTI data	▲	▲	▲	▲	▲		▲				▽	✗
✗	✗	▽	▲	▲	✗		Patterns	✗	▲	▲				✗	▲				
	✗			✗	✗	✗	Identity	▲	▲		✗			▲					
				✗	✗	✗	Course of Action	✗	▽		▲			▲		✗	✗		
	✗		✗	✗	✗	✗	Versioning	✗	✗		▲	▲	✗	✗				▽	▽
	✗		✗	✗	✗	✗	Author	✗	✗		✗			✗					
			✗	✗	✗	✗	Confidence, count	✗	✗		✗	▲	▽	▲				▽	▽
			✗	✗	✗	✗	Markings	✗	✗		✗		▲	▲	✗			▽	▽
	✗		✗	✗	✗	✗	Artifacts	▲	▲	▲		▲		✗	✗				

Table 8. Typical use case and example CTI.

Typical Use Case	Example CTI
Email Blocklist	Simple block based on sender email address, domain or IP
Spam/Email Filter	Complex block based on sender IP, domain, email address, mail content, attachments, links, etc.
Firewall/Router ACL	IP address, port, may use connection rate (DDoS) or mask/simple patterns
NIDS	Complex, source/destination, addresses, URLs, file content, Malware IoC, Source reputation, etc.
HIDS/SIEM	Complex, source/destination, addresses, URLs, file content, Malware IoC, Source reputation, system IoCs (registry, files, paths).
Malware Analysis	Complex, known sources, poor reputation, email, file content, etc.
Human, SOC, DB	Complex dataset to build threat picture and analyse threats.

In Table 7, the formats and languages are graded on how well the test criteria have been met as per the following key: a blank means that the criterion or feature is neither met nor supported; the

'▽' symbol means that the feature is partially supported and some but not all criteria are met; the '✖' symbol means that the criteria are met or the feature is supported in a satisfactory manner; and the '▲' symbol means that the requirement criteria and feature requirements are exceeded. Table 8 below describes some very typical example use cases and examples of the types of CTI that those use cases may consume.

From the analysis of the various use cases, CTI formats and sampled feeds, it became clear that some were better suited at representing CTI for a given use case, e.g., due to being simpler or richer. This is illustrated in Table 9, where the available formats and languages are correlated against the security use cases according to the information that is given in Table 7.

Table 9. Formats and languages suitability per use case.

Formats and Languages	Typical Use Case						
	Email Blocklist	Spam/Email Filter	Firewall/Router ACL	NIDS	HIDS/SIEM	Malware Analysis	Human, SOC, DB
STIX 1.x	0.67	**0.74**	0.50	0.70	**0.72**	0.70	0.74
STIX 2.x	0.67	0.68	0.50	0.61	0.66	**0.70**	0.65
MAEC	0.67	0.63	**0.67**	0.78	0.66	**0.70**	0.71
CVRF	0.00	0.26	0.00	0.26	0.45	0.48	0.45
IDEA	0.83	0.68	**0.67**	0.87	**0.72**	0.63	0.77
CIF (platform API)	0.67	0.26	0.33	0.22	0.24	0.26	0.23
MISP (platform API)	0.67	0.68	**0.67**	0.65	0.69	0.67	0.71
Snort/Suricata rules	0.50	0.42	**0.67**	0.52	0.52	0.41	0.48
DNSBL	**1.00**	0.37	**0.67**	0.30	0.28	0.26	0.29
RSS	0.83	0.58	0.50	0.48	0.41	0.30	0.42
Text CSV	0.83	0.58	0.50	0.52	0.41	0.26	0.39
Text list	**1.00**	0.42	**0.67**	0.30	0.24	0.22	0.26

For each use case, the format or language achieving the highest suitability score is shown in boldface, with the score ranging from 0 (lowest) to 1 (highest). The expression used for computing the suitability score $s(f, u)$ of any format or language f against some use case u is given by

$$s(f, u) = \frac{1}{n(u)} \#\{a \in c(u) : f(a) \text{ covers } u(a)\}$$

where the set $c(u)$ is comprised of the criteria/features being applicable for the use case u, whose number is $n(u)$. $f(a)$ and $u(a)$ denote the level at which the criterion/feature a is supported and required, respectively. The ordering '▽', '✖', '▲' of the symbols in increasing support of features allows us to determine if the needs of a particular use case are being met. Let us take the *email blocklist* use case as an example, that is we have $u =$ "email blocklist" in the above expression. According to Table 7, this use case requires the features

$$c(u) = \{\text{Blocklist, IPv4 address, IPv6 address, Email address, Domain, Complexity}\}$$

and hence $n(u) = 6$. It is immediately seen in Table 7 that STIX 1.x protocol can adequately support only four out six features and therefore for $f =$ "STIX 1.x" we get $s(f, u) = \frac{4}{6} = 0.67$, which is also depicted in Table 9. Regarding the two features not counted for in STIX 1.x, namely *Blocklists* and *Complexity*, we see that the former is not supported while the latter implies that the protocol is unnecessarily over complex in the way that the information is provided (as stated in the assessment criteria of Table 6). It is interesting to note that the IDEA format (followed by STIX 1.x and MISP) is found to be the most suitable for the majority of the use cases considered, whereas it is located among the next most suitable formats and languages for the remaining ones—something that clearly justifies its design goals. On the other hand, Table 9 shows that the use case of "Firewall/Router ACL" is the one that most formats and languages can largely support.

The direction of the information flow is also a factor in the original design and the use of several of the formats were examined. Table 10 shows the flow direction and the formats noted as most suitable.

Table 10. Format suitability.

Direction	Suitable Format
From sensor/detection, (probe, IDS, log, alert, honeypot, etc.) to CTI collection or aggregation system.	IDEA, MAEC, text (device specific), CSV, custom JSON, proprietary, etc.
Between or extraction from CTI collection or aggregation systems.	STIX, MISP, MAEC, CVRF, CSV, custom JSON.
From CTI collection or aggregation systems to consuming cyber protective systems or devices.	CSV, IDS rules, Text blocklist.

CTI data from sensors or detection mechanisms tend to be specific to the source type or detection mechanism used. IDEA is a custom format designed to transport CTI from sensors to a central system. MAEC is quite popular with honeypot providers. CTI collection and aggregation systems, or extraction of data from them, are best suited to the formats that can provide the best fit for the data being shared or extracted. Such examples are a simple CSV for bulk IP data; CVRF for vulnerabilities; and STIX, MISP and custom JSON formats for a rich representation of CTI. The format used to distribute CTI to cyber protective systems or devices needs to be one that can be directly consumed, e.g., IDS rule sets, IP/domain lists, MD5 signatures, etc. When examining the suitability of the various formats and given the original use case and design criteria for the formats, the results are as we expected; this does not make any one format better than any other, it depends on the use and the requirements.

6. Conclusions

Through research and analysis, it quickly became apparent that the quantity of CTI sources and formats is vast. As noted above, more than half of the threat intelligence feeds sampled from these sources were either retransmitted or of unknown origin. The support for STIX is apparent in many platforms and the consensus from the research would suggest it has industry and community support. However, its use is not widespread and often poorly implemented. The trend is to use API or platform-specific formats that are a better fit with the given use case.

The question of which format to use depends on the use case; the creation, coding and use of custom JSON formats is a quick and simple way to meet requirements of a specific use case, or there may be a preference to adhere to existing standards. Our recommendation would be to use the best fit; the evidence from the research has shown that even the producers and key supporters of standards still produce their own, lightweight, custom JSON formats, regardless the time scales, processes and ratification needed by standards.

Our recommendations on the distribution and sharing of CTI is to follow the best practice, where applicable, with the common descriptors and conventions in the language. It was found that relying on the IDEA format (and possibly MISP or STIX) might constitute a best practice for the majority of the security use cases considered due to its ability in meeting their CTI needs. In addition, most of the formats are capable of supporting access control services being offered by means of a firewall or router.

Many of the issues we encountered with the quality and the distribution of CTI could be reduced by including the origin and freshness/timestamp data in feeds, keeping threat data complete. Clearly, the vast number of CTI sources offer an opportunity for further research into assessing and improving the quality of CTI feeds. Where resources are constrained, e.g., in IoT devices, better association between the threat and target surface could provide focused CTI able to more effectively protect these devices.

Author Contributions: A.R., S.S., and N.K. contributed equally. The authors read and approved the final manuscript as well as the authors order. All authors have read and agreed to the published version of the manuscript.

Funding: This project has received funding from the European Union's Horizon 2020 research and innovation programme under grant agreement No. 786698. This work reflects authors' view and the agency is not responsible for any use that may be made of the information it contains.

Conflicts of Interest: The authors declare that they have no conflict of interest.

References

1. Roberts, S.J.; Brown, R. *Intelligence–Driven Incident Response*; O'Reilly Media: Sevastopol, CA, USA, 2017.
2. Menges, F.; Sperl, C.; Pernul, G. Unifying cyber threat intelligence. In *Trust, Privacy and Security in Digital Business (TrustBus), Lecture Notes in Computer Science*; Springer: Berlin, Germany, 2019; Volume 11711, pp. 161–175.
3. Poputa–Clean, P. SANS Institute, Automated Defense, Using Threat Intelligence to Augment Security. Available online: https://www.sans.org/reading--room/whitepapers/threats/automated--defense--threat--intelligence--augment--35692 (accessed on 3 April 2020).
4. Appala, S.; Cam–Winget, N.; McGrew, D.A.; Verma, J. An actionable threat intelligence system using a publish–subscribe communications model. In Proceedings of the 2nd ACM Workshop on Information Sharing and Collaborative Security, Denver, CO, USA, 12–16 October 2015; pp. 61–70.
5. Wagner, T.D. Cyber Threat Intelligence for "Things". In Proceedings of the 2019 International Conference on Cyber Situational Awareness, Data Analytics and Assessment (Cyber SA), Oxford, UK, 3–4 June 2019; pp. 1–2.
6. MITRE Corp. Making Security Measurable. 2018. Available online: https://msm.mitre.org/ (accessed on 3 April 2020).
7. Barnum, S. Standardizing cyber threat intelligence information with the Structured Threat Information eXpression (STIX). 2014. Available online: http://www.standardscoordination.org/sites/default/files/docs/STIX_Whitepaper_v1.1.pdf (accessed on 3 April 2020).
8. Connolly, J.; Davidson, M.; Richard, M.; Skorupka, C. Trusted Automated eXchange of Indicator Information (TAXII™). 2012. Available online: http://taxii.mitre.org/about/documents/Introduction_to_TAXII_White_Paper_November_2012.pdf (accessed on 3 April 2020).
9. OASIS Open Introduction to TAXII. 2018. Available online: https://oasis--open.github.io/cti--documentation/taxii/intro.html (accessed on 3 April 2020).
10. MITRE Corp. Cyber Observable eXpression (CybOX™) Archive Website. 2017. Available online: http://cyboxproject.github.io/ (accessed on 3 April 2020).
11. Sauerwein, C.; Sillaber, C.; Mussmann, A.; Breu, R. Threat Intelligence Sharing Platforms: An Exploratory Study of Software Vendors and Research Perspectives. In Proceedings of the 13th International Conference on Wirtschaftsinformatik, St. Gallen, Switzerland, 12–15 February 2017.
12. Zrahia, A. Threat intelligence sharing between cybersecurity vendors: Network, dyadic, and agent views. *J. Cybersecur.* **2018**, *4*, 1–16. [CrossRef]
13. Brown, S.; Gommers, J.; Serrano, O. From Cyber Security Information Sharing to Threat Management. In Proceedings of the 2nd ACM Workshop on Information Sharing and Collaborative Security, Denver, CO, USA, 12–16 October 2015; pp. 43–49.
14. Liu, R.; Zhao, Z.; Sun, C.; Yang, X.; Gong, X.; Zhang, J. A Research and Analysis Method of Open Source Threat Intelligence Data. In Proceedings of the 3rd International Conference of Pioneering Computer Scientists, Engineers and Educators (ICPCSEE), Changsha, China, 22–24 September 2017; Part I, Communications in Computer and Information Science. Springer: Berlin, Germany, 2017; Volume 727, pp. 352–363.
15. Sauerwein, C.; Pekaric, I.; Felderer, M.; Breu, R. An analysis and classification of public information security data sources used in research and practice. *Comput. Secur.* **2019**, *82*, 140–155. [CrossRef]
16. Abu, M.; Selamat, S.; Ariffin, A.; Yusof, R. Cyber Threat Intelligence—Issue and Challenges. *Indones. J. Electr. Eng. Comput. Sci.* **2018**, *10*, 371–379.
17. Pala, A.; Zhuang, J. Information sharing in cybersecurity: A review. *Decis. Anal.* **2019**, *16*, 1–25. [CrossRef]
18. Tounsi, W.; Rais, H. A survey on technical threat intelligence in the age of sophisticated cyber attacks. *Comput. Secur.* **2018**, *72*, 212–233. [CrossRef]
19. Menges, F.; Pernul, G. A comparative analysis of incident reporting formats. *Comput. Secur.* **2018**, *73*, 87–101. [CrossRef]
20. Mavroeidis, V.; Bromander, S. Cyber threat intelligence model: An evaluation of taxonomies, sharing standards, and ontologies within cyber threat intelligence. In Proceedings of the 2017 European Intelligence and Security Informatics Conference (EISIC), Athens, Greece, 11–13 September 2017; pp. 91–98.

21. Burger, E.W.; Goodman, M.D.; Kampanakis, P.; Zhu, K.A. Taxonomy model for cyber threat intelligence information exchange technologies. In Proceedings of the ACM Workshop on Information Sharing & Collaborative Security (WISCS), Scottsdale, AZ, USA, 3 November 2014; pp. 51–60. [CrossRef]
22. Asgarli, E.; Burger, E. Semantic ontologies for cyber threat sharing standards. In Proceedings of the 2016 IEEE Symposium on Technologies for Homeland Security (HST), Waltham, MA, USA, 10–11 May 2016; pp. 1–6.
23. Serrano, O.; Dandurand, L.; Brown, S. On the Design of a Cyber Security Data Sharing System. In Proceedings of the 2014 ACM Workshop on Information Sharing & Collaborative Security, Scottsdale, AZ, USA, 3 November 2014; pp. 61–69.
24. Sullivan, C.; Burger, E. "In the public interest": The privacy implications of international business-to-business sharing of cyber-threat intelligence. *Comput. Law Secur. Rev.* **2017**, *33*, 14–29. [CrossRef]
25. Wagner, T.D.; Mahbub, K.; Palomar, E.; Abdallah, A.E. Cyber threat intelligence sharing: Survey and research directions. *Comput. Secur.* **2019**, *87*, 101589. [CrossRef]
26. Zibak, A.; Simpson, A. Cyber threat information sharing: Perceived benefits and barriers. In Proceedings of the 14th International Conference on Availability, Reliability and Security, Canterbury, UK, 26–29 August 2019; pp. 1–9. [CrossRef]
27. Wagner, C.; Dulaunoy, A.; Wagener, G.; Iklody, A. MISP: The Design and Implementation of a Collaborative Threat Intelligence Sharing Platform. In Proceedings of the 2016 ACM on Workshop on Information Sharing and Collaborative Security, Vienna, Austria, 24 October 2016. [CrossRef]
28. Skopik, F. *Collaborative Cyber Threat Intelligence: Detecting and Responding to Advanced Cyber Attacks at National Level*; CRC Press: Boca Raton, FL, USA, 2018.
29. Farnham, G. *Tools and Standards for Cyber Threat Intelligence Projects*; SANS Institute InfoSec Reading Room: Bethesda, MA, USA, 2013.
30. Friedman, J.; Bouchard, M. *Definitive Guide to Cyber Threat Intelligence*; CyberEdge: Annapolis, MD, USA, 2015.
31. EclecticIQ. Cabby—TAXII Client Implementation. 2018. Available online: https://github.com/EclecticIQ/cabby (accessed on 3 April 2020).
32. OASIS Open. OASIS TC Open Repository: TAXII 2 Client Library Written in Python. 2018. Available online: https://github.com/oasis-open/cti-taxii-client (accessed on 3 April 2020).
33. MITRE Corp. The MITRE Corporation. 2018. Available online: https://www.mitre.org/ (accessed on 3 April 2020).
34. MITRE Corp. About MAEC. 2018. Available online: http://maecproject.github.io/about-maec/ (accessed on 3 April 2020).
35. OASIS Open. Introduction to STIX. 2018. Available online: https://oasis-open.github.io/cti-documentation/ (accessed on 3 April 2020).
36. OASIS. Introduction to STIX. 2018. Available online: https://oasis-open.github.io/cti-documentation/stix/intro (accessed on 3 April 2020).
37. OASIS. OASIS CTI CybOX Subcommittee. 2018. Available online: https://www.oasis-open.org/committees/tc_home.php?wg_abbrev=cti-cybox (accessed on 3 April 2020).
38. OASIS. OASIS Cyber Threat Intelligence (CTI) TC. 2017. Available online: https://www.oasis-open.org/committees/tc_home.php?wg_abbrev=cti (accessed on 3 April 2020).
39. MITRE Corp. CVE—Common Vulnerabilities and Exposures. 2018. Available online: http://cve.mitre.org/index.html (accessed on 3 April 2020).
40. OASIS Open. CSAF Common Vulnerability Reporting Framework (CVRF) Version 1.2. 2017. Available online: https://docs.oasis-open.org/csaf/csaf-cvrf/v1.2/cs01/csaf-cvrf-v1.2-cs01.html (accessed on 3 April 2020).
41. CESNET. Intrusion Detection Extensible Alert. 2018. Available online: https://www.cesnet.cz/en/index (accessed on 3 April 2020).
42. CIRCL. Malware Information Sharing Platform MISP—A Threat Sharing Platform. 2018. Available online: https://www.circl.lu/services/misp-malware-information-sharing-platform/ (accessed on 3 April 2020).
43. CSIRT Gadgets LLC. CSIRT Wiki, Getting Started—Welcome to the CSIRTG–EX Software Development Kit. 2018. Available online: https://github.com/csirtgadgets/csirtg/wiki (accessed on 3 April 2020).
44. Cisco. Snort. 2018. Available online: https://snort.org/ (accessed on 3 April 2020).
45. OISF. Suricata Open Source IDS / IPS / NSM engine. 2018. Available online: https://suricata-ids.org/ (accessed on 3 April 2020).

46. Spamhaus. Understanding DNSBL Filtering. 2018. Available online: https://www.spamhaus.org/whitepapers/dnsbl_function/ (accessed on 3 April 2020).
47. Winer, D. RSS 2.0 Specification. Available online: https://cyber.harvard.edu/rss/rss.html (accessed on 3 April 2020).
48. FireEye, Inc. Free Security Software—IOC Tools (Indicator of Compromise). Available online: https://www.fireeye.com/services/freeware.html (accessed on 3 April 2020).
49. Mandiant. GitHub Repository. Available online: https://github.com/mandiant (accessed on 3 April 2020).
50. Danyliw, R. Internet Engineering Task Force (IETF), RFC 7970. The Incident Object Description Exchange Format Version 2. Available online: https://tools.ietf.org/html/rfc7970 (accessed on 3 April 2020).
51. Lookingglass. Welcome to the OpenTPX Project! Available online: https://opentpx.org/ (accessed on 3 April 2020).
52. Cisco Security Alerts. Available online: https://tools.cisco.com/security/center/cvrf_20.xml. (accessed on 3 April 2020).
53. Oracle Security & Patch Update Advisories. Available online: http://www.oracle.com/ocom/groups/public/@otn/documents/webcontent/1932662.xml. (accessed on 3 April 2020).
54. Red Hat Security Advisories. Available online: https://www.redhat.com/security/data/cvrf/ (accessed on 3 April 2020).
55. Malc0de Database. Available online: http://malc0de.com/database/ (accessed on 3 April 2020).
56. NC4 Soltra. Connecting to PickupSTIX. Available online: https://www.soltra.com/en/documentation/ctx--soltra--edge/connecting--to--pickupstix/ (accessed on 3 April 2020).
57. Abuse.Ch. Ransomware Tracker. 2016. Available online: https://ransomwaretracker.abuse.ch/tracker/ (accessed on 3 April 2020).
58. NC4/Soltra LLC, PickUpStix. Available online: https://www.soltra.com/en/documentation/ctx--soltra--edge/connecting--to--pickupstix/ (accessed on 3 April 2020).
59. Anomali, Limo—Free Intel Feed. Available online: https://www.anomali.com/platform/limo (accessed on 3 April 2020).
60. Leach, P.; Mealling, M.; Salz, R. RFC4122, A Universally Unique IDentifier (UUID) URN Namespace. Available online: https://tools.ietf.org/html/rfc4122 (accessed on 3 April 2020).

© 2020 by the authors. Licensee MDPI, Basel, Switzerland. This article is an open access article distributed under the terms and conditions of the Creative Commons Attribution (CC BY) license (http://creativecommons.org/licenses/by/4.0/).

Review

Systematic Review and Quantitative Comparison of Cyberattack Scenario Detection and Projection

Ivan Kovačević *, Stjepan Groš and Karlo Slovenec

University of Zagreb, Faculty of Electrical Engineering and Computing, Unska 3, HR-10000 Zagreb, Croatia; stjepan.gros@fer.hr (S.G.); karlo.slovenec@fer.hr (K.S.)
* Correspondence: ivan.kovacevic@fer.hr

Received: 15 September 2020; Accepted: 15 October 2020; Published: 19 October 2020

Abstract: Intrusion Detection Systems (IDSs) automatically analyze event logs and network traffic in order to detect malicious activity and policy violations. Because IDSs have a large number of false positives and false negatives and the technical nature of their alerts requires a lot of manual analysis, the researchers proposed approaches that automate the analysis of alerts to detect large-scale attacks and predict the attacker's next steps. Unfortunately, many such approaches use unique datasets and success metrics, making comparison difficult. This survey provides an overview of the state of the art in detecting and projecting cyberattack scenarios, with a focus on evaluation and the corresponding metrics. Representative papers are collected while using Google Scholar and Scopus searches. Mutually comparable success metrics are calculated and several comparison tables are provided. Our results show that commonly used metrics are saturated on popular datasets and cannot assess the practical usability of the approaches. In addition, approaches with knowledge bases require constant maintenance, while data mining and ML approaches depend on the quality of available datasets, which, at the time of writing, are not representative enough to provide general knowledge regarding attack scenarios, so more emphasis needs to be placed on researching the behavior of attackers.

Keywords: targeted attacks; attack scenario; intrusion detection; alert correlation; cyber situational awareness; attack projection

1. Introduction

The disappointing truth about cyber security today is that targeted cyberattacks, which are usually motivated by financial gain, often go undetected for weeks, even months [1], allowing for attackers to do a lot of damage. If attacks were discovered at an early stage, the damage they cause could be significantly reduced. At the time of writing, attacks are primarily detected using network and host intrusion detection systems (NIDS and HIDS, respectively). Network intrusion detection systems monitor network traffic, while host intrusion detection system monitor different data sources on a host (logs, network traffic, system parameters). They both try to detect incidents primarily using rules that describe what incidents are, with more and more tools appearing on the market trying to use anomaly detection. No matter the approach, those tools generate a lot of noise, visible in high amount of detections and false alarms, and often provide multiple detections of the same attack. Finally, they do not provide sufficient insights regarding security-related events they detect, resulting in attacks being overlooked by the security personnel [2]. The main problem is that attackers, while trying to achieve their goals by going through an attack scenario, often cause a lot of low-level incidents to be detected. We define an attack scenario as a partially ordered collection of attack steps or actions, such as information gathering, exploitation, malware installation, and credential misuse, which particular attackers perform in their attempt to compromise an organization. Examples of several attack scenarios can be found in [3]. IDSs can not detect such scenarios on their own, because they observe individual events and network connections, and can potentially detect attack techniques, technical errors, and anomalies they contain. An attack scenario, on the other hand, combines multiple attack techniques to reach the goals of the attacker, and usually results in multiple IDS alerts. In order to overcome those issues, approaches appeared that try to gain higher-level insights by combining multiple low-level incidents into larger attack scenarios that caused them. They work with alerts that have already been generated by an IDS component. Furthermore, when certain attack scenario is detected, it is—at least theoretically—possible to predict its future phases. Predicting future phases of attacks is usually called attack projection.

As researchers use a plethora of different success metrics and datasets to evaluate their approaches, it is not immediately clear how to compare their results and draw conclusions about them. In many cases, articles avoid using previously defined metrics and only report results using their own metric, making their results biased. The main goal of this survey is to enumerate and describe frequently used datasets and success metrics, and provide expanded comparison tables in an attempt to reduce the bias of published work in this research area.

Overall, this survey presents an overview of cyberattack scenarios' detection and projection research and its state-of-the-art, together with an overview of the underlying concepts and practical problems. Because of this focus on high-level attack scenarios, low-level intrusion and anomaly detection methods are outside of the scope of this survey. Unlike other related surveys, this survey makes a more thorough effort of comparing the results of the various approaches and the available evaluation metrics, so the main contributions of this survey are a unified comparison of results of relevant papers and the evaluation metrics in this research area.

The structure of the paper is as follows. Section 2 gives some preliminaries by describing the goals of attack scenario detection and projection in the context of cyber situational awareness. The following section, Section 3 gives an overview of novel existing surveys in this area, which constitute related work. In Section 4, relevant evaluation metrics for alert correlation and attack scenario detection are discussed and compared. Section 5 presents some of the commonly used datasets, as well as some novel datasets that show potential, with their pros and cons. After that, Section 6 describes our methods and gives a narrative overview of selected research papers published in this area, beginning from the early, pioneering approaches, and finishing with the state-of-the-art at the time of writing. The central contribution of this paper is given in Section 7, where common evaluation measures are calculated and displayed on several comparison tables. Finally, the paper gives a discussion in Section 8, and closes with the conclusion in Section 9, followed by a list of references.

2. Preliminaries: Achieving Cyber Situational Awareness

Cyber situational awareness (CSA) is the application of the model of situational awareness (SA), elaborated in the military domain by Endsley [4], in the context of information systems. The original SA model consists of three levels: (i) perception, (ii) comprehension, and (iii) projection, and it considers SA separately from decision-making and performance outcome, as shown in Figure 1. Perception focuses on gathering elements that contribute to the current situation, upon which comprehension then builds a meaningful representation of the situation. Finally, once the current situation has been comprehended, projection aims to foresee the possible future developments of the situation and its elements [5]. Because humans have a very limited attention and working memory capacity, Endsley suggested that systems should be designed to automate some SA tasks, and fuse the numerous and diverse elements of the situation into a small and manageable number of meaningful objects [4]. This would allow human attention to shift towards the most relevant aspects of the situation, consequently improving the decision-making process.

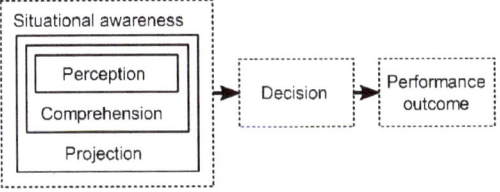

Figure 1. Endsley's model of situational awareness.

The following three subsections describe some aspects of CSA that are closely related to the scope of this paper: (A) attack detection, (B) attack projection, and (C) the problem of uncertainty. Attack detection includes perception and comprehension, while attack projection includes the projection level of CSA. The problem of uncertainty refers to uncertainty that is present in all CSA tasks that can distort CSA and lead to misinterpretation of the current situation.

2.1. Attack Detection

The primary means of performing attack detection is via intrusion detection systems, which are in CSA considered to be a simple sensors, belonging to the perception level of CSA [6]. The first big problem with intrusion detection systems in practice is the high number of generated low-level alerts making their analysis hard [2] and prone to errors [6]. IDS systems also suffer from large false positive (FP) and false negative (FN) error rates [7,8].

The second problem is the difficulty of comprehending the alerts' meaning and context, vital for making decisions upon which appropriate actions are taken, which is, IDSs often fail to lead to the detection of sophisticated attacks [9]. To make matters worse, the investigation of numerous false alerts, as well as the resulting quarantining of suspicious computer systems, can end up being very expensive for organizations [9]. Because IDSs generate alarms for individual events, these alarms can be grouped and analyzed to produce high-level information that is more comprehensible for personnel [6]. For example, one attack could result in numerous IDS alarms, and it would be beneficial to group those alarms according to attacks they probably resulted from [5].

Tadda and Salerno [5] analyzed the application of SA principles to the cyber domain, specifically in the detection of multi-stage attacks. They formulated the goal of CSA in cyber defense as the creation of attack tracks. Attack tracks are hypotheses of complex multi-stage attacks containing all of the supporting evidence extracted from the system's data. Essentially, events and alerts are grouped and correlated according to a potential multi-stage attack scenario they may be resulting from.

2.2. Attack Projection

The highest level of CSA is attack projection, as summarized by Yang et al. [10]. Attack projection aims to predict the future phases of an attack by combining the observed malicious activities with the knowledge about the system's configuration and vulnerabilities, and the attacker's behavior patterns. In theory, taking a proactive approach and predicting the next moves of an attacker should be possible [11]. Problems that make attack projection hard in practice are sensor inaccuracy, attack obfuscation, development of new attack strategies over time, and a lack of datasets required for comprehensive evaluation [10]. Sensor inaccuracy refers to FP and FN errors, while the obfuscation and development of new attacks stem from the attacker actively trying to conceal the attack [10].

A recent survey of attack projection was made by Husák et al. [11]. In addition to attack projection, they also surveyed related tasks, specifically intention recognition, intrusion prediction, and network security situation forecasting. The task of intention recognition, as described by Ahmed and Zaman [12], focuses on estimating the ultimate goal of an attack, and as such is closely related to attack projection. Intrusion prediction [13] attempts to predict future intrusions based on the knowledge of past intrusions combined with various additional data, while network security situation prediction [14] tries to forecast the overall future tendency of the risk of the network being attacked. The tasks of intrusion prediction and network security situation forecasting are outside of the scope of this survey.

2.3. The Problem of Uncertainty

The greatest challenge in CSA, which manifests in all CSA tasks, is managing uncertainty [15], which can distort SA and mislead the interpretation of the situation [16]. One type of uncertainty stems from the lacking knowledge about existing vulnerabilities and the chances of their exploitation, which also includes the security awareness of the organization's employees, depending on which they could be more or less susceptible to social engineering [15]. Li et al. [15] refer to this type of uncertainty as static uncertainty. In contrast, dynamic uncertainty occurs as a consequence of attackers actively trying to avoid detection, and manifests itself during the intrusion, making the attack very hard to detect and trace, as well as after the intrusion has ended during analysis and forensics. They argue that cyber defense systems must employ approaches that can deal with the all-present uncertainty, such as e.g., Bayesian networks, and that deterministic approaches, such as attack graphs, are not well suited for practical use in intrusion detection and response.

3. Related Work

There are several relevant surveys regarding cyber-attack detection and prediction. In this paper, we have chosen three recent surveys that we perceived as being of high quality, as related work. Table 1 gives an overview of those surveys with regard to their scope, classes of methods and systems they proposed, presented insights to evaluate efficiency, and some of the open research problems that their authors identified. We also elaborate the differences between our survey and four other surveys later in this section. It is important to note that surveys of low-level intrusion and anomaly detection methods that do not deal with high-level scenarios of attacks are outside the scope of this survey.

Salah et al. [17] proposed a classification of various products that implement alert correlation capabilities, such as IDS, Security Information, and Event Management (SIEM) systems, and network management tools, with regard to the number of different information sources that they use, type of application, correlation method, and their architecture. Depending on the architecture, they can be either centralized, decentralized, or hierarchical. Information sources discriminate between systems using single and multiple source types. Correlation methods are given in Table 1. Information source types that they reviewed are the following:

- alert database containing alerts from low-level sensors such as Network Intrusion Detection Systems (NIDS) or Host Intrusion Detection Systems (HIDS);
- topology information, such as hosts, connectivity, etc.;
- vulnerabilities database, e.g., Common Vulnerabilities and Exposures (CVE) database;
- trouble ticketing system information;
- ontology database, with domain semantics that can be used for automatic inference;
- cases database, containing rules for associating alerts with common and known problems; and,

- knowledge representation, rules and models either put together by experts or inferred from datasets, which are used for correlating alerts.

Finally, they proposed an alert correlation process model that consists of four phases: (*i*) alert preprocessing, (*ii*) alert reduction, (*iii*) alert correlation, and (*iv*) alert prioritization; and made an analysis of various products with regard to the proposed dimensions. Our survey differs from this survey in that, while they focus on correlation systems' architecture, we focus on success metrics and compare the approaches' results.

Hubballi et al. [18] focused on the approaches for reducing false positive alerts, and gave a broader view and classification than the survey mentioned earlier. This survey also gave an overview of the use of the various techniques in commercial SIEM systems together with their performance scores. This survey focuses on approach classification and determining the pros and cons of various classes, and unlike our survey, does not deal with results of their evaluation.

Husák et al. [11] has the closest scope to the scope of this paper. They analyze attack projection, attack prediction, intention recognition, and network security situation forecasting. In contrast to the previously listed surveys [17,18], this survey gives a summary of the results from the published approaches, but, unlike our survey, does not examine those results and limits itself to a small set of metrics. Additionally, Husák et al. [11] analyze several novel proposals that we omitted here, most of which are either out of scope of our paper or use their own datasets or evafluation. The readers are encouraged to refer to Tables 2–5 in that survey for an overview of evaluation details for papers that are not covered here. Their survey also includes attack projection related research and concludes that it has seen much less research than attack detection [11]. Husák states that the main research trend in this area is using methods, such as e.g., data mining, in order to discover new attack patterns from datasets. It is also worth mentioning that Husák in [19] already did some work consistent with that trend. Finally, one of their conclusions was that approaches published in modern papers support real-time operation, while the older ones did not focus on it as much.

There was also a recent survey by Navarro et al. [20], which focused on problems of reproducibility and replicability, and the properties of datasets the approaches used. Although they covered a large number of papers and datasets, they did not deal with evaluation metrics and results. An older survey by Mirheidari et al. [21] examined the properties of different classes of algorithms used in alert correlation, such as their memory requirements and ability to identify false positive detections and previously not seen attacks, but did not examine and compare individual approaches and their evaluation methods.

We also found several surveys regarding collaborative intrusion detection systems (CIDSs), such as [22,23], but these primarily focus on correlation and integration of alerts from distributed intrusion detection sensors with the goal of detecting coordinated distributed attacks, rather than reconstructing attack scenarios.

Although the surveys differ in scope and depth, a large portion of each focuses on the same set of problems common to attack detection and projection. Various authors came to similar conclusions about the articles published in this area, especially with regard to the open research questions. The recurring themes are the lack of proper evaluation methods and success measures, low quality of available datasets, bad performance in practical operation, scalability issues, as well as the detection of unknown attacks. Problems with performance in practical application and the discovery of unknown attacks could stem from the flaws in the way that the attacks and attackers have been researched and modeled, and further research on attackers' behavior could result in better models and improvements in the aforementioned areas.

Table 1. Overview of analyzed surveys in attack detection and prediction.

	Salah et al. [17]	Hubballi and Suryanarayanan [18]	Husák et al. [11]
Year	2013	2014	2018
Scope	Alert correlation	False alert minimization (a subset of more specific methods)	Attack projection Attack intention recognition Attack prediction Network security situation forecasting
Classes of models / methods / techniques	**Similarity based methods** 　Attribute based 　Temporal based 　Sequential based 　Pre-post conditions 　Graphs 　Codebook 　Markov models 　Bayesian networks 　Neural networks 　Others **Case based methods** 　Expert based 　　Expert rules 　　Pre-defined scenarios **Inferred knowledge**	**Signature enhancement** **Stateful signatures** **Vulnerability signatures** **Alarm Mining** 　Clustering 　Classification 　Neural Network 　Frequent pattern mining **Alarm Correlation** 　Multi-Step 　Knowledge based 　Complementary evidence based 　Causal relation based 　Fusion based 　Attack Graph based 　Rule based **Alarm Verification** **Flow analysis** **Alarm Prioritization** **Hybrid methods**	**Discrete Models** 　Graph models 　　Attack Graphs 　　Bayesian Networks 　　Markov Models 　　Game Theoretical models **Continuous Models** 　Time Series 　Grey Models **Machine Learning and Data Mining** 　Machine Learning 　　Neural Networks 　　SVM 　　(etc.) 　Data Mining **Other Approaches** 　(e.g., similarity based correlation)
Notes regarding evaluation	-	-	Evaluation in surveyed proposals: 　Proof-of-concept 　Testbed 　Live data (e.g., honeynet) 　Public Dataset (DARPA, etc.) 　Custom Dataset 　Virtual Attacks 　Comparison with other algorithms
Identified open research problems	Lack of standard strategies for evaluating performance and effectiveness Development of scalable system architectures Improving FP reduction Development of new strategies for coping with unseen types of attacks	Evaluation on a common dataset Analysis of approach performance Standardization of formats and reports Evaluation of systems' real time operation Adapting to changes in the environment	Methods relying on attack models require continuous model maintenance Low quality of evaluation datasets Low prediction accuracy and usability in practice Evaluation of methods on a common dataset with common and meaningful metrics Discovering novel attacks and security paradigms

This survey selects representative papers regarding attack scenario detection and projection, and it focuses on presenting their crucial points and practical aspects. Our methods are described in Section 6.1. Unlike the other related surveys identified in this section, this survey makes a more thorough effort of comparing the results of the various approaches and their evaluation methods in the context of situational awareness. This also includes the comparison of approaches over several datasets and success metrics, which was not provided by any of the other surveys reviewed in this section. Finally, our survey contains several recent approaches that were not included the surveys above. The details of our contributions are further discussed in Section 7.

4. Evaluation Metrics

Metrics for cyberattack scenario detection are needed in order to compare the success of various approaches. Because the problem is not a simple matter of TP and TN errors, like in low-level intrusion detection, each of them measures the success in terms of achieving a goal and poses a certain research challenge. This section gives an overview of some of the proposed metrics for cyberattack scenario detection. Similarly to FP and FN errors in low-level intrusion detection, Tadda and Salerno [5,24] defined success metrics in the context of systems providing high-level CSA. These metrics are divided into confidence, purity, cost utility, timeliness, and effectiveness metrics, and share similarities with success metrics used in data science. All of them rely on the notion of attack tracks described earlier. However, the survey of the representative papers has shown that, out of those metrics, only the confidence metrics, namely precision, are somewhat used in practice. This can be seen in Table 2, where the metrics usage frequency among the surveyed articles is shown.

Table 2. Usage of evaluation metrics in the surveyed papers.

Metric	Used in # Approaches	References
Reduction (8)	6	[25–30]
R_C (5)	4	[31–38]
R_S (6)	4	[31–38]
Episode Reduction (9)	2	[39,40]
QoA (7)	1	[41]
Recall (1)	0	-
Precision (2)	0	-
Fragmentation (3)	0	-
Misassociation (4)	0	-

Confidence metrics score the detection of real attack tracks. Tadda and Salerno define four confidence metrics: recall (1), precision (2), fragmentation (3), and mis-association (4).

$$\text{Recall} = \frac{\#\text{Correct Detections}}{\#\text{Known Attack Tracks}} \qquad (1)$$

$$\text{Precision} = \frac{\#\text{Correct Detections}}{\#\text{Detected Attack Tracks}} \qquad (2)$$

$$\text{Fragmentation} = \frac{\#\text{Fragments}}{\#\text{Known Attack Tracks}} \qquad (3)$$

$$\text{mis-association} = \frac{\#\text{Detections neither Correct nor Fragments}}{\#\text{Detected Attack Tracks}} \qquad (4)$$

In expressions (1)–(4), known attack tracks (KAT) are the attack tracks labeled in the dataset as the ground truth, detected attack tracks (DAT) are those detected by the system, correct detections (CD) are detected attack tracks that are at the same time known attack tracks, and fragments (F) are detections that should have been detected as a part of a larger attack track.

Ning et al. [34] instead used completeness (5) and soundness (6) measures for alert correlation.

$$R_C = \frac{\#\text{Correctly Correlated Alerts}}{\#\text{Related Alerts}} \qquad (5)$$

$$R_S = \frac{\#\text{Correctly Correlated Alerts}}{\#\text{Correlated Alerts}} \qquad (6)$$

In expressions (5) and (6), correlated alerts (CA) are alerts that the system classified as part of the attack tracks, related alerts (RA) are alerts that belong to the real attack tracks in the dataset, and correctly correlated alerts (CCA) are alerts classified into correct attack tracks [34].

Yu and Frinckle [41] used a metric that aims at scoring the quality of the alerts that are produced by alert correlation systems from the viewpoint of the security analyst, called the Quality of Alers (QoA) (7).

$$\text{QoA} = 100\% - \frac{\text{FP} + \text{FN} + \text{RT}}{\text{TA}} \qquad (7)$$

In expression (7), FP is the number of the false positive alerts generated by the system, FN is the number of the false negatives, RT is the number of repeated true alerts, and TA is the number of real intrusive actions [41]. To the best of our knowledge, only Yu and Frinckle use this metric, but it nonetheless captures important information regarding the evaluated system, and calculating it for other papers is possible. It is worth noting that this measure is high if the number of errors produced by the system is low, but also indicates that the number of repeated true alerts should be low.

Some authors, such as [27], aim at minimizing the number of alerts, and measure the amount of reduction between the original number of alerts and the number of new, often high-level alerts, which result from the alert correlation systems. Although this type of reduction measurement (8) is common among most of the surveyed papers, some of them, such as [39,40], use another variant, called episode reduction, abbreviated here as ER (9).

$$\text{Reduction} = 100\% - \frac{\text{Number of Output Alerts}}{\text{Number of Input Alerts}} \qquad (8)$$

$$\text{ER} = 100\% - \frac{\text{Number of Episodes With Reduction}}{\text{Number of Input Episodes}} \qquad (9)$$

In [39], episodes are mined collections of alerts that could potentially represent multi-step attack scenarios. Episode reduction in those papers refers to filtering out meaningless episodes and benign episodes. In addition to episode reduction, the papers [39,40] also use other episode-related metrics, but as they are highly dependent on the methodology and circumstances, they cannot be directly used for comparison with other papers.

Another evaluation metric used in attack projection, such as in the approach that was proposed by Fava et al. [42], is the ratio of correct predictions of the next scenario steps. Farhadi et al. [43] went for a more loose approach, in which they also measured the ratio of predictions in which the real next attack scenario step was one of the two predicted as most probable. Approaches that include an implementation, such as e.g., Wang et al. [44], often benchmark performance in terms of latencies on expected real-time load. In-depth comparison of attack projection metrics and evaluation results will be left for future work.

Some metrics above are representations of other metrics used in artificial intelligence, with a difference in their semantics. R_C (5), for example, is equivalent to another commonly used metric, the true positive rate (TPR), with the main difference being that R_C is specialized for alerts and attack tracks. The equivalence becomes evident when CCA in (5) is replaced with TP, and RA with (TP+FN), which results in the TPR formula. Precision defined in (2) is a variant of the commonly used precision metrics, which is evident when CD in (2) is replaced with TP, and DAT with (TP+FP), resulting in the common precision metric. R_S is also equivalent to precision, which again gets obvious when CCA in (6) is replaced with TP, and CA with (TP+FP). The difference between (2) and (6) is that (2) measures precision of attack tracks, while (6) measures the precision of assigning alerts to attack tracks.

Because most surveyed papers contain information regarding their alert and attack counts, it is possible to calculate completeness, soundness, and QoA metrics that can be used for comparison over a common dataset, regardless of the metrics used in the original papers. It is also possible to compare the reduction metrics, but these metrics lack deeper insight and should be used alongside other criteria. Any special considerations regarding success metrics will be discussed alongside the datasets and approaches in question in Sections 5 and 6.

5. Evaluation Datasets

Measuring the performance and effectiveness of different approaches the key is to have good datasets. When different methods are evaluated against the same dataset then they can be easily compared while using the same success metrics. Yet, the nature of the attacks is such that a single dataset is not sufficient and there are a number of different datasets available each with their own advantages and disadvantages. Different attack and normal data used in different datasets make a comparison between methods tested on different datasets difficult and not worth doing. To make things more complicated, a number of papers use custom datasets that are not used by anyone else. In the following subsections, we provide an overview of some commonly used datasets across the reviewed papers, as well as some new datasets that appeared recently.

5.1. Darpa Intrusion Detection Evaluation Datasets

The first widely used dataset for IDS evaluation is DARPA1998, published by the MIT Lincoln laboratory in 1998. This dataset was originally used to evaluate intrusion detection systems in the Air Force Research Laboratory (AFRL). The dataset was generated using a simulated network and contains a mix of normal network activities and attacks. It is divided into a training set containing seven weeks of traffic and a test set of two weeks of traffic (MIT Lincoln Laboratory, 2019). All of the attacks in the dataset are labeled. The 300 simulated attacks that appear in the dataset are described in detail by Kendall [45] (Kendall, 1999). In 1999, a newer and improved version of the dataset was created, called the DARPA 1999 [46].

In the aftermath of the Wisconsin Evaluation Rethink workshop in 2000, and the subsequent meetings, the MIT Lincoln Laboratory produced a new dataset containing two DDoS attack scenarios, called the DARPA 2000 [3]. The first scenario, LLDOS 1.0, features a novice attacker performing a DDoS attack in several phases, beginning with the initial probing of the network from the internet, and ending with the abuse of an existing service to perform a DDoS attack. The second scenario, LLDOS 2.0.2, features a slightly stealthier attacker performing the same attack using a malware. As the two datasets hold information about attack scenarios, they are popular for evaluation of tasks, such as alert correlation and attack projection.

5.2. Def Con Ctf Datasets

Alongside annual DEF CON conference [47], a popular international capture the flag (iCTF) cybersecurity competition is held. After each competition, the organizers publish the collected network packet captures containing attack traffic, which researchers can use to analyze attacks and discover attack patterns. The packet captures are published on the DEF CON CTF web site [48]. A notable property of CTF datasets is that the vast majority of their traffic is attack traffic, as opposed to benign traffic in other datasets [49].

5.3. Cyber Treasure Hunt Dataset

Vigna [50] organized several cybersecurity exercises in order to evaluate students, one of which was a treasure hunt competition. The students were split into teams and tried to attack a simulated payroll system by following provided step-by-step instructions with the goal of performing malicious money transfers [50]. The students were required to document the details of the attacks, and network traffic was collected, resulting in a dataset containing labeled multi-step attack scenarios [51].

5.4. Unb Datasets

Because of the significant changes in network usage patterns since the age of DARPA datasets [49], as well as the criticism surrounding those datasets [52,53], the Canadian Institute of Cybersecurity (CIC) created several novel IDS/IPS datasets (University of New Brunswick—Canadian Institute for Cybersecurity, n.d.) that are still not much used. The first dataset ISCXIDS2012 was created in 2012, using simulated agents developed to mimic previously observed user behavior patterns in the Institute. Shiravi et al. [54] provided the detailed description of this dataset, as well as the guidelines and requirements for creating valid and realistic datasets. The main requirements can be summarized, as follows: (i) network traffic should not be modified after the capture has finished, (ii) the data should be labeled automatically during its creation, (iii) the capture should contain the entire network interactions and diverse intrusions, and (iv) the simulated data must eliminate the need for later sanitization and anonymization.

In a similar manner, after Gharib et al. [49] identified a more complete set of requirements for realistic intrusion detection datasets, the same institute created a newer dataset in 2017, called CICIDS2017. The details regarding this dataset are explained by Sharafaldin et al. [55].

5.5. Other Datasets

The previous datasets are either heavily used by researchers, or they show a good promise as is the case with CICIDS2017. There is a number of other datasets used less frequently than the previous ones, but still interesting enough to be mentioned. One example of such datasets are the DARPA Grand Challenge Problem (GCP) datasets, which are used in four surveyed articles, but to the best of our knowledge are not publicly available.

Some datasets are collected while using honeypots. Honeypots are network decoys that are used to lure attackers and observe their actions [56], whose entire network traffic is, by definition, malicious. Thus, logged honeypot traffic can be used to research attackers and attack patterns. The next source of attack data comes from simulated attack scenarios in which a number of researchers pretended to be attackers and perform attacks, resulting in custom datasets. In both cases, honeypots and simulated attacks, it is frequently the case that the authors do not publish much details of the attacks that happened nor datasets used, hindering the reproducibility of their research.

An overview of several other notable intrusion detection datasets was given by Gharib et al. [49]. Besides the already mentioned, they reference the CAIDA, LBNL, CDX, Kyoto, Twente, UMASS, and ADFA datasets, each with its advantages and weaknesses. A plethora of additional datasets is available throughout the internet, e.g., packet captures [57]. In the authors' opinion, these datasets are usually highly specific and, in most cases, not well fit for intrusion detection evaluation and learning.

5.6. Criticism of Existing Datasets

McHugh gave an extensive critique of the DARPA datasets in 2000 [52], mainly focusing on the suitability of its data and the simulated environment used to create it for intrusion detection evaluation. He argues that the dataset contains biases, with the major ones stemming from the employed attack taxonomy, which concentrates on the attacker's viewpoint, ignoring the viewpoint of the IDS, and the simplistic false positive based receiver operator curves (ROC) used to present results, making them a nonrealistic general measure of success. Specifically, DARPA '98 was synthesized without deeper insights regarding the causes of real-word false positive alerts and lacks realistic statistical properties [52]. Further analysis performed by Brown et al. [53] confirmed that its statistical properties significantly differed from those expected in a real network.

Gharib et al. [49] analyzed and compared 11 commonly used datasets in 2016 using the dimensions they proposed, and came to the conclusion that all of them had severe weaknesses when it comes to traffic completeness. Most of them failed at either providing diverse attack examples, a sufficient variety of protocols, adequate data labeling, a good feature set, etc. For detailed information, readers are encouraged to examine Table 3 in their paper [49]. The authors also state that datasets should be produced in a dynamic manner, so they can be periodically updated to reflect changes in attack and network traffic patterns over time. Finally, they proposed a weighted ranking framework for the datasets based on the proposed dimensions.

6. Survey of Developments in Attack Scenario Detection and Projection

This section provides an overview of the developments in attack scenario detection and projection, beginning with early approaches in the 1990s, towards the state of the art approaches at the time of writing. The overview only examines the most important points of the articles, as perceived by the authors of this survey.

6.1. Methods

Surveyed articles were chosen in the following manner:

1. Three Google Scholar and Scopus searches were performed, using the following keywords: "attack scenario detection alert correlation", "attack scenario reconstruction alert correlation", and "attack scenario projection alert correlation".
2. A table containing 300 articles was populated while using the top 100 articles returned by each Google Scholar search.
3. We sorted the table by four criteria, each time listing the first 25 results whose titles and abstracts fell within the scope of the survey. The first criterion was the total number of citations, in order to find the most influential papers. The second criterion was the number of citations divided by the age of the paper, in order to find more recent papers that gained traction. The third criterion was the rank that the Google Scholar search yielded, used to measure the relevance of the paper. Finally, as we

noticed the lack of newer papers on the list, we added a fourth criteria, which filtered out all papers older than 2015, and sorted the remaining papers using their Google Scholar ranks. At this point, the list contained 100 article mentions.

4. All 23 articles ([25,26,31–34,38,58–73]) that appeared more than once on the list were considered for inclusion in the survey. Additionally, [35,36] were included, because, although each of them appeared only once on the list, they were in essence representing the same approach. Porras et al. [58] was excluded because it primarily concerns with mission impact and elimination of duplicate and false positive alerts, and not attack scenarios. Hossain et al. [59] were excluded, because they primarily focus on detecting low-level attacks on a single host. Pietraszek et al. [71] was excluded, because they primarily focused on reducing false-positive IDS alerts.
5. We included additional 8 articles ([19,28,39,40,42,43,74,75]) that were cited in the surveys that were covered in Section 3 and seemed to be relevant.
6. Additional 15 articles ([8,37,44,76–87]) were included because their results were cited by other articles already included at this point.
7. The steps described above were repeated using Scopus searches with the same search terms. While the first search returned the target 100 results, subsequent searches only found a handful of articles. In this manner, additional 9 articles ([29,30,88–94]) were found and considered for inclusion in our survey. Alhaj et al. [92] was excluded because their goal is correlating alerts into individual attack steps, but they do not reconstruct the high-level attack scenarios, effectively performing alert aggregation.
8. Finally, four articles were included on an individual basis. Huang et al. [95] was included, because it considers operational planning of attacks. Yu and Frincke [41] was included, because it introduced an expressive metrics, Quality of Alerts (7). GhasemiGol and Ghaemi-Bafghi [27], and Albanese et al. [96] were included due to using approaches whose types were underrepresented.

The annual distribution of surveyed methods and papers is shown in Figure 2. Interestingly, we noticed that older methods (i.e., [31–34]) were published while using a larger number of papers, with later papers extending and updating the originally presented method, while newer methods (i.e., [73]) are usually published using one or two papers. This can be seen as the difference between the numbers of papers and methods in Figure 2, especially noticeable between the years 2001 and 2004. It can also be observed that scientific production in this field peaked in the early 2000s, followed by research at a slower pace thereafter.

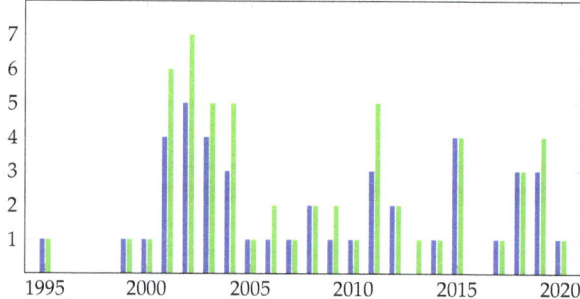

Figure 2. Annual distribution of surveyed methods (blue) and papers (green).

It should be noted that this method contains biases, as it only covers sources that are written in English and discriminates against articles on the basis of quantitative measures rather than their actual quality. Each article we surveyed is biased because it evaluates its approach using only a subset of datasets and success metrics. Our main goal in this survey is to try to expand this subset and quantify their individual bias. In Section 7, we collected the original counts of alerts from the papers and calculated commonly used success metrics described in Section 4 in order to provide a more complete comparison of the approaches' results.

The approaches are split into several subsections according to their similarities. On the highest level are three subsections: State transition analysis technique (Section 6.2), Event and/or Alert Correlation Relying on Expert Knowledge (Section 6.3), and Event and/or Alert Correlation Relying on Data Mining and Machine Learning (Section 6.4). The first deals with older approaches, based on explicit attack scenario signatures, the second with approaches that do not describe explicit attack scenarios, but use some sort of rules to build attack scenarios, and the third with approaches that use machine learning, often combined with rules, to build attack scenarios.

6.2. State Transition Analysis Technique

Early research efforts in attack detection aimed to develop easy to understand and efficient ways of describing attack signatures independently of the underlying audit data formats, and flexible enough to cover the majority of possible attack scenarios. Several of these relied on describing low-level attack scenarios while using the state transition analysis technique (STAT). STAT defines a state machine with an initial state, transition states, a target state, and signature actions that trigger transitions between states [8]. Once event logs are transformed into a state transition diagram, the diagram can be compared with transition diagrams for known attacks. Approaches, such as these work directly with audit data, such as event logs, as opposed to latter approaches working on top of alerts generated from IDSs.

Ilgun et al. [8] proposed USTAT, a HIDS that is based on STAT, with the U in the abbreviation indicating that it is a UNIX-based STAT prototype. In USTAT, low-level attack scenarios are described using state machines that end in compromised states. States are logical expressions parametrized by the system's attributes. Directed edges, called transitions, connect states that start with state being a precondition and end with a state being post-condition. Transitions are triggered when signatures of possibly malicious activities are detected, causing the state machine to enter a new state describing the expected resulting changes to the environment. As audit data enters USTAT, it performs state transitions, which are then analyzed by its inference engine. If the inference engine recognizes a described attack sequence, USTAT acts as an IPS, taking the action that is specified by its rules. USTAT was only able to recognize changes in known attributes in the system, ignoring any outer entities, such as users, and they could describe and recognize only explicitly specified scenario variant patterns inside the collected audit data. It was validated and compared to other tools while using a list of defined requirements for such tools.

Eckmann et al. [86] proposed a STAT based attack detection language called STAT language (STATL), a STAT based attack detection language, which is independent of the underlying IDS and provides performance enhancements over USTAT. The language compiles into rule formats for supported HIDSs and NIDSs, such as e.g., USTAT, NetSTAT, and WinSTAT, extended with applicable USTAT plugins. The authors validated STATL's expressiveness by describing 48 different attack scenarios for various IDSs. They successfully described all of those 48 attack scenarios while using STATL.

6.3. Event and/or Alert Correlation Relying on Expert Knowledge

Research soon focused on intrusion detection on a higher level. Instead of analyzing audit data and raising low-level alerts, some researchers began combining the already generated alerts and additional data from various sources into plausible larger-scale attack scenarios. Huang et al. [95] made an analogy of intrusion detection and response to military operations. In military operations, commanders need to have some hypothesis regarding the adversary's operational and strategic goals to decide on a proper reaction to his actions discovered on the tactical level, such as e.g., troop movements. In cybersecurity, low-level alerts, which could likely be FP or represent irrelevant attacks, do not provide the bigger picture and the higher-level insight that is needed to make decisions. Nevertheless, they can be combined to uncover possible attack strategies on the operational and strategic levels, which provide a deeper understanding of the current system state. Additionally, sensors that are based on different rulesets and detection types, especially if distributed over different parts of the network, offer diverse viewpoints on the information system, and should be used in a complementary manner [63].

6.3.1. Alert Correlation Based on Alert Attributes

Huang et al. [95] proposed a collaborative server-client IDS focused on the attacker's higher-level strategies, mimicking military operations. The server collects alerts from the clients and hypothesizes about the possible large-scale attack strategy. Clients are informed of the hypothesis and they build a goal tree based on domain knowledge to check whether the detected alerts could prove it. Goal trees hold a hierarchical representation of an actor's goal with the combination of sub-goals the actor used to achieve it. Finally, they send the evidence and detected low-level goals to the server, so it can adjust and possibly form a new hypothesis. The authors identified that further work needs to be done in detecting the attacker's decoy attacks and adding probabilities to the goal tree.

Valdes and Skinner [25,85] proposed a probabilistic alert correlation approach based on attribute similarity. A similarity function and a threshold are defined for each alert attribute, as well as the situation-specific similarity expectation, correcting that similarity for some specific situations, such as e.g., IP spoofing. Similar alerts from various sensors are grouped into meta-alerts, containing lists of original attribute values. Evaluation was performed while using recorded live data, where the number of generated alerts was reduced seven times, and by performing experiments over a simulated e-commerce network, where the approach managed to reconstruct the basic steps of the attack scenario as meta-alerts.

Morin et al. [62] proposed a formal model for cyber security, called M2D2. The model introduces relations between information system characteristics, vulnerabilities, security tools (e.g., IDS), and events. Network topology is described using a hypergraph in which nodes represent network interfaces, while edges, which can connect more than two nodes, represent networks. Data is mostly populated from available databases, such as e.g., ICAT vulnerability database, but parts such as topology information must be entered manually. The model can be used for aggregating alerts and determining causality relations between them. An improved version, M4D4 [80], added support for various new network device types, and describes a higher-level attack nomenclature using graphs. At the time of publishing, the authors were working on a proof-of-concept M4D4 Prolog reasoning engine, integrated into an information system to aid in alert correlation. Because M2D2 and M4D4 were developed to describe networks ad attacks, the authors validated the models by describing several situations and use cases.

Morin and Debar [61] combined M2D2 [62] with the formalism of chronicles [97] in order to describe benign and malicious scenarios as chronicles using expert rules with temporal constraints. Benign chronicles contain false positive alerts, while malicious chronicles contain alerts and events related to recognized attack scenarios. M2D2 correlates alerts with chronicles into high-level alerts and security operators are presented with high-level alerts and alerts that do not belong to a chronicle, resulting in an alert reduction. The approach was validated while using recorded logs of alerts from a real network.

Qin and Lee [68] correlate alerts into attack scenarios that are based on the Granger Causality Test (GCT). This test aims to check whether lags between events in a time series could indicate a causality relationship between them. The authors construct ordered sequences of alerts, called hyper-alerts, based on feature similarity and prioritize them using a manually configured Bayesian Network. Prioritized hyper-alerts are presented to security analysts. When they choose to investigate a hyper-alert, the approach searches for other causality related hyper-alerts via GCT, potentially revealing neighboring steps of the attack scenario. The approach was evaluated while using Grand Challenge Problem (GCP) v3.1. and DEFCON 9 datasets. On the GCP dataset, they scored reduction, completeness, and soundness of 91%, 94%, and 87%, respectively, while, on the DEFCON 9 dataset, they achieved alert reduction of 99.7%. They identified missed IDS alerts as the main limiting factor of their approach. In [69], the authors improved alert prioritization by including a knowledge base containing information regarding the surrounding network assets, and added an auxiliary method of alert correlation based on a knowledge base containing known attack plans.

Valeur et al. [26] correlated alerts and used alert verification to identify failed attacks. e.g., if an attempt of a service exploit over a machine is detected, and that machine does not run or expose the targeted service, then that attack was a failure. This is done actively, while using a vulnerability scanner script executed each time an exploit alert is reported. Unfortunately, active alert verification has drawbacks regarding system performance and can produce alerts as a side effect. Related alerts are merged into a hierarchical tree of meta-alerts, alerts on a higher level of abstraction, according to common attribute values and a sliding time-window. The next stage of the proposed system is attack focus recognition, aiming to discover hosts that are often sources or destinations of attacks. Finally, high-level multistage attack scenarios are detected by matching chains of alerts to predefined attack scenarios described while using STATL, and displayed as graphs to the operators. Additional tools that are offered by the system are impact analysis, which follows dependencies within the system and tries to determine the full impact of an attack, and alert prioritization, assigning priority values to alerts. Evaluation of FP alert reduction was performed on several datasets, including DARPA, Honeypot, DEF CON CTF, and the Treasure Hunt dataset, as well as live data from the Rome Air Force Research Laboratory's networks.

Barzegar and Shajari [38] converted each alert into an ontology according to its attributes. Alerts were considered to be neighboring steps of the attack if the ontology similarity according to their proposed measure was larger than a set treshold. The approach was evaluated using DARPA 2000 LLDOS 1.0 and MACCDC 2012 datasets. On DARPA 2000, it correlated alerts into four groups, three corresponding to parts of the attack scenario and one to false positive alerts. They provided completeness and soundness measures over only a subset of this dataset, and did not address the fragmentation of the reconstructed scenario, so we excluded those results from Section 7. The depictions of their latency results lack concrete time units and they seem to grow at least polynomially with the number of alerts, so we do not consider this approach to be applicable in real time.

6.3.2. Preconditions and Post-Conditions

As noted by Cheung et al. [67], low-level intrusion detection approaches relying on signatures, such as e.g., STATL, lack the flexibility to be used in order to describe high-level scenarios in a way that could support all their equivalent variants. They concluded that this lack of flexibility is addressed using the precondition and post-condition approach, in which scenarios are assembled from individual attacks by matching their preconditions and post-conditions.

Cuppens and Ortalo [81] proposed a language for describing cyberattacks, called LAMBDA. The language represents attacks using attacker actions defined through their preconditions and post-conditions, as well as their detection actions, describing steps needed to detect the action, and verification actions, describing the steps needed to confirm that the attack succeeded. The later can e.g., refer to determining that a vulnerable service is running on the target machine. In addition, the language defines event combination operators in order to specify ordering of events required for an action to be recognized, including e.g., parallel and sequential ordering. Attack actions can be combined if post-conditions of one action satisfy parts of the preconditions of the other. Cuppens and Miege [60] later used LAMBDA in the MIRADOR project to automatically generate a set of correlation rules from the described attacks. LAMBDA was validated by describing various attacks and situations.

Templeton and Levitt [82] introduced the model JIGSAW, which models attack scenarios while using attack actions with required capabilities, conceptually the same as preconditions, and provided capabilities, similar to post-conditions. Capabilities are atomic objects that describe the state of the system, and alongside technical attributes describe the attacker's knowledge about the system and policy statements. They proposed use cases for the model in tasks, such as e.g., FP reduction, attack detection, and attack projection. Finally, they identified attack generation, formal theory of attacks, and attack nondeterminism as future research problems, and stated that models based on signatures and exploits do not generalize well. Similarly to LAMBDA, they validated JIGSAW by describing several examples of attacks.

Cuppens et al. [83] distance themselves from earlier works using explicit scenario specification, and concentrate on using plans and objectives to detect attack scenarios. Action preconditions and post-conditions are still defined while using LAMBDA, but they are not combined according to a pre-defined scenario. Instead, actions are treated as possible stepping-stones that lead the system to a state in which some given security properties have been compromised. The authors added a concept of virtual alerts, as placeholders to support possible attacker's actions that were not detected, but are nonetheless required for the scenario. The article validates the proposed approach by describing example attack scenarios.

Ning et al. [31,32,34] and Ning and Xu [33] developed an alert correlation model and several alert analysis tools. The model shares similarities with JIGSAW and LAMBDA, since it describes attack scenarios while using the same precondition and post-condition paradigm, but it also presents a concrete alert correlation method. Alerts are mapped one-on-one to entities, called hyper-alerts, which are then correlated using partial preconditions and post-conditions that are specified by experts. Only a partial match of the conditions is required, because alerts corresponding to some preconditions required for later scenario steps could be missing (e.g., FN alerts). Correlation results in connected directed acyclic graphs (DAGs) of hyper-alerts representing possible attack scenarios. The authors later added concepts supporting attack strategy detection, by defining a measure of similarity of the previously generated DAG with known scenarios that are defined in the form of attack strategy graphs, and developed the toolkit TIAA for alert analysis. Attack strategy graphs are automatically built by generalizing already seen hyper-alert correlation graphs and extracting rules. Finally, they evaluated the method while using the DARPA 2000 dataset, and reported completeness scores between 62.5% and 94.74% and soundness scores between 92.3% and 100%, depending on the subset of the dataset.

Cheung et al. [67] proposed the Correlated Attack Modeling Language (CAML), based on the precondition post-condition paradigm. Unlike previous similar work, they primarily focused on modularity, and developed the language to describe modular scenarios. The main idea is that higher-level scenarios can be executed while using a combination of various lower-level scenarios, with the lowest level being individual attack steps. They validated a prototype by compiling the language to P-BEST expert rules [98], and then tested them on the DARPA Grand Challenge Problem (GCP) v2.0 dataset.

Ning et al. [66] argued that a major drawback of earlier precondition and post-condition based approaches is that they have problems with handling missed alerts. They proposed several techniques that combine the hyper-alert correlation graph generated while using approaches, such as Ning's earlier work with graphs generated by techniques relying on alert attribute similarity and audit trails to complete the correlation graph. Two graphs are combined while using a set of proposed rules. They validated the proposed techniques while using the DARPA 2000 dataset, on which only one of the attacks had miscorrelated alerts.

Saad and Traore [35,36] introduced an ontology-based technique for semantically clustering alerts into attack scenarios. The proposed ontology contains various alert attributes and context information. For clustering, the semantic relevance between two alerts is calculated as the normalized sum of the weights of ontology relations occurring between them, resulting in an alert correlation graph (ACG). The ACG is then clustered into subgraphs that represent possible attack scenarios. Finally, causality relations between alerts of a possible attack scenario are determined using prerequisites and consequences of attacks stored in the ontology, resulting in an attack scenario graph. Evaluation was performed on the DARPA 2000 dataset, which resulted in soundness and completeness between 96% and 100%. As no details about counts of various alert classes are available in the paper, it was not possible to verify or revise these scores. The authors intended to continue improving the ontology relations, use machine learning to reduce FP alerts, and research attack-pattern similarity. They considered the latter very challenging, because the available datasets were very limited and outdated.

Milajerdi et al. [73] developed a system that contains a high-level host-based IDS component and an alert correlation system that builds high-level attack scenarios. The IDS component is based on high-level signatures that detect the attacker's tactics, techniques, and procedures (TTPs) from system events. Detected TTPs are then correlated into high level scenario graphs (HSGs) using prerequisites based on the Mandiant Attack Lifecycle Model [99] and information flows between them. After additional preprocessing, HSGs are classified using estimated overall TTP severity scores into malicious attack scenarios and benign episodes. The approach was evaluated while using recorded traffic from a DARPA-organized red team exercise [100] and a live red team engagement over a simulated enterprise network. In both instances, it scored very high accuracy and recall rates in attack scenario detection. For the first experiment, both rates were 100%.

6.3.3. Attack Graphs and Trees

Goldman et al. [84] and Geib and Goldman [75] proposed an IDS extension system, called SCYLLARUS, for attack plan recognition and projection. The system has a hierarchical plan library in which higher-level plans can be decomposed into a partially ordered series of lower-level plans. Top-level plans are complete attack scenarios, executed with concrete attacker goals. The system analyses IDS alerts, according to them, tries to determine the likelihood of a plan being executed, and provides information about probable next targets according to that plan for taking proactive measures. They evaluated the prototype of the first version of SCYLLARUS using a custom attack simulation dataset.

Hughes and Sheyner [74] describe attack scenario graphs, graphs that are based on vulnerabilities and exploits representing stages of possible attack scenarios reaching a target asset, and the possible usage of such graphs in a model for threat mitigation. Attack scenario graphs can be used in order to predict possible future steps of attacks. They validated the model by demonstrating several usage examples.

Noel et al. [79] used attack graphs that were constructed offline to correlate events into attack scenarios in real time. Events are matched to exploits on the attack graph, and a new weighted graph, the event graph, is constructed with edge weights representing event distances on the original attack graph. Event distances are used to calculate the measure of correlation of the observed sequence of events to the scenario represented by the attack graph. The approach was validated while using several practical experiments. The first experiment proved that the approach could filter out attacker's actions based on preconditions and post-conditions. The second experiment, which contained seven attack scenarios mixed into a custom dataset, was used to tune and analyze hyper-parameters. Finally, the third experiment, in which they mixed 10,000 attack-scenario and other event-related alerts into a custom dataset, analyzed the system's performance when confronted with larger number of alerts. The authors claimed that all 10,000 alerts were correctly identified by the system, but did not provide details regarding the alert counts.

Wang et al. [44,65] concentrated on addressing performance issues of attack graphs in real-time application. They proposed the queue-graph data structure, which only correlates the latest instances of alert types, to increase alert correlation performance, and performed aggregation over indistinguishable transitive alerts to produce a more compact graph. The approach also hypothesizes about missing alerts and predicts possible future attack steps. Alert correlation was validated using the DARPA 2000 dataset, but, besides a high-level overview and the claims of it being satisfactory, no additional results were presented [44]. Real-time performance was validated using the Treasure Hunt dataset [44].

Liu et al. [37] use attack patterns, described using proposed models called attack path graphs (APGs), in order to describe the attack phases an attacker has to make to reach his goal. The high-level attack pattern graph starts with a probing phase, continues with the scanning and intrusion phases, and ends with the goal phase. Alerts are correlated while using the edges of the APG for each source and target host, and an attack graph is constructed by aggregating APGs based on the attackers and victims. The approach was evaluated using the DARPA 2000 dataset, with a reported completeness measure of 100% and soundness measure ranging from 98.3% to 100%. Using the original counts from the article [37], we calculated the measures from scratch, resulting in lower scores than provided in the original paper. The revised scores are available in Section 7 and the explanation in Section 8.

Roschke et al. [29] improved upon the work in [65] by using explicit attack graph-based correlation and considering a larger number of alerts with the same type when building attack scenarios. Explicit correlation aids in forensic analysis, while the consideration of multiple alert type instances enables the detection of a larger number of similar attack scenarios. The method was evaluated over a custom dataset containing recorded university network traffic and simulated attacks. It managed to detect the simulated attack scenario and reduced the number of reported alerts by at least 95.58%.

Ahmadinejad et al. [91] developed a hybrid approach that combines attack graphs that were constructed for the target network and alert feature similarity. Alerts for known attacks are correlated using the attack graph, with hypothesized unknown attack steps inserted into the graph in situations when the attack sequence on the graph is inconsistent, while alerts representing unknown attacks are grouped according to their feature similarity into collections that the authors refer to as hyper-alerts. Evaluation was performed over the DARPA 2000 dataset and several other synthetic datasets. The methods successfully reconstructed the attack scenario while providing sufficient performance for real-time operation. Although the results on the DARPA 2000 dataset indicate the accuracy and recall of 100%, the authors never explicitly state this and do not provide sufficient quantitative success metrics, so we only managed to calculate the alert reduction of 93.58%.

Albanese et al. [89] proposed a system that correlates alerts into attack scenarios while using a generalized dependency graph (GDG) and a probabilistic temporal attack graph (PTAG). The GDG describes the dependencies between IT resources and services provided by the organization, while the PTAG is an attack graph that contains temporal distributions for the delays between consecutive attack steps. The construction of these graphs requires a considerable amount of manual work. After the attack scenarios have been constructed, the system attempts to predict the most probable next attack steps using probabilities from the PTAG and their overall impacts on the system calculated from the GDG. Evaluation over synthetic attack graphs has shown that the system is able to perform in real time.

A later paper by Albanese et al. [96] presented a method for identifying sequences of events or alerts that are poorly explained by existing attack patterns in the knowledge base, and could potentially represent novel attacks. The method conceptually resembles anomaly detection and it is applicable to both IDS signatures and alert correlation, where high-level attack patterns are described using i.e., attack graphs.

An evaluation on the IDS level was performed using two days of recorded traffic. However, the use case with high-level attack scenarios was not evaluated, hence the minus (−) sign in Section 7.

6.4. Event and/or Alert Correlation Relying on Data Mining and Machine Learning

Many of the surveyed later works criticized approaches relying on pre-defined knowledge and parameters, because they only worked well with pre-defined scenarios and could not detect previously unknown attacks. Recent research in this area aims at employing machine learning and data mining, often with specialized and pre-configured models, to learn new attack patterns from datasets and recorded network traffic.

Dain and Cunningham [77,78] defined attack scenarios as ordered sequences of alerts, in which new alerts can be added if they satisfied a similarity criteria. A new scenario was defined if an alert was not similar to any of the existing scenarios. The approach they called the heuristic approach, as compared the attributes of the new alert with the last alert in each scenario using a non-linear function, which was trained on the labeled DEF CON 8 CTF dataset using the square error minimization technique in MATLAB, and correctly joined 88.81% of alerts into scenarios on the test set. They compared this results to the naïve approach, which only compared the IP addresses of the results, but this approach scored better on the DEF CON 8 CTF dataset, by correctly assigning 93.91% of alerts. We can conclude that this was probably a result of the properties of this specific dataset. Another proposed technique was based on data mining. It considered previous three alerts in the scenario for comparison, and used some additional features. The best results were obtained while using decision trees, which correctly joined 99.99% of alerts to scenarios. Unfortunately, that particular dataset was rarely used, so these results cannot be directly compared to other methods, as visible in Section 7.

Zhu and Ghorbani [70] used a multi-layer perceptron (MLP) and a support vector machine (SVM), trained together with a pre-built alert correlation matrix (ACM) extracted from the training set, in order to learn a classifier to decide whether two alerts should be grouped in the same attack scenario or not. Validation was performed using the DARPA 2000 dataset. They extracted attack graphs from the predicted alert correlation scores, and concluded that both models managed to group a subset of alerts into meaningful scenarios. Although most of the discovered attack scenario's steps resemble that of Ning et al. [34], the system failed to correlate the final DoS attack in the dataset due to spoofed IP addresses [70]. No quantitative results that could be compared to other approaches were provided in the article.

Yu and Frincke [41] described alerts as observations of hidden actions of attackers, and proposed to model attack scenarios using Hidden Colored Petri nets (HCPN). Action preconditions and post-conditions, as represented by input and output arcs in the HCPN, were taken from the work of Ning et al. [31–34]. HCPN colors represent actors, transitions represent actions, and places represent the system's resources. Based on this model, they provided an algorithm to determine the next most probable steps of the attacker. Training and evaluation was performed on the DARPA 2000 dataset, resulting in a TP rate of next step prediction of 93.3%. They focused on a novel measure of success they introduced, Quality of Alerts (QoA) (7) [41], as the principal success metrics. We used the provided alert class counts and definitions from this article to check their QoA scores and calculate completeness and soundness scores. The results and comparison can be found in Section 7. In addition to DARPA 2000, the authors used the DARPA Grand Chellenge Problem dataset, reaching a QoA score of 93%.

Fava et al. [42] used Variable-Length Markov Models (VLMM) in order to discover scenarios, or more precisely the attacker's behavior patterns, and to project the most probable next action of the attacker. The proposed methodology assigns symbols to alerts belonging to attack tracks based on their attributes and the chosen alphabet, and uses the generated symbol sequences in order to train VLMMs later used for projection. Evaluation was performed using a custom-built dataset, reaching a correct prediction rate of 90%.

Sadoddin and Ghorbani [28] proposed a modified version of the FP_Growth algorithm, based on structured patterns, for attack scenario pattern mining. Alerts are aggregated according to their sources and targets, and a Frequent Structure Pattern Tree (FSP_Tree) is incrementally built and frequent patterns are extracted. The method was evaluated while using the DARPA 2000 dataset, synthetically generated patterns, and real alert data from the Fred-eZone. The DARPA 2000 LLDOS 1.0 attack scenario was recognized as two scenarios, because of the IP spoofing used in the last step, and there were two more unlabeled scenarios detected inside the background LLDOS 1.0 traffic. Only alert reduction was scored, were the number of alerts was reduced from 922 to 29, resulting in a reduction ratio of 96,85%. In the second experiment, synthetic patterns were used to test performance, which seemed to support real-time application. Finally, by applying the system on real network traffic, they revealed patterns of a large number of various bot attacks.

Ren et al. [76] used Bayesian networks to group alerts that belong to the same scenario together. The proposed approach has an offline phase, in which features contributing to the correlation of alerts are identified, and an online phase, in which alerts are probabilistically correlated using the previously identified features. Before the offline phase is performed, features that represent the generalization hierarchy of existing features, e.g., IP address ranges, are added. Probabilities can be periodically recalculated to capture new correlation patterns. It is important to note that the correlation is not done on basis of alerts, but on the basis of the alert's types. Evaluation was performed while using the DARPA 2000 dataset, correctly correlating 96.5% of alert pairs into scenarios, and a honeynet dataset labeled by experts, on which it correctly correlated 93.2% of alert pairs before recalculation, and 96.1%, after recalculation. It was possible to calculate completeness and soundness measures using the information provided in the article. Section 7 shows the results.

Soleimani and Ghorbani [39] proposed a correlation system relying on decision trees (DTs) to discover malicious attack scenarios. Alerts are first aggregated into hyper-alerts, in the context of the paper, groups of similar alerts. Hyper-alerts are then combined into episodes, sequences of hyper-alerts of various lengths. A DT classifier is used to determine whether the detected episodes are malicious, and adds malicious episodes into a model episode tree, a data structure used to store attack scenario patterns. The DT classifier is initially trained on labeled data. If the DT classifies episodes as benign, rules based on alert attributes are used to assess whether those episodes could be potential attack scenarios and, if so, they are presented to the admin who can approve them to be used to update the classifier. After malicious episodes have been detected, the model episode tree is used to calculate the probabilities of potential next hyper-alerts, and a prediction of next probable attacker's steps is made. The detection and episode filtering was evaluated using the DARPA 2000 dataset and a recorded real worm attack from an ISP. After characteristic attack scenarios have been learned, the scenario detection rate scored between 94.61% and 99.78%, and episode reduction for episodes of length 6 reached 99.94%.

Fredj [90] extends attack graphs with probabilities of attacks, and refers to this type of graphs as Markov chain attack graphs (MCAG). The MCAG is constructed from a database of organization's hosts and network information, including firewall policies, and statistical information from a dataset or online data containing attack traffic. Alerts are correlated using the MCAG and clustered into attack scenarios by finding their final states, which do not have outgoing edges to other states. The probabilities allow for predicting the next probable steps in the scenario. Evaluation was performed on the DEF CON CTF 17 dataset, where it correlated attacks by 97 attackers into 85 attack scenarios.

GhasemiGol and Ghaemi-Bafghi [27] used alerts partial entropy (APE) to identify alerts containing common information in order to produce hyper-alerts and attack scenarios. Their central idea is that alerts corresponding to same attacks will have more common information than alerts that correspond to different attacks. Alerts are clustered using the density-based spatial clustering of applications with noise (DBSCAN) algorithm based on the measure of partial entropy. DBSCAN was originaly proposed for spatial databases, and performs clustering on a basis of density [101]. After clustering, additional features describing alert generalization hierarchies are added, and only a specified number of alerts containing the largest amount of information in a cluster are chosen as its representatives. The approach was evaluated on the DARPA 2000 dataset, where it reduced the original 34,819 alerts to seven hyper-alerts, corresponding to major events in the scenario. Although the main semantics of the DARPA 2000 attacks were preserved, details regarding alert classification rates are not available in the paper. Finally, they argue that the resulting hyper-alerts could be further filtered by editing the thresholds for entropy, but that would also cause information loss.

Ramaki et al. [40] proposed the real-time episode correlation algorithm (RTECA), which correlates alerts into scenarios using a causal correlation matrix (CCM), estimated during the algorithm's offline phase using an alert dataset, and it predicts the next probable steps of the detected scenario. Alerts are aggregated into hyper-alerts, defined as alert aggregates, based on their attack type and a given time window. Variable-length hyper-alert sequences are then converted to directed acyclic graphs (DAGs), called episodes, which are, during the online phase of the algorithm, classified as either critical or benign, depending on the strength of the correlation between their hyper-alerts calculated using the CCM. Critical episodes are reported as potential attack scenarios, while the weakly correlated benign episodes are analyzed to learn potential new scenarios and update the CCM accordingly. CCM's values are estimated offline while using distinct similarity functions over some of the hyper-alert's attributes, and statistical properties calculated from a stored attack tree. The attack tree constitutes of correlated hyper-alerts from all identified critical episodes, together with frequencies of those episodes used to calculate conditional hyper-alert probabilities. These probabilities can be used to predict the most probable next steps of an attack scenario. Evaluation was performed using the DARPA 2000 LLDOS 1.0 dataset and DARPA Grand Challenge Problem dataset 3.1. On DARPA 2000, it successfully identified the main steps of the scenario, with an episode reduction score of 99.95% for episodes of length 4. Evaluation on the Grand Challenge Problem (GCP) v3.1 dataset successfully identified previously unknown attack patterns and correctly projected between 68.5% and 99.93% of next scenario steps, with an episode reduction score of 99.95%.

A later paper by Ramaki et al. [88] presents a method relying on Bayesian attack graphs (BAGs) for alert correlation and attack scenario projection. The BAG is an attack graph that contains conditional probabilities for attack steps when some preceding attack steps have been observed. This probabilities are estimated off-line using a dataset of recorded alerts. Evaluation was performed on the DARPA 2000 LLDOS 1.0 scenario, where it successfully discovered the entire attack scenario with an alert reduction of 99.34% and projected the next steps of the attack with an accuracy between 92.3% and 99.2%. The authors claim that the entire attack sequence was correctly reconstructed, which is equivalent to the R_C (5) score of 100%.

Holgado et al. [72] used Hidden Markov Models (HMMs) with pre-defined attack stages to learn and predict steps of attacks. After pre-processing, alerts are labeled with keywords and severity scores obtained from the related CVE [102] records, to handle previously unknown types of alerts automatically. The approach was trained and evaluated while using DARPA 2000 LLDOS1.0 and a custom-made dataset containing scripted DDoS attacks. It was sucessful in identifying phases of the attack and calculated the risk of reaching the final phase for each alert. Although the authors provided a graphical representation of their results for DARPA 2000 that suggests a success rate of almost 100%, they did not provide enough numerical information to calculate comparable metrics.

Husák and Kašpar [19] used sequential rule mining on alerts from the SABU alert sharing platform [103] to learn attack sequence rules, and evaluated their usability in attack scenario projection. The evaluation showed that the observed attack patterns change quickly, but they are nonetheless stable enough to be usable during several days. This paper was more focused on the properties of the dataset than the method itself.

Ghafir et al. [87] developed a machine-learning based IDS technology, called MLAPT, whose low-level detection modules are outside the scope of this survey. It correlates alerts of its detection modules into larger-scale attacks using pre-defined attack steps, and pre-defined grouping and correlation rules. Upon discovery of partial or full attack scenarios, it generates high-level alerts for the security team. The final component of MLAPT is an attack projection module that predicts next steps of partial attack scenarios. It relies on several ML algorithms that were trained over a dataset of attack scenarios the system detected in a period of six months. MLAPT was evaluated while using three proprietary datasets and simulated attack scenarios inside the campus network. It achieved a total R_C of 81.8%, R_S of 73.4%, and successfully predicted the next steps, depending on the classifier in 68 to 85% of cases. In [64], the authors improved the attack scenario projection component with a Hidden Markov Model (HMM). The improved version was evaluated using two synthetic datasets, with a projection success rate of about 40% for projection from the first step of a scenario, increasing to 100% for projection from the middle of a scenario.

Zhang et al. [30] used alert attributes and Markov Chains (MC) to estimate alert causal probabilities from a dataset. Similar alerts were aggregated into hyper-alerts and then grouped according to the IP addresses they contained. A small set of alert attributes, such as the source IP address, was used to estimate the probability that alerts in an alert sequence are consequences of previous alerts. Using the mined statistics and alerts that involve multiple hosts, the method creates an attack graph that represents attack steps in the network. Evaluation was performed using the DARPA 2000 dataset. The method successfully reconstructed large parts of both attack scenarios, and led to an alert reduction of 96%. Unfortunately, detailed evaluation results needed to calculate other metrics are not available. Because this approach is based on mining rules from a dataset, it is likely to give the best results when detecting repetitive attacks, such as those that are characteristic for computer worms, and will likely not perform as well when facing novel attack scenarios.

Zhang et al. [93] define rules and algorithms for grouping alerts into attack scenarios and reasoning about the next steps of the attacker. Alerts are first aggregated into intrusion actions with regard to their features. Intrusion actions are then added to sessions representing communication between pairs of hosts. The criteria for splitting sessions are unusually large time gaps between intrusion actions. Finally, sessions are grouped into attack scenarios if they share common intrusion actions. Historical data is used offline to estimate correlation probabilities of sessions, and this probabilities are used to predict the next steps of the attack. The method was evaluated on the DARPA LLDOS 1.0 scenario and the CICIDS 2017 dataset. It has shown a promising prediction accuracy larger than 90%, and it managed to reconstruct the attack scenarios with alert reduction of 83.12% on the LLDOS 1.0 scenario. Unfortunately, the reconstructed graphs of attack scenarios are large and difficult to analyze, and the available quantitative results were insufficient for calculating other success metrics.

Hu et al. [94] correlate alerts into attack sequences using network-specific attack graphs constructed with the MulVAL Toolkit [104]. The creation of the attack graph requires knowledge about the target network and existing vulnerabilities. Generated attack sequences are then clustered using a sequence similarity metric into attack scenarios. The method takes into account that attack steps may have been missed by the IDS, including both unknown and failed attacks. Validation is first preformed in an experimental setting, and later whlie using the DEF CON CTF 23 dataset. In both cases, the method successfully reconstructed the attack scenarios. Unfortunately, quantitative information on the results was not published.

7. Results

The main contribution of this survey is a unified comparison of results of the surveyed papers. A basic overview of the papers is shown in Table 3, with the following columns:

1. Model gives a basic description of the concepts which the method or tool is based on. Some of the articles propose attack description languages, such as STATL, and are therefore labeled as Language. Similarity stands for feature similarity and refers to models which compare alert features, often using pre-defined formulas for similarity. Articles relying on attack graphs and attack trees are labeled as Graph. Most papers either use a knowledge base (KB) populated by experts, or machine learning (ML) or data mining (DM) to extract attack scenarios. In some cases, names of data mining algorithms or machine learning models are provided. Hidden Markov model is abbreviated as HMM, Similarity as Sim., Decision Trees as DT, and Markov chain as MCh.
2. Domain knowledge refers to the amount of domain knowledge that needs to be provided to the model to be fully functional. Since languages can be used in all sorts of systems, based on KB, AI, or hybrid models, the field is left empty. One star means that there is only minimal domain knowledge entry needed, in most cases about IP ranges in the network. Three stars signify that the system needs regular maintenance of a knowledge base to operate, while two stars are between the two extremes.
3. Level of evaluation describes whether the authors validated their approaches on datasets without quantitative success measurement, denoted by V, evaluated them on datasets using quantitative success metrics, denoted by E, or validated them using examples, use-cases, and/or formal proofs, denoted by F.
4. Real-time refers to the ability of the proposed system to work in a real online environment. It is important to note that empty fields do not necessarily mean that the system is not applicable, but that the question of such use was not considered in the paper. The papers which support real-time operation are marked with a plus sign in this column. Question marks denote that the paper left an impression that it should support real-time usage, but it did not contain clear evidence for it.

The next seven columns indicate whether a particular dataset was used in the paper. Datasets that have been used in a paper either for validation or for evaluation are marked with a plus sign or with the version of the dataset. If the approach was evaluated over at least one of the selected datasets, and it contains evaluation results, it is denoted by E. The asterisk in the final column for [93] signifies that they used CICIDS 2017 rather than a custom dataset.

Analysis of the papers has shown that four datasets were used for evaluation in a consistent manner more than once, namely DARPA 2000, Cyber Treasure Hunt, GCP 3.1, and DEF CON 9 datasets. The results for those datasets are listed in Tables 4–6, with some additionally depicted in Figure 3. In most cases, we used original data from the papers (alert counts, false positives, false negatives, etc.) in order to calculate success metrics from scratch, according to the definitions in Section 4. Figures that were not given in the paper and are purely our own work are underlined, and the results that significantly differ from the ones stated in the papers are written in bold, with original values written in parenthesis.

Ning et al. calculated the completeness and soundness over subsets of the LLDOS 1.0 scenario, so the results in the table are also not available in their original articles and, hence, underlined. We encountered a similar situation with Liu et al., but our results on the dataset subset basis differed from those that are mentioned in the paper. Finally, in cases where original information was missing from the papers, we transcribed results directly from the papers. Such results are denoted with *(P)* in the table. Ramaki et al. [88] is a special case, because we converted their claim of detecting the entire attack scenario into the R_C score of 100%.

Table 3. Overview of the basic properties and evaluation details of the surveyed papers.

	Details					Datasets Used					
Papers	Year	Models	Domain knowledge	Level of evaluation	Real-time	Cyber Treasure Hunt	DARPA 1999	DARPA 2000	DARPA GCP (ver.)	DEFCON (#)	LiveD- data	Custom / experiment
Ilgun et al. [8]	1995	Language		V								+
Huang et al. [95]	1999	Language		-								
Cuppens and Ortalo [81]	2000	Language		F								
Valdes and Skinner [25]	2001	Sim.	*	E							+	+
Templeton and Levitt [82]	2001	Language		F								
Goldman et al. [75,84]	2001	KB, Graph	***	V								+
Dain and Cunningham [77]	2001	ML, DM	*	E	+					8		
Ning et al. [31–34]	2002	KB	***	E				+		8		
Eckmann et al. [86]	2002	Language		V								+
Morin et al. [62,80]	2002	KB	***	F								
Morin and Debar [61]	2003	KB	***	V	?						+	
Cuppens and Miege [60]	2002	KB	***	V								+
Cuppens et al. [83]	2002	KB	***	F								
Cheung et al. [67]	2003	KB	***	V	+				2.0			
Hughes and Sheyner [74]	2003	KB, Graph	***	F								
Qin and Lee [68,69]	2003	Sim., KB	**	E					3.1	9		
Valeur et al. [26]	2004	Sim.	*	E		+	+	+		9	+	+
Ning et al. [66]	2004	Sim., KB	**	V				+				
Noel et al. [79]	2004	KB, Graph	***	V	?							+
Wang et al. [44,65]	2005	Sim., Graph	**	V	+	+		+			+	
Zhu and Ghorbani [70]	2006	ML Ensemble	*	V	?			+				
Yu and Frincke [41]	2007	KB, HCPN	**	E	+			+	?			
Liu et al. [37]	2008	KB, Graph	**	E	?			+				
Fava et al. [42]	2008	Markov models	*	E	+							+
Sadoddin and Ghorbani [28]	2009	DM: FSP_Growth	*	E	?			+			+	+
Ren et al. [76]	2010	Sim., ML	*	E	+			+			+	
Roschke et al. [29]	2011	Graph	**	E	?							+
Ahmadinejad et al. [91]	2011	Graph, Sim.	**	E	+			+				+
Albanese et al. [89]	2011	KB, Graph	***	E	+							+
Saad and Traore [35,36]	2012	KB, Graph	**	E		+		+				
Soleimani and Ghorbani [39]	2012	DM, ML: DT	*	E	+			+				+
Albanese et al. [96]	2014	KB, Rules	***	-	+							
Fredj [90]	2015	KB, Graph	**	E	+					17		
GhasemiGol and Ghaemi-Bafghi [27]	2015	ML	*	E				+				
Ramaki et al. [40]	2015	DM on streams	**	E	+			+	3.1			+
Ramaki et al. [88]	2015	KB, Graph, ML	**	E	+			+				
Holgado et al. [72]	2017	KB, ML: HMM	**	V	+			+				+
Barzegar and Shajari [38]	2018	Sim.	*	E				+				+
Husák and Kašpar [11]	2018	DM: TopKRules	*	V	?						+	
Ghafir et al. [64,87]	2018	KB, ML: HMM	**	E	+						+	+
Milajerdi et al. [73]	2019	KB, Graph	***	E	+						+	+
Zhang et al. [30]	2019	ML: MCh	*	E	+			+				
Zhang et al. [93]	2019	ML, Rules	*	E	+			+				*
Hu et al. [94]	2020	ML, Graph	**	V						23		+

Table 4. Selected success metrics calculated on the basis of results on the DARPA 2000 dataset.

Papers	Year	Scenario LLDOS 1.0		
		R_S (6)	R_C (5)	QoA (7)
Saad and Traore [35,36]	2012	(P) 99.70%	(P) 100.00%	
Ramaki et al. [88]	2015		(P) 100.00%	
Ren et al. [76]	2010	87.14%	96.83%	
Ning et al. [31–34]	2002	94.06%	94.06%	85.91%
Yu and Frinckle [41]	2007	100.00%	93.75%	93.75%
Liu et al. [37]	2008	(100%) 86.27%	(99.02%) 87.13%	

Table 5. Alert reduction calculated on the basis of results on the DARPA 2000 dataset.

Papers	Year	LLDOS 1.0		Reduction on LLDOS 1.0 and LLDOS 2.0.2
		Reduction (8)	Episode Reduction (9)	
GhasemiGol and Ghaemi-Bafghi [27]	2015	99.98%		
Ramaki et al. [88]	2015	99.34%		
Yu and Frincke [41]	2007	98.35%		
Sadoddin and Ghorbani [28]	2009	(96%) 96.85%		
Zhang et al. [30]	2019	96.21%		(P) 96%
Ning et al., Ning and Xu [31–34]	2002	94.43%		95.64%
Ahmadinejad et al. [91]	2011	93.58%		
Zhang et al. [93]	2019	83.12%		
Valeur et al. [26]	2004			53.00%
Ramaki et al. [40]	2015		(P) 99.95%	
Soleimani and Ghorbani [39]	2012		(P) 99.94%	

Table 6. Alert reduction results on Cyber Treasure Hunt (CTH), GCP 3.1, and DEF CON 9 datasets.

Papers	Year	CTH	GCP 3.1	DEF CON 9
Saad and Traore [35,36]	2012	99.999%		
Ramaki et al. [40]	2015		92.00%	
Qin and Lee [68,69]	2003		91.37%	99.66%
Valeur et al. [26]	2004	99.962%		96.81%

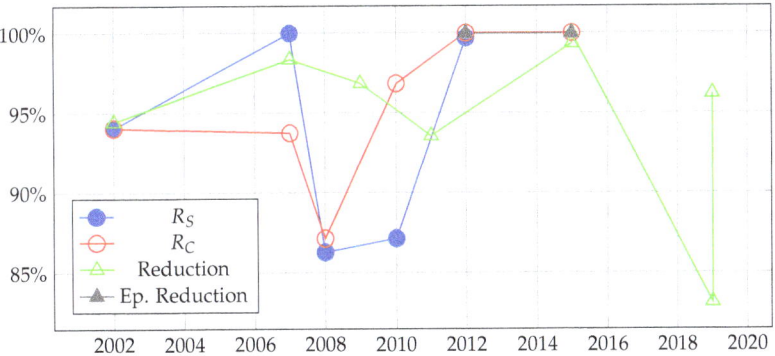

Figure 3. Scores achieved (percentages) on the DARPA 2000 LLDOS 1.0 scenario over years.

8. Discussion

We calculated new metrics for 9 articles, and verified others where possible, as can be seen in Tables 4–6. Soundness and completeness measures were not consistent for Liu et al. [37], where our calculations yielded lower values than reported in the original paper. We found that they may have made a typing error when presenting their correlation results. While calculating, they may have copied the number of correctly correlated results from the inside network in LLDOS 1.0 (44 in the paper), to the correctly correlated alert count from the DMZ portion of the scenario. If we assume that the latter number was actually 57 and not 44, the calculated results from the paper written in parenthesis are the correct values. In contrast, the calculation of episode reduction for Sadoddin and Ghorbani resulted in a slightly larger value than that presented in the paper, because the authors, possibly intentionally, made a rounding error. Because the reduction metrics shown in Tables 5 and 6 depend heavily on the characteristics of initial alerts raised by the IDS, the fact that many authors use different IDS systems with different configurations makes them an imprecise comparison criteria.

In Table 4, we chose to sort the papers according to calculated scores for completeness, R_C, because it reflects the absence of false negative errors, which are, in practice, much more dangerous than false positive errors. We expect that approaches exhibiting false negative errors in a controlled and sterile setting, such as a dataset, where the adversary is essentially a simple automaton, will be fairly easy to fool with novel and modified attack patterns when facing intelligent adversaries in real environments. Interestingly, if newer methods perform worse than older methods on some old datasets, they could still potentially perform better on contemporary attacks that are not represented in theese old datasets.

Knowledge-based approaches and approaches that learn attack patterns from datasets, are both types of signature recognition systems. As such, they must be constantly manually updated with new high-level attack patterns described by rules and/or new datasets in order to correlate alerts correctly. As a step in this direction, the authors of UNB datasets continuously publish new and updated datasets, as described in Section 5.4. Unfortunately, this procedure makes the comparison of attack detection approaches difficult, since older approaches would have to be re-evaluated each time that a new dataset is published in order to be comparable. A possible solution is to evaluate new approaches while using a larger number of datasets, including both older popular datasets and novel ones. To compensate for the inconsistent use of metrics, approaches should aim to publish detailed information on the correlation results, so that different evaluation metrics can be calculated independently if necessary. As shown in Tables 4 and 5, 4 approaches did not provide the underlying counts of the alerts and only published their final results.

Another problem in detecting high-level attacks is missed alarms, i.e., situations where IDS did not trigger alarms for an attack. This usually happens, because the IDS does not have the appropriate signatures. Ning et al. [66] addressed this by using domain knowledge to create hypotheses for missed alerts and later validated these hypotheses while using raw audit data, while Wang et al. [44] handled this using alarm similarity. As some of the surveyed papers suggest, one possible solution to this problem would be to integrate different data sources, such as anomaly detector alarms and event logs, and another would be to incorporate more knowledge about attackers and their high-level targets into the approach. As for the first solution, Sommer and Paxson [105] argue that a serious disadvantage of anomaly detectors based on machine learning is that ML is not suitable for meaningful outlier detection. To expand on this, the fact that an event is an outlier has little to do with the fact that it is malicious, and most of the outliers can be traced back to software and policy changes and benign network errors, resulting in a large proportion of anomalies being false positives. The second solution deals with the attacker's operations, tactics, and objectives, and may provide better quality, but it relies on large amounts of domain knowledge about attackers and continuous manual work. To quote all too often quoted Sun Tzu: "If you know others and know yourself, you will not be imperiled in a hundred battles". An example of a high-level model that is based on attacker tactics is the MITRE ATT&CK framework [106], which we encourage all readers to review.

Most of the metrics collected in this survey measure the technical characteristics of the approaches, rather than their impact on productivity in a security operations center (SOC). The only metric in the surveyed literature that focuses on user experience is QoA (7), along with its weighted variant described in [41]. Together with the CSA metrics that were proposed in [5], it is largely ignored in the surveyed papers. Most commercial systems today rely on human operators, combine subscription-based domain knowledge with anomaly detection to find attacks, and provide analytical capabilities for further manual validation and investigation.

Finally, approaches since 2004 mostly support real-time operation, and would be sufficent for detecting scripted attack scenarios of computer worms and low-profile attackers. Some limited projection of attacks is also possible. However, they are far from detecting and projecting attack scenarios of high-profile intelligent adversaries, such as Advanced Persistent Threats (APTs), and a lot of research in applications of AI is needed to cope with them.

9. Conclusions

This paper gave an overview of notable advances in high-level attack scenario detection and projection. It explained the context, reasoning, and practical problems that encouraged research in this area, and followed the evolution of proposals from the pioneering approaches to the state-of-the-art at the time of writing. Approaches relying on manually defined domain knowledge require constant maintenance and are, in most cases, not very flexible, while data mining and ML approaches highly depend on the quality of the available datasets. In this survey, emphasis was put on comparing the papers by calculating comparable measures of success on two most commonly used datasets, namely the DARPA 2000 dataset and the Treasure Hunt dataset. We suggest that future approaches evaluate on a larger number of datasets, including both the popular older ones, and novel ones, and focus more on the value they provide to the security personell. At the time of writing, the available datasets do not seem representative enough to enable the extraction of knowledge regarding attackers and attack scenarios aside from bot attacks. The tasks of attack scenario detection and projection have proven to be very challenging, and they require further research.

Author Contributions: Conceptualization, I.K. and S.G.; methodology, I.K.; validation, I.K.; formal analysis, I.K.; investigation, I.K. and K.S.; data curation, I.K.; Writing—Original draft preparation, I.K.; Writing—Review and editing, I.K., S.G., and K.S.; visualization, I.K. and K.S.; supervision, S.G.; project administration, S.G.; funding acquisition, S.G. All authors have read and agreed to the published version of the manuscript.

Funding: This research was carried out as a part of *Cyber Conflict Simulator* research and development project co-financed by the EU under the KK.01.2.1.01.0054 grant agreement.

Conflicts of Interest: The authors declare no conflict of interest. The funders had no role in the design of the study, in the collection, analyses, and interpretation of data, in the writing of the manuscript, and in the decision to publish the results.

Reference

1. Verizon. 2020 Data Breach Investigations Report (DBIR). 2020. Available online: https://enterprise.verizon.com/resources/reports/dbir/2020/introduction/ (accessed on 1 September 2020).
2. Bhatt, S.; Manadhata, P.K.; Zomlot, L. The operational role of security information and event management systems. *IEEE Secur. Priv.* **2014**, *12*, 35–41. [CrossRef]
3. MIT Lincoln Laboratory. 2000 DARPA Intrusion Detection Scenario Specific Datasets. Available online: https://www.ll.mit.edu/r-d/datasets/2000-darpa-intrusion-detection-scenario-specific-datasets (accessed on 11 August 2020).

4. Endsley, M.R. Design and evaluation for situation awareness enhancement. In *Proceedings of the Human Factors Society Annual Meeting*; Sage: Los Angeles, CA, USA, 1988; pp. 97–101.
5. Tadda, G.P.; Salerno, J.S. Overview of cyber situation awareness. In *Cyber Situational Awareness*; Springer: Boston, MA, USA, 2010; pp. 15–35.
6. Barford, P.; Dacier, M.; Dietterich, T.G.; Fredrikson, M.; Giffin, J.; Jajodia, S.; Jha, S.; Li, J.; Liu, P.; Ning, P.; et al. Cyber SA: Situational awareness for cyber defense. In *Cyber Situational Awareness*; Springer: Boston, MA, USA, 2010; pp. 3–13.
7. Axelsson, S. The base-rate fallacy and the difficulty of intrusion detection. *ACM Trans. Inf. Syst. Secur. (TISSEC)* **2000**, *3*, 186–205. [CrossRef]
8. Ilgun, K.; Kemmerer, R.A.; Porras, P.A. State transition analysis: A rule-based intrusion detection approach. *IEEE Trans. Softw. Eng.* **1995**, *21*, 181–199. [CrossRef]
9. Bass, T. Intrusion detection systems and multisensor data fusion. *Commun. ACM* **2000**, *43*, 99–105. [CrossRef]
10. Yang, S.J.; Du, H.; Holsopple, J.; Sudit, M. Attack projection. In *Cyber Defense and Situational Awareness*; Springer: New York, NY, USA, 2014; pp. 239–261.
11. Husák, M.; Komárková, J.; Bou-Harb, E.; Čeleda, P. Survey of attack projection, prediction, and forecasting in cyber security. *IEEE Commun. Surv. Tutor.* **2018**, *21*, 640–660. [CrossRef]
12. Ahmed, A.A.; Zaman, N.A.K. Attack Intention Recognition: A Review. *IJ Netw. Secur.* **2017**, *19*, 244–250.
13. Abdlhamed, M.; Kifayat, K.; Shi, Q.; Hurst, W. Intrusion prediction systems. In *Information Fusion for Cyber-Security Analytics*; Springer International Publishing: Cham, Switzerland, 2017; pp. 155–174.
14. Leau, Y.B.; Manickam, S. Network security situation prediction: A review and discussion. In *International Conference on Soft Computing, Intelligence Systems, and Information Technology*; Springer: Berlin/Heidelberg, Germany, 2015; pp. 424–435.
15. Li, J.; Ou, X.; Rajagopalan, R. Uncertainty and risk management in cyber situational awareness. In *Cyber Situational Awareness*; Springer: Boston, MA, USA, 2010; pp. 51–68.
16. Endsley, M.R.; Connors, E.S. Foundation and challenges. In *Cyber Defense and Situational Awareness*; Springer: New York, NY, USA, 2014; pp. 7–27.
17. Salah, S.; Maciá-Fernández, G.; DíAz-Verdejo, J.E. A model-based survey of alert correlation techniques. *Comput. Netw.* **2013**, *57*, 1289–1317. [CrossRef]
18. Hubballi, N.; Suryanarayanan, V. False alarm minimization techniques in signature-based intrusion detection systems: A survey. *Comput. Commun.* **2014**, *49*, 1–17. [CrossRef]
19. Husák, M.; Kašpar, J. Towards predicting cyber attacks using information exchange and data mining. In Proceedings of the 2018 14th International Wireless Communications & Mobile Computing Conference (IWCMC), Limassol, Cyprus, 25–29 June 2018; pp. 536–541.
20. Navarro, J.; Deruyver, A.; Parrend, P. A systematic survey on multi-step attack detection. *Comput. Secur.* **2018**, *76*, 214–249. [CrossRef]
21. Mirheidari, S.A.; Arshad, S.; Jalili, R. Alert correlation algorithms: A survey and taxonomy. In *Cyberspace Safety and Security*; Springer: New York, NY, USA, 2013; pp. 183–197.
22. Zhou, C.V.; Leckie, C.; Karunasekera, S. A survey of coordinated attacks and collaborative intrusion detection. *Comput. Secur.* **2010**, *29*, 124–140. [CrossRef]
23. Vasilomanolakis, E.; Karuppayah, S.; Mühlhäuser, M.; Fischer, M. Taxonomy and survey of collaborative intrusion detection. *ACM Comput. Surv. (CSUR)* **2015**, *47*, 1–33. [CrossRef]
24. Salerno, J. Measuring situation assessment performance through the activities of interest score. In Proceedings of the 2008 11th International Conference on Information Fusion, Cologne, Germany, 30 June–3 July 2008; pp. 1–8.
25. Valdes, A.; Skinner, K. Probabilistic alert correlation. In *International Workshop on Recent Advances in Intrusion Detection*; Springer: Berlin/Heidelberg, Germany, 2001; pp. 54–68.
26. Valeur, F.; Vigna, G.; Kruegel, C.; Kemmerer, R.A. Comprehensive approach to intrusion detection alert correlation. *IEEE Trans. Dependable Secur. Comput.* **2004**, *1*, 146–169. [CrossRef]
27. GhasemiGol, M.; Ghaemi-Bafghi, A. E-correlator: An entropy-based alert correlation system. *Secur. Commun. Netw.* **2015**, *8*, 822–836. [CrossRef]

28. Sadoddin, R.; Ghorbani, A.A. An incremental frequent structure mining framework for real-time alert correlation. *Comput. Secur.* **2009**, *28*, 153–173. [CrossRef]
29. Roschke, S.; Cheng, F.; Meinel, C. A new alert correlation algorithm based on attack graph. In *Computational Intelligence in Security for Information Systems*; Springer: Berlin/Heidelberg, Germany, 2011; pp. 58–67.
30. Zhang, Y.; Zhao, S.; Zhang, J. RTMA: Real time mining algorithm for multi-step attack scenarios reconstruction. In Proceedings of the 2019 IEEE 21st International Conference on High Performance Computing and Communications; IEEE 17th International Conference on Smart City; IEEE 5th International Conference on Data Science and Systems (HPCC/SmartCity/DSS), Zhangjiajie, China, 10–12 August 2019; pp. 2103–2110.
31. Ning, P.; Cui, Y.; Reeves, D.S. Constructing attack scenarios through correlation of intrusion alerts. In Proceedings of the 9th ACM Conference on Computer and Communications Security, Washington, DC, USA, 18–22 November 2002; pp. 245–254.
32. Ning, P.; Cui, Y.; Reeves, D.S. Analyzing intensive intrusion alerts via correlation. In *International Workshop on Recent Advances in Intrusion Detection*; Springer: Berlin/Heidelberg, Germany, 2002; pp. 74–94.
33. Ning, P.; Xu, D. Learning attack strategies from intrusion alerts. In Proceedings of the 10th ACM Conference on Computer and Communications Security, New York, NY, USA, 27–31 October 2003; pp. 200–209.
34. Ning, P.; Cui, Y.; Reeves, D.S.; Xu, D. Techniques and tools for analyzing intrusion alerts. *ACM Trans. Inf. Syst. Secur. (TISSEC)* **2004**, *7*, 274–318. [CrossRef]
35. Saad, S.; Traore, I. Extracting attack scenarios using intrusion semantics. In *International Symposium on Foundations and Practice of Security*; Springer: Berlin/Heidelberg, Germany, 2012; pp. 278–292.
36. Saad, S.; Traore, I. Semantic aware attack scenarios reconstruction. *J. Inf. Secur. Appl.* **2013**, *18*, 53–67. [CrossRef]
37. Liu, Z.; Wang, C.; Chen, S. Correlating multi-step attack and constructing attack scenarios based on attack pattern modeling. In Proceedings of the 2008 International Conference on Information Security and Assurance (ISA 2008), Busan, Korea, 24–26 April 2008; pp. 214–219.
38. Barzegar, M.; Shajari, M. Attack scenario reconstruction using intrusion semantics. *Expert Syst. Appl.* **2018**, *108*, 119–133. [CrossRef]
39. Soleimani, M.; Ghorbani, A.A. Multi-layer episode filtering for the multi-step attack detection. *Comput. Commun.* **2012**, *35*, 1368–1379. [CrossRef]
40. Ramaki, A.A.; Amini, M.; Atani, R.E. RTECA: Real time episode correlation algorithm for multi-step attack scenarios detection. *Comput. Secur.* **2015**, *49*, 206–219. [CrossRef]
41. Yu, D.; Frincke, D. Improving the quality of alerts and predicting intruder's next goal with Hidden Colored Petri-Net. *Comput. Netw.* **2007**, *51*, 632–654. [CrossRef]
42. Fava, D.S.; Byers, S.R.; Yang, S.J. Projecting cyberattacks through variable-length markov models. *IEEE Trans. Inf. Forensics Secur.* **2008**, *3*, 359–369. [CrossRef]
43. Farhadi, H.; AmirHaeri, M.; Khansari, M. Alert correlation and prediction using data mining and HMM. *ISeCure ISC Int. J. Inf. Secur.* **2011**, *3*, 77–101.
44. Wang, L.; Liu, A.; Jajodia, S. An efficient and unified approach to correlating, hypothesizing, and predicting intrusion alerts. In *European Symposium on Research in Computer Security*; Springer: Berlin/Heidelberg, Germany, 2005; pp. 247–266.
45. Kendall, K.K.R. A Database of Computer Attacks for the Evaluation of Intrusion Detection Systems. Ph.D. Thesis, Massachusetts Institute of Technology, Cambridge, MA, USA, 1999.
46. MIT Lincoln Laboratory. 1999 DARPA Intrusion Detection Evaluation Dataset. 1999. Available online: https://www.ll.mit.edu/r-d/datasets/1999-darpa-intrusion-detection-evaluation-dataset (accessed on 11 August 2020).
47. DEF CON. DEF CON. Available online: https://www.defcon.org (accessed on 26 August 2020).
48. DEF CON. DEF CON CTF Archive. Available online: https://www.defcon.org/html/links/dc-ctf.html (accessed on 11 August 2020).
49. Gharib, A.; Sharafaldin, I.; Lashkari, A.H.; Ghorbani, A.A. An evaluation framework for intrusion detection dataset. In Proceedings of the 2016 International Conference on Information Science and Security (ICISS), Pattaya, Thailand, 19–22 December 2016; pp. 1–6.
50. Vigna, G. Teaching hands-on network security: Testbeds and live exercises. *J. Inf. Warf.* **2003**, *2*, 8–24.

51. Vigna, G.; Eckmann, S.T.; Kemmerer, R.A. Attack languages. In Proceedings of the IEEE Information Survivability Workshop, Boston, MA, USA, 24–26 October 2000.
52. McHugh, J. Testing intrusion detection systems: A critique of the 1998 and 1999 darpa intrusion detection system evaluations as performed by lincoln laboratory. *ACM Trans. Inf. Syst. Secur. (TISSEC)* **2000**, *3*, 262–294. [CrossRef]
53. Brown, C.; Cowperthwaite, A.; Hijazi, A.; Somayaji, A. Analysis of the 1999 darpa/lincoln laboratory ids evaluation data with netadhict. In Proceedings of the 2009 IEEE Symposium on Computational Intelligence for Security and Defense Applications, Ottawa, ON, Canada, 8–10 July 2009; pp. 1–7.
54. Shiravi, A.; Shiravi, H.; Tavallaee, M.; Ghorbani, A.A. Toward developing a systematic approach to generate benchmark datasets for intrusion detection. *Comput. Secur.* **2012**, *31*, 357–374. [CrossRef]
55. Sharafaldin, I.; Lashkari, A.H.; Ghorbani, A.A. Toward Generating a New Intrusion Detection Dataset and Intrusion Traffic Characterization. In Proceedings of the 4th International Conference on Information Systems Security and Privacy (ICISSP 2018), Funchal, Portugal, 22–24 January 2018; pp. 108–116.
56. Provos, N. A Virtual Honeypot Framework. In Proceedings of the USENIX Security Symposium, San Diego, CA, USA, 9–13 August 2004; pp. 1–14.
57. NETRESEC AB. Publicly Available PCAP Files. 2015. Available online: https://www.netresec.com/?page=PcapFiles (accessed on 13 May 2020).
58. Porras, P.A.; Fong, M.W.; Valdes, A. A mission-impact-based approach to INFOSEC alarm correlation. In *International Workshop on Recent Advances in Intrusion Detection*; Springer: Berlin/Heidelberg, Germany, 2002; pp. 95–114.
59. Hossain, M.N.; Milajerdi, S.M.; Wang, J.; Eshete, B.; Gjomemo, R.; Sekar, R.; Stoller, S.; Venkatakrishnan, V. SLEUTH: Real-time attack scenario reconstruction from COTS audit data. In Proceedings of the 26th USENIX Security Symposium (USENIX Security 17), Vancouver, BC, Canada, 16–18 August 2017; pp. 487–504.
60. Cuppens, F.; Miege, A. Alert correlation in a cooperative intrusion detection framework. In Proceedings of the 2002 IEEE Symposium on Security and Privacy, Berkeley, CA, USA, 12–15 May 2002; pp. 202–215.
61. Morin, B.; Debar, H. Correlation of intrusion symptoms: An application of chronicles. In *International Workshop on Recent Advances in Intrusion Detection*; Springer: Berlin/Heidelberg, Germany, 2003; pp. 94–112.
62. Morin, B.; Mé, L.; Debar, H.; Ducassé, M. M2D2: A formal data model for IDS alert correlation. In *International Workshop on Recent Advances in Intrusion Detection*; Springer: Berlin/Heidelberg, Germany, 2002; pp. 115–137.
63. Cuppens, F. Managing alerts in a multi-intrusion detection environment. In Proceedings of the Seventeenth Annual Computer Security Applications Conference, New Orleans, LA, USA, 10–14 December 2001.
64. Ghafir, I.; Kyriakopoulos, K.G.; Lambotharan, S.; Aparicio-Navarro, F.J.; AsSadhan, B.; BinSalleeh, H.; Diab, D.M. Hidden Markov models and alert correlations for the prediction of advanced persistent threats. *IEEE Access* **2019**, *7*, 99508–99520. [CrossRef]
65. Wang, L.; Liu, A.; Jajodia, S. Using attack graphs for correlating, hypothesizing, and predicting intrusion alerts. *Comput. Commun.* **2006**, *29*, 2917–2933. [CrossRef]
66. Ning, P.; Xu, D.; Healey, C.G.; Amant, R.S. Building Attack Scenarios through Integration of Complementary Alert Correlation Method. 2004. Available online: http://citeseerx.ist.psu.edu/viewdoc/summary?doi=10.1.1.60.4412 (accessed on 19 October 2020).
67. Cheung, S.; Lindqvist, U.; Fong, M.W. Modeling multistep cyber attacks for scenario recognition. In Proceedings of the DARPA Information Survivability Conference and Exposition, Washington, DC, USA, 22–24 April 2003; pp. 284–292.
68. Qin, X.; Lee, W. Statistical causality analysis of infosec alert data. In *International Workshop on Recent Advances in Intrusion Detection*; Springer: Berlin/Heidelberg, Germany, 2003; pp. 73–93.
69. Qin, X.; Lee, W. Attack plan recognition and prediction using causal networks. In Proceedings of the 20th Annual Computer Security Applications Conference, Tucson, AZ, USA, 6–10 December 2004; pp. 370–379.
70. Zhu, B.; Ghorbani, A.A. Alert correlation for extracting attack strategies. *IJ Netw. Secur.* **2006**, *3*, 244–258.
71. Pietraszek, T.; Tanner, A. Data mining and machine learning—Towards reducing false positives in intrusion detection. *Inf. Secur. Tech. Rep.* **2005**, *10*, 169–183. [CrossRef]

72. Holgado, P.; Villagrá, V.A.; Vazquez, L. Real-time multistep attack prediction based on hidden markov models. *IEEE Trans. Dependable Secur. Comput.* **2017**, *17*, 134–147.
73. Milajerdi, S.M.; Gjomemo, R.; Eshete, B.; Sekar, R.; Venkatakrishnan, V. Holmes: Real-time apt detection through correlation of suspicious information flows. In Proceedings of the 2019 IEEE Symposium on Security and Privacy (SP), San Francisco, CA, USA, 19–23 May 2019; pp. 1137–1152.
74. Hughes, T.; Sheyner, O. Attack scenario graphs for computer network threat analysis and prediction. *Complexity* **2003**, *9*, 15–18. [CrossRef]
75. Geib, C.W.; Goldman, R.P. Plan recognition in intrusion detection systems. In Proceedings of the DARPA Information Survivability Conference and Exposition II. DISCEX'01, Anaheim, CA, USA, 12–14 June 2001; pp. 46–55.
76. Ren, H.; Stakhanova, N.; Ghorbani, A.A. An online adaptive approach to alert correlation. In *International Conference on Detection of Intrusions and Malware, and Vulnerability Assessment*; Springer: Berlin/Heidelberg, Germany, 2010; pp. 153–172.
77. Dain, O.M.; Cunningham, R.K. Building scenarios from a heterogeneous alert stream. In *Proceedings of the 2001 IEEE Workshop on Information Assurance and Security*; United States Military Academy: West Point, NY, USA, 2001.
78. Dain, O.; Cunningham, R.K. Fusing a heterogeneous alert stream into scenarios. In *Applications of Data Mining in Computer Security*; Springer: Boston, MA, USA, 2002; pp. 103–122.
79. Noel, S.; Robertson, E.; Jajodia, S. Correlating intrusion events and building attack scenarios through attack graph distances. In Proceedings of the 20th Annual Computer Security Applications Conference, Tucson, AZ, USA, 6–10 December 2004; pp. 350–359.
80. Morin, B.; Mé, L.; Debar, H.; Ducassé, M. A logic-based model to support alert correlation in intrusion detection. *Inf. Fusion* **2009**, *10*, 285–299. [CrossRef]
81. Cuppens, F.; Ortalo, R. Lambda: A language to model a database for detection of attacks. In *International Workshop on Recent Advances in Intrusion Detection*; Springer: Berlin/Heidelberg, Germany, 2000; pp. 197–216.
82. Templeton, S.J.; Levitt, K. A requires/provides model for computer attacks. In *NSPW '00: Proceedings of the 2000 Workshop on New Security Paradigms*; Association for Computing Machinery: New York, NY, USA, 2001; pp. 31–38.
83. Cuppens, F.; Autrel, F.; Miege, A.; Benferhat, S. Correlation in an intrusion detection process. In Proceedings of the Internet Security Communication Workshop, Tunis, Tunisia, 19–21 September 2002; pp. 153–172.
84. Goldman, R.P.; Heimerdinger, W.; Harp, S.A.; Geib, C.W.; Thomas, V.; Carter, R.L. Information modeling for intrusion report aggregation. In Proceedings of the DARPA Information Survivability Conference and Exposition II. DISCEX'01, Anaheim, CA, USA, 12–14 June 2001; pp. 329–342.
85. Valdes, A.D.J.; Skinner, K. Probabilistic Alert Correlation. U.S. Patent 7,917,393, 29 March 2011.
86. Eckmann, S.T.; Vigna, G.; Kemmerer, R.A. STATL: An attack language for state-based intrusion detection. *J. Comput. Secur.* **2002**, *10*, 71–103. [CrossRef]
87. Ghafir, I.; Hammoudeh, M.; Prenosil, V.; Han, L.; Hegarty, R.; Rabie, K.; Aparicio-Navarro, F.J. Detection of advanced persistent threat using machine-learning correlation analysis. *Future Gener. Comput. Syst.* **2018**, *89*, 349–359. [CrossRef]
88. Ramaki, A.A.; Khosravi-Farmad, M.; Bafghi, A.G. Real time alert correlation and prediction using Bayesian networks. In Proceedings of the 2015 12th International Iranian Society of Cryptology Conference on Information Security and Cryptology (ISCISC), Rasht, Iran, 8–10 September 2015; pp. 98–103.
89. Albanese, M.; Jajodia, S.; Pugliese, A.; Subrahmanian, V. Scalable analysis of attack scenarios. In *European Symposium on Research in Computer Security*; Springer: Berlin/Heidelberg, Germany, 2011; pp. 416–433.
90. Fredj, O.B. A realistic graph-based alert correlation system. *Secur. Commun. Netw.* **2015**, *8*, 2477–2493. [CrossRef]
91. Ahmadinejad, S.H.; Jalili, S.; Abadi, M. A hybrid model for correlating alerts of known and unknown attack scenarios and updating attack graphs. *Comput. Netw.* **2011**, *55*, 2221–2240. [CrossRef]
92. Alhaj, T.A.; Siraj, M.M.; Zainal, A.; Elshoush, H.T.; Elhaj, F. Feature selection using information gain for improved structural-based alert correlation. *PLoS ONE* **2016**, *11*, e0166017. [CrossRef]

93. Zhang, K.; Zhao, F.; Luo, S.; Xin, Y.; Zhu, H. An intrusion action-based IDS alert correlation analysis and prediction framework. *IEEE Access* **2019**, *7*, 150540–150551. [CrossRef]
94. Hu, H.; Liu, J.; Zhang, Y.; Liu, Y.; Xu, X.; Huang, J. Attack scenario reconstruction approach using attack graph and alert data mining. *J. Inf. Secur. Appl.* **2020**, *54*, 102522. [CrossRef]
95. Huang, M.Y.; Jasper, R.J.; Wicks, T.M. A large scale distributed intrusion detection framework based on attack strategy analysis. *Comput. Netw.* **1999**, *31*, 2465–2475. [CrossRef]
96. Albanese, M.; Erbacher, R.F.; Jajodia, S.; Molinaro, C.; Persia, F.; Picariello, A.; Sperlì, G.; Subrahmanian, V. Recognizing unexplained behavior in network traffic. In *Network Science and Cybersecurity*; Springer: New York, NY, USA, 2014; pp. 39–62.
97. Dousson, C. Extending and unifying chronicle representation with event counters. In Proceedings of the 15th European Conference on Artificial Intelligence, Lyon, France, 21–26 July 2002; IOS Press: Amsterdam, The Netherlands, 2002; pp. 257–261.
98. Lindqvist, U.; Porras, P.A. Detecting computer and network misuse through the production-based expert system toolset (P-BEST). In Proceedings of the 1999 IEEE Symposium on Security and Privacy (Cat. No. 99CB36344), Oakland, CA, USA, 14 May 1999; pp. 146–161.
99. Mandiant, A. Exposing One of China's Cyber Espionage Units. 2013. Available online: https://www.fireeye.com/content/dam/fireeye-www/services/pdfs/mandiant-apt1-report.pdf (accessed on 13 May 2020).
100. Keromytis, A.D. Transparent Computing Engagement 3 Data Release. 2018. Available online: https://github.com/darpa-i2o/Transparent-Computing (accessed on 8 October 2020).
101. Ester, M.; Kriegel, H.P.; Sander, J.; Xu, X. *A Density-Based Algorithm for Discovering Clusters in Large Spatial Databases with Noise*; Kdd: Seattle, WA, USA, 1996; pp. 226–231.
102. *Common Vulnerabilities and Exposures (CVE)*; The MITRE Corporation: McLean, VA, USA; Available online: https://cve.mitre.org/ (accessed on 1 September 2020).
103. CESNET. 1998 DARPA Intrusion Detection Evaluation Dataset. 2016. Available online: https://sabu.cesnet.cz/en/start (accessed on 21 August 2020).
104. Ou, X.; Govindavajhala, S.; Appel, A.W. MulVAL: A Logic-Based Network Security Analyzer. In Proceedings of the USENIX Security Symposium, Baltimore, MD, USA, 31 July–5 August 2005; pp. 113–128.
105. Sommer, R.; Paxson, V. Outside the closed world: On using machine learning for network intrusion detection. In Proceedings of the 2010 IEEE Symposium on Security and Privacy, Oakland, CA, USA, 16–19 May 2010; pp. 305–316.
106. Strom, B.E.; Applebaum, A.; Miller, D.P.; Nickels, K.C.; Pennington, A.G.; Thomas, C.B. *Mitre Att&ck: Design and Philosophy*; Technical Report; The MITRE Corporation: McLean, VA, USA, 2018. Available online: https://www.mitre.org/sites/default/files/publications/pr-18-0944-11-mitre-attack-design-and-philosophy.pdf (accessed on 1 September 2020).

Publisher's Note: MDPI stays neutral with regard to jurisdictional claims in published maps and institutional affiliations.

© 2020 by the authors. Licensee MDPI, Basel, Switzerland. This article is an open access article distributed under the terms and conditions of the Creative Commons Attribution (CC BY) license (http://creativecommons.org/licenses/by/4.0/).

Article

Resilience Evaluation of Multi-Path Routing against Network Attacks and Failures

Hyok An [1], Yoonjong Na [1], Heejo Lee [1,*] and Adrian Perrig [2]

[1] Department of Computer Science and Engineering, Korea University, Seoul 02841, Korea; anhyok@korea.ac.kr (H.A.); nooryyaa@korea.ac.kr (Y.N.)
[2] Department of Computer Science, ETH Zurich, 8092 Zurich, Switzerland; adrian.perrig@inf.ethz.ch
* Correspondence: heejo@korea.ac.kr

Abstract: The current state of security and availability of the Internet is far from being commensurate with its importance. The number and strength of DDoS attacks conducted at the network layer have been steadily increasing. However, the single path (SP) routing used in today's Internet lacks a mitigation scheme to rapidly recover from network attacks or link failure. In case of a link failure occurs, it can take several minutes until failover. In contrast, multi-path routing can take advantage of multiple alternative paths and rapidly switch to another working path. According to the level of available path control, we classify the multi-path routing into two types, first-hop multi-path (FMP) and multi-hop multi-path (MMP) routing. Although FMP routing supported by networks, such as SD-WAN, shows marginal improvements over the current SP routing of the Internet, MMP routing supported by a global Internet architecture provides strong improvement under network attacks and link failure. MMP routing enables changing to alternate paths to mitigate the network problem in other hops, which cannot be controlled by FMP routing. To show this comparison with practical outcome, we evaluate network performance in terms of latency and loss rate to show that MMP routing can mitigate Internet hazards and provide high availability on global networks by 18 participating ASes in six countries. Our evaluation of global networks shows that, if network attacks or failures occur in other autonomous systems (ASes) that FMP routing cannot avoid, it is feasible to deal with such problems by switching to alternative paths by using MMP routing. When the global evaluation is under a transit-link DDoS attack, the loss rates of FMP that pass the transit-link are affected significantly by a transit-link DDoS attack, but the other alternative MMP paths show stable status under the DDoS attack with proper operation.

Keywords: network security; multi-path routing; high availability; Internet-scale evaluation

Citation: An, H.; Na, Y.; Lee, H.; Perrig, A. Resilience Evaluation of Multi-Path Routing against Network Attacks and Failures. *Electronics* **2021**, *10*, 1240. https://doi.org/10.3390/electronics10111240

Academic Editor: Victor A. Villagrá

Received: 15 April 2021
Accepted: 19 May 2021
Published: 24 May 2021

Publisher's Note: MDPI stays neutral with regard to jurisdictional claims in published maps and institutional affiliations.

Copyright: © 2021 by the authors. Licensee MDPI, Basel, Switzerland. This article is an open access article distributed under the terms and conditions of the Creative Commons Attribution (CC BY) license (https://creativecommons.org/licenses/by/4.0/).

1. Introduction

1.1. Motivations

The Internet was not designed to maintain high availability in the face of malicious activities. Recent patches to improve Internet security and availability have been constrained by the current Internet architecture design. As network-based attacks on availability continue to increase each year [1], an improvised Internet architecture should offer availability and security based on its design, provide incentives for deployment, and consider economic, political, and legal issues at the design stage [2].

Currently, the Internet provides single path (SP) routing, which lacks a rapid mitigation mechanism to counter Internet hazards, such as network congestion, link failures, or DDoS attacks. Under network congestion or DDoS attacks, the performance of the Internet, such as latency or loss rate, is highly affected, and, in the case of a link failure, it can take several minutes until failover [3]. Fault recovery of the border gateway protocol (BGP) at times takes several minutes before the routes converge to a consistent form [4]. Path outages even lead to significant disruptions in communication, which may last tens of minutes or longer [5–7]. However, if such a network congestion or network failure occurs from a link that an autonomous system (AS) cannot handle, sophisticated operations or cooperation among AS administrators will be required. Detailed routing information is maintained only within a single AS. The information shared with other providers and ASes is heavily filtered and summarized using BGP running at the border routers between ASes [8].

1.2. Multi-Path Routing Approaches

Many approaches have been proposed to solve the problem from network attacks and failures using concepts of multiple paths. A link failure can occur in several ways, such as through natural hazards or by a DDoS attack. In multihoming [9], a network is connected to multiple providers and uses its own range of addresses. If one of the links fails, the protocol recognizes the failure and reconfigures its routing tables. Although multihoming provides a way to connect multiple providers, it is not a multi-path routing but rather an SP routing because we cannot select the connection to the providers for every communication instantly. Resilient overlay networks (RON) [10] are a remedy for potential problems, such as BGP's fault or path outages, because BGP hides information about traffic conditions and topological details in the interests of scalability and policy enforcement. RON can often find paths between its nodes, even while wide-area Internet routing protocols, such as BGP, cannot. This RON approach is an overlay network that selects different waypoints, but it cannot select the actual network path. Multi-path TCP (MPTCP) is the de-facto standard multi-path transport protocol [11]. MPTCT can only make use of a single path per local network interface and, therefore, does not permit path choice as in MMP.

Software Defined WANs (SD-WAN) is a multi-path approach that can mitigate these problems by providing multiple routing paths from the first hop. However, if network congestion or a link failure occurs in other hops through a hazard or adversary that a victim AS cannot control, the SD-WAN will be unable to deal with the problem as it the SD-WAN has no control over other such hops, which can be described as the first type of multi-path routing, first-hop multi-path (FMP) routing. Figure 1 contrasts fully disjoint and partially disjoint paths. In Figure 1a, source AS *Src* has two links—one to next AS A_1, and the other to AS A_2. The AS following A_1 and A_2 is AS *B*. Source AS *Src* has two paths to destination AS *Dst*. However, the two paths use the same links between AS *B* to AS *Dst*. These paths can be disconnected by only one link, that is, link B-C or C-Dst. Figure 1b shows the non-overlapping paths from *Src* to *Dst*. As we can see, we cannot guarantee that the given paths are non-overlapped if the paths are established only by source AS's link selection. The second type of multi-path routing is multi-hop multi-path (MMP) routing, which can take advantage of controlling the paths available between a source and a destination. MMP can change to an alternative path to mitigate the problem, if hazards or attacks occur in other hops that a victim AS cannot control. However, to implement MMP on the current Internet architecture, a redesign of the Internet architecture is required.

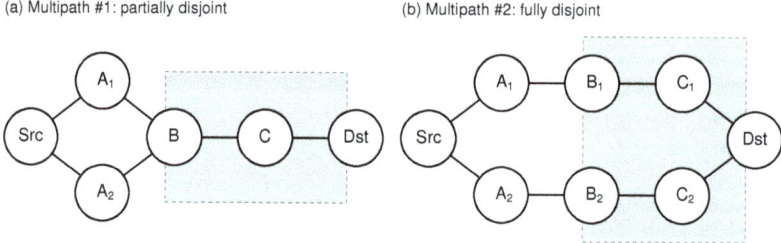

Figure 1. Source AS *Src* has two links to next hop ASes A_1 or A_2 but (**a**) shows partially disjoint paths that the multi-paths between *Src* and *Dst* uses same links from AS *B* to AS *Dst*, which can be disconnected by a single link failure, whereas (**b**) shows fully disjoint paths that cannot be disconnected by a single link failure.

1.3. Objectives

We classify multi-path routing to be of two types—FMP and MMP—based on path control and verify that the effectiveness of multi-path with real networks deployed globally, especially MMP routing, shows more resilience than FMP used currently. MMP routing enables changing to alternate paths to mitigate the network problem in other hops, which cannot be controlled by FMP routing. In this study, we evaluated the network performance in terms of latency and loss rate [12].

These results indicate that MMP can mitigate a DDoS attack of a transit-link and provide high availability on a global network using scalability, control, and isolation on next-generation networks (SCION) [13,14]. SCIONLab [15,16] is a global research network used to test SCION. We can join the SCION network using our own computation resources and set up and run our own ASes. An AS will actively participate in routing in a SCIONLab network and enable realistic experimentation using the unique properties of the SCION architecture. We conducted an evaluation of multi-path selection on the MMP and compared it with FMP. For a global evaluation, we used a SCIONlab network with nodes from six countries, that is, the United States, Switzerland, Germany, the Republic of Korea, Singapore, and Japan, and considered 18 SCION ASes. For a multi-path test, we assume that there is an attacker attempting to consume the entire bandwidth for communication. To compare MMP with FMP, a situation is assumed in which heavy traffic occurs among the ASes affecting both links configured for SD-WAN, but which cannot be controlled by it.

1.4. Contributions

The contributions of our study are threefold:

1. We classify multi-path routing to be of two types—FMP and MMP—based on path control. Because SP routing lacks rapid mitigation mechanisms to counter Internet hazards, FMP routing can mitigate these problems by selecting a connection link to the first hop beyond its control. However, it cannot deal with a problem in another hop. To mitigate this, MMP routing can take advantage of controlling all paths available between a source and a destination.
2. We verify that the effectiveness of multi-path with real networks deployed globally, especially MMP routing, shows more resilience than FMP used currently. Using the global network, we show that FMP mitigates certain cases of network problems, but MMP gains impressively lower loss rate by comparison.
3. When the global evaluation is under a transit-link DDoS attack, the loss rates of FMP that pass the transit-link are affected significantly by a transit-link DDoS attack, but the other alternative MMP paths show stable status under the DDoS attack with proper operation under 1% loss rates. Meanwhile, the impact to loss rate of FMP is over 40% by a bandwidth of 700 Mbps.

2. Background and Related Work

2.1. Link Failure and Transit-Link DDoS Attack

A link failure can occur in several ways, such as through natural hazards or a DDoS attack. Rerouting is one solution for such a case. Unlike a traditional DDoS attack, Comelt [17] or CrossFire [18] have shown that, by attacking only the core links, an adversary can effectively degrade the victim's network connectivity or cause it to fail using only a small number of resources. Such DDoS attacks are difficult for a victim to detect because their targets are usually several hops away from the victim's AS and are typically not controllable by the victim. However, if a DDoS attack or a link failure occurs from a link that a victim AS has no control over, rerouting from a traditional Internet environment would be a challenge.

2.2. Multihoming

In multihoming [9], a network is connected to multiple providers and uses its own range of addresses. The edge routers of the network communicate with the providers using a dynamic routing protocol, typically BGP. BGP announces its own network address range to all providers. If one of the links fails, the protocol recognizes the failure and re-configures its routing tables. Multihoming requires the use of an address space that is accepted by all providers and causes an increase in the global routing table. Although multihoming provides a way to connect multiple providers, it is not a multi-path routing but rather an SP routing because we cannot immediately select the connection to the providers for every communication.

2.3. Multi-Path Routing

Multi-path TCP (MPTCP) is the de-facto standard multi-path transport protocol [11]. MPTCT can only make use of a single path per local network interface and, thus, does not permit path choise as in MMP. The partial-reliability extension of the stream control transmission protocol (PR-SCTP) offers a primitive the possibility to define a lifetime parameter [19,20]. Liu et al. [21] used linear programming to evaluate multi-path routing from a traffic engineering perspective. Although PR-SCTP offers multihoming capabilities, additional IP addresses are used as a backup in the case of a failure; thus, PR-SCTP is not a fully multi-path protocol and does not address the problem that we describe in this paper [22].

In Reference [23], the authors consider the problem of routing data over multiple disjoint paths and propose a framework for multi-path routing in mobile ad hoc networking (MANET). In Reference [24], they develop an analytical framework for evaluating multi-path routing, and Reference [25] shows the Secure Message Transmission (SMT) protocol, which safeguards the data transmission against arbitrary malicious behavior of other nodes in MANET. Recently, flying ad hoc networks (FANETs) are rapidly proliferating and leading the emergence of the Internet of Drones and its applications. Reference [26] proposes a jamming-resilient multi-path routing protocol, also called JarmRout so that intentional jamming and disruption or isolated and localized failures do not interrupt the overall network performance of FANETs. The JarmRout is designed based on a combination of three major schemes, which are link quality scheme, traffic load scheme, and spatial distance scheme, to select maximally spatial node-disjoint multiple paths with high link quality and light traffic load to deliver the data packets from an source to destination nodes. In our paper, we focus on multi-path for the global network, not the local network.

2.4. Cognitive Packet Network

The Cognitive Packet Network (CPN) is a Quality of Service (QoS)-driven routing protocol, in which each flow specifies the QoS metric (e.g., delay, loss, jitter, or other composite metrics) that it wishes to optimize [27]. The work describes an experimental system that implements multi-path routing. In Reference [28], the authors show how such multi-path routing schemes can be used to mitigate and block network attacks, and in Reference [29] the dynamic recovery and QoS robustness of multi-path routing schemes is demonstrated in the presence of attacks. Reference [30] addresses one of the challenges of multi-path policies in that they create transients in multi-hop systems. In the recent papers by Reference [31,32], the use of multi-path routing to mitigate and recover from attacks with the help of software defined networking (SDN) controllers is discussed, and demonstrated with experimental results. The work in Reference [33] details how SDN controllers can be exploited together with their dynamic ability to vary paths for given connections, to optimize the access to Fog services. The SDN system in the test-bed was extended using the "cognitive packet routing algorithm" [34]. In our paper, we conider and discuss SD-WAN as WAN of SDN on multi-path for the global network.

2.5. SD-WAN: Software Defined WANs

SDN continues to attract both industry and research communities as a modern paradigm for the management and operation of computer networks. SDN is changing how computer networks are managed by providing a centralized network management. SDN separates the control plane of the network with different devices providing a centralized view of the network from the data plane [35]. Conceptually, SDN supports large-scale network control with simple operations and is a new method for preventing network-based attacks [36].

SD-WAN is the combination of a group of network technologies (overlay, performance routing, and firewalling) that uses software to make a WAN more intelligent, flexible, and easier to manage. The key benefits of a SD-WAN are flexibility and intelligence. SD-WAN is an overlay that enables the use of different underlays, i.e., the Internet, MPLS, VPLS, and LTE. For customers building a SD-WAN over the Internet, they can choose their best partner in a location-by-location manner. This also means that customers can mix the underlays and inherit their key benefits, such as the reliability of MPLS and the low cost of Internet access. SD-WAN solutions have been designed from scratch with the application being the key object to manipulate. Thus, monitoring, prioritization, and routing use applicative parameters in their policies and are fully orchestrated. SD-WAN policies, parameters, and topologies are centrally managed in real-time.

The quality of SD-WAN depends on the underlays: for multinational customers looking for savings on their access budget, and thus considering the low-cost last-mile of the Internet, the user experience may be impacted. Because SD-WAN cannot control what happens between selected Internet service providers (ISPs), customers may even lose connections between sites for minutes or even hours. Furthermore, SD-WAN solutions are not interoperable, and customers are vendor-locked. SD-WAN is a growth relay for equipment vendors, system integrators, and service providers and is a way for the cloud service providers to get closer to their customers. Typically, SD-WANs are operated using intra-domain routing protocols.

2.6. Resilient Overlay Networks

RON [10] is a remedy for potential problems, such as BGP's fault or path outages, because BGP hides information about traffic conditions and topological details in the interests of scalability and policy enforcement. Distributed applications layer a "resilient overlay network" over the underlying Internet routing substrate. The nodes comprising a RON reside in a variety of routing domains and cooperate with each other to forward data on behalf of any pair of communicating nodes in the RON. Because ASes are independently administrated and configured, and routing domains rarely share interior links, they generally fail independently of each other. As a result, if the underlying topology has physical path redundancy, RON can often find paths between its nodes, even when wide-area routing Internet protocols, such as BGP, cannot. This RON approach is an overlay network which selects different way-points, but it cannot select the actual network path.

2.7. BGP Poisoning

BGP poisoning is a rerouting technique used in BGP. By using BGP poisoning, an AS can change the routing path that it has no control over. Instead of multi-path routing, BGP poisoning can be another solution for hazards or attacks occurring in an AS that the administrator has no control. For example, Nyx uses BGP poisoning as a countermeasure for a link failure, and LIFEGUARD uses BGP poisoning to recover from a routing failure [37,38]. However, BGP poisoning is not a standard technique. BGP poisoning can also be used as an offensive method, such as BGP hijacking [39].

2.8. SCION Multi-Path Routing Architecture

SCION is the first clean-slate Internet architecture designed to provide route control, failure isolation, and explicit trust information for end-to-end communication. SCION organizes existing ASes into groups of independent routing planes, called isolation domains (ISD), which interconnect to provide global connectivity. Isolation domains provide natural isolation of the routing failures and misconfigurations, give endpoints strong control for both inbound and outbound traffic, provide meaningful and enforceable trust, and enable scalable routing updates with a high path freshness. As a result, the SCION architecture provides strong resilience and security properties as an intrinsic consequence of its design. In addition to high security, SCION also provides a scalable routing infrastructure, and efficient packet forwarding.

As a path-based architecture, SCION end hosts learn about available network path segments, and combine them into end-to-end paths that are carried in the packet headers. Path construction embedded cryptographic mechanisms are constrained to the route policies of the ISPs and receivers, offering a path choice to all parties, i.e., senders, receivers, and ISPs. This approach enables path-aware communication, an emerging trend in networking. These features also enable multi-path communication, which is an important approach for high availability, rapid failover in the case of network failures, increased end-to-end bandwidth, dynamic traffic optimization, and resilience to DDoS attacks [14]. SCION enables a native network-wide multi-path routing, allowing traffic differentiation based on latency and the bandwidth requirements, i.e., latency-critical traffic can be sent over the satellites, and the remaining traffic can be forwarded on a terrestrial network, thus improving the cost-effectiveness of the system [40].

3. Problem Statement

Network congestion occurs owing to traffic-based bandwidth consumption. Congestion is caused by malicious attackers or flash crowds [10] of innocent users. Under network congestion, that is, when the performance of the Internet, including the latency or loss rate, is highly affected by a hazard or when a link failure occurs, it can take several minutes or more until failover [3]. Fault recovery of BGP at times takes several minutes before the routes converge to a consistent form [4]. Path outages even lead to significant disruptions in communication lasting tens of minutes or longer [5–7].

3.1. Single Path or Multi-Path Routing

Multihoming, SD-WAN and SCION appear to be multi-path approaches when viewed we simply look at them as a topology because it is feasible to make multiple connections to other ASes using them. However, if we consider them in terms of routing, each approach operates differently. For clarity, we define here the meanings of SP, FMP, and MMP, including their major differences from the routing mechanism of multihoming, SD-WAN, and SCION.

3.1.1. Single Path (SP)

SP routing is provided by the current Internet architecture. It lacks rapid mitigation mechanisms to counter Internet hazards, such as network congestion or link failures. For example, multihoming is connected to multiple providers, and the edge routers of the network communicate with the providers using a dynamic routing protocol. If one of the links fails, the protocol re-configures its routing tables for a new single path.

3.1.2. First-Hop Multi-Path (FMP)

FMP routing can mitigate these problems by selecting a connection link to the first hop. However, if network congestion or a link failure occurs in another hop, it cannot deal with the problem because the SD-WAN does not have the authority to control the path in the other hops. For example, SD-WAN is an overlay with the ability to use different underlays, that is, the Internet, MPLS, VPLS, and LTE. Because SD-WAN cannot control all that is happening between the selected internet service providers (ISPs), customers may even lose connections to sites for minutes or even hours.

3.1.3. Multi-Hop Multi-Path (MMP)

MMP routing can control all paths available between a source and a destination. This is a crucial approach to high availability, rapid failover in the case of network failures, increased end-to-end bandwidth, dynamic traffic optimization, and resilience to

DDoS attacks. SCION enables native network-wide multi-path routing, allowing traffic differentiation based on latency and bandwidth requirements.

3.2. Types of Multi-Path Routing: FMP and MMP

Although FMP and MMP are not directly comparable, SD-WAN offers application-aware properties (DPI, encryption, and firewalling) that can also be used with SCION. However, SCION offers a super-set of SD-WAN properties at the underlay level: enhanced path selection, trust in the network, and real end-to-end control (not only the first hop). As the core difference, MMP is a public global Internet infrastructure with end-to-end control, whereas FMP providers typically build private networks within the first hop (see Figure 2). An MMP customer can obtain a connection from any ISP offering MMP, whereas an FMP customer is typically locked to one provider (Though there exist certain alliances, it is definitely not an open network). Typically, SD-WANs are operated using intra-domain routing protocols.

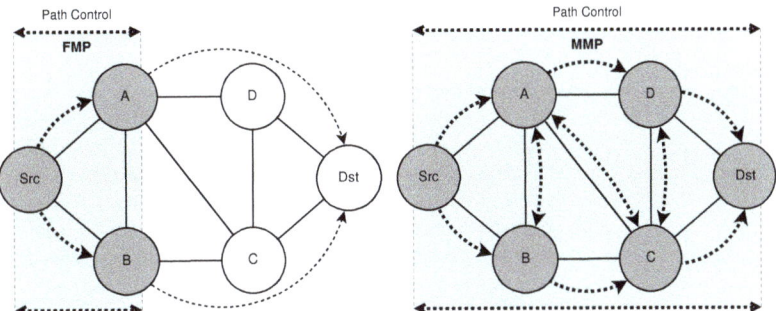

Figure 2. Source AS *Src* can choose a link between/among first hops to mitigate a link congestion/failure; however, FMP cannot choose a path after the first hops even if there are network hazards before the destination *Dst*. MMP can control all ASes between *Src* and *Dst*, allowing a path to be chosen by considering the link status.

It is, therefore, possible to distinguish FMP for one customer controlled using SD-WAN from MMP for any customer in a network dynamically controlled through a public infrastructure, such as SCION. SCION has a much more powerful multi-path routing. For instance, consider the case of a customer with an AS who has a regular Internet connection and an SD-WAN connection. The SD-WAN system can then send a packet over one of two connections. However, a single SCION Internet connection can enable 10 or more paths (depending on the topology) between the source and destination. The path p_n list without the routing loop in Figure 2 is shown below.

- p_0: $Src \to A \to B \to C \to Dst$
- p_1: $Src \to A \to B \to C \to D \to Dst$
- p_2: $Src \to A \to C \to Dst$
- p_3: $Src \to A \to C \to D \to Dst$
- p_4: $Src \to A \to D \to Dst$
- p_5: $Src \to A \to D \to C \to Dst$
- p_6: $Src \to B \to A \to C \to Dst$
- p_7: $Src \to B \to A \to C \to D \to Dst$
- p_8: $Src \to B \to A \to D \to Dst$
- p_9: $Src \to B \to A \to D \to C \to Dst$
- p_{10}: $Src \to B \to C \to Dst$
- p_{11}: $Src \to B \to C \to D \to Dst$
- p_{12}: $Src \to B \to C \to A \to D \to Dst$

3.3. Path Establishment against Link Failures on SCION

Since SCION forwarding paths are static, they collapse when one of the links fails. Link failures are handled by a three-pronged approach that typically masks link failures without any outage to the application and rapidly re-establishes fresh working paths [13]:

- Beaconing periodically establishes new working paths.
- SCION control message protocol (SCMP) (SCION-equivalent of ICMP) is used for path-segment revocation. Failed links result in the rapid erasure of affected path segments from path servers.
- SCION end hosts use multi-path communication by default, thus masking link failures to an application with another working path. As multi-path communication can increase availability (even in environments with very limited path choice [10]), SCION beacon servers actively attempt to create disjoint paths, SCION path servers make an effort to select and announce disjoint paths, and end hosts compose path segments to achieve maximum resilience to path failure. Consequently, we expect that most link failures in SCION will be unnoticed by the application, unlike the frequent (although mostly brief) outages in the currently available Internet [38,41].

4. Evaluation

Figure 3 shows a part of the world map (Background world map designed by Layerace/Freepik.) with the nodes used to measure the effectiveness of FMP and MMP on SCIONLab, which deploys SCION on research networks. The core ASes are placed at Carnegie Mellon University in the United States, Magdeburg in Germany, SWISSCOM (SCMN) in Switzerland, NUS in Singapore, and KISTI Daejeon in Korea. We evaluate network performance in terms of latency and loss rate [12].

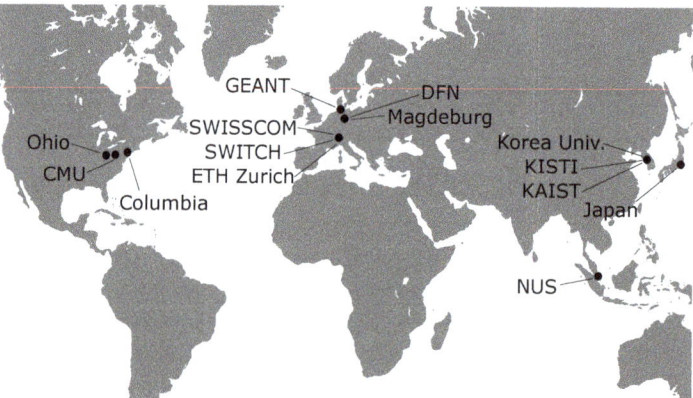

Figure 3. The nodes used for the evaluation with the 6 participating countries including 18 participating SCION ASes. Some dots have several nodes. The nodes *Ohio* and *Japan* are deployed on Amazon Web Services (AWS).

To evaluate latency and loss rate, we employ the SCION ICMP (SCMP) testing tool and SCION bandwidth testing tool provided by SCIONLab [15]. We generate normal/heavy traffic adopting the routing policies of SP, FMP, and MMP on the SCION protocol. The results of the evaluation show proper operation for each routing policy. Under the attack using the heavy traffic, the background ongoing traffic of the network is very small compared to the heavy traffic, and it is enough to be ignored to see the results. This evaluation does not cover the whole cases of the multi-path but it shows the point of comparison for the main characteristics among SP, FMP, and MMP routing.

Initially, for comparison with SP, FMP, and MMP, we select six SCION ASes between Korea and Switzerland. Next, to test the effectiveness of MMP at the Internet-scale during a transit-link DDoS attack, we test using a larger topology with 6 countries and 18 SCION ASes participating in it. During this test, we select four core ASes (the United States, Germany, Switzerland and Korea) from four different countries and set the transit-link between the core ASes as direct path selected countries as shown in Figure 4. We then use these transit-links and measure the loss rate of the paths between the countries.

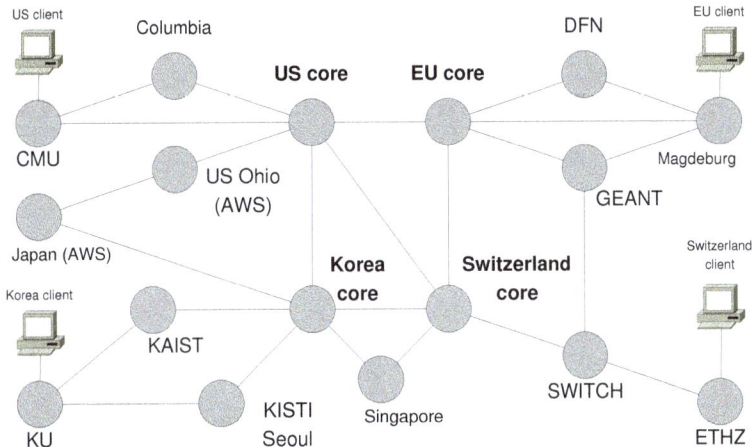

Figure 4. Global topology using SCIONLab for a real-world network evaluation. The core ASes are placed at Carnegie Mellon University (CMU) in the United States (US), Magdeburg in Germany, SWISSCOM (SCMN) in Switzerland, NUS in Singapore, and KISTI Daejeon in Korea. The links between the core ASes are operated as a transit-link for communication. All links are adopted as multi-path to show the effectiveness of MMP. The ETHZ node includes two ASes.

As shown in Figure 5, six ASes are involved between the source *Src* and destination *Dst* for real Internet evaluation between Switzerland and Korea. *A, B, C, D, E,* and *F* ASes indicate KU, KISTI Seoul, KAIST, KISTI Daejeon (KR core AS), CMU (US core AS), and SCMN (Switzerland core AS), respectively. There are two links between *F* and *Dst*. After then, we use given global networks to show effectiveness of multi-path routing under high latency.

4.1. Assumptions

We assume that a normal user from the source node *Src* is communicating with the destination node *Dst*. We assume that the usual latency or loss rate of the default routing path is lower than other alternative paths. For a multi-path test, we assume that an attacker consumes the bandwidth of the path that the victim is currently using, making it difficult for the victim to communicate by degrading the network performance, such as through latency increase or packet loss. In this case, we assume that the attacker knows the path setting of the router because in a legacy router protocol, the path setting is pre-configured with a single path. In addition, we assume that an attacker is a passive attacker that cannot recognize when the victim has changed its path.

4.2. Comparison between SP, FMP and MMP Routing

An evaluation of utilization of multiple available MMP paths is also applied to compare the performance using two and four paths, which are written as a multi-path above. In Figure 5, the path $p_0 = \{Src \to A \to B \to D \to F \nearrow Dst\}$ between *Src* and *Dst* shows under heavy traffic of the communication.

- $p_0 = \{Src \to A \to B \to D \to F \nearrow Dst\}$: A low-latency path that will be under heavy traffic of communication.

- $p(SP) = \{Src \to A \to B \to D \to E \to F \nearrow Dst\}$: A high-latency path that will initially be used as a single path.
- $p(FMP) = \{Src \to A \to C \to D \to F \searrow Dst\}$: A low-latency path that will be partially affected by an attack.
- $p(MMP) = \{Src \to A \to C \to D \to E \to F \searrow Dst\}$: A high-latency path that will mostly be unaffected by an attack.

As aforementioned, FMP cannot deal with network congestion or a link failure if the problem occurs in other hops that FMP cannot control. However, because MMP can take advantage of a path that an FMP, such as SD-WAN, cannot control, MMP can mitigate the problem if such a situation occurs. Figure 5 shows the topology uses for an evaluation towards comparison between FMP and MMP under such a situation. We assume that AS A, which controls the source, uses SD-WAN, and, because an SD-WAN configuration through cooperation with other ASes is difficult, we also assume that the SD-WAN does not control the path passing other AS hops. We use $p(SP)$, $p(FMP)$, and $p(MMP)$ as the path of SP, FMP, and MMP routing.

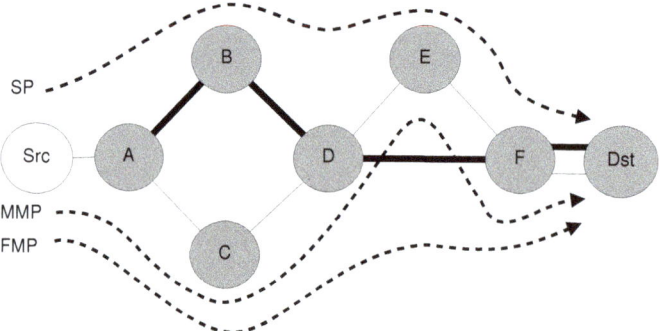

Figure 5. Topology used for comparison between MMP and FMP. The solid bold line indicates heavy traffic. The scenario is a case in which the traffic congestion affects both paths of FMP and MMP, whereas the alternative path found by MMP does not.

If traffic congestion occurs in path $\{A \to B \to D\}$, path $\{D \to F\}$, and path $\{F \nearrow Dst\}$, then both links configured for SP and FMP will suffer a degradation in network performance even though SD-WAN switches its link from a SP path $p(SP)$ to an FMP path $p(FMP)$, or vice versa. However, when it comes to $p(MMP)$, it can be used as an alternative option because MMP can control the path outside the source AS. Because path $\{A \to C \to D\}$, path $\{D \to E \to F\}$, and path $\{F \searrow Dst\}$ are unaffected by traffic congestion, using this $p(MMP)$, we can mitigate the network performance degradation.

Figure 6 and Table 1 show the results of the network performance of each path $p(x)$ when x is SP, FMP or MMP, under traffic congestion. If traffic congestion occurs on the $p(x)$ mentioned previously, $p(SP)$ and $p(FMP)$, which are options for SD-WAN, will suffer from degraded network performance with an average increase in latency of approximately 46% (SP: 270 ms to 396 ms) and 31% (FMP: 318 ms to 419 ms), and a loss rate of 4.7% and 4%, respectively. However, for the $p(MMP)$, which is an alternative path that can be configured using MMP, the network suffers from almost no network performance degradation.

Table 1. Average latency and loss rate with FMP paths and MMP path during attack.

Path	Loss Rate	Avg. Latency
$p(SP)$	4.7%	396.66 ms
$p(FMP)$	4%	419.19 ms
$p(MMP)$	0%	318.73 ms

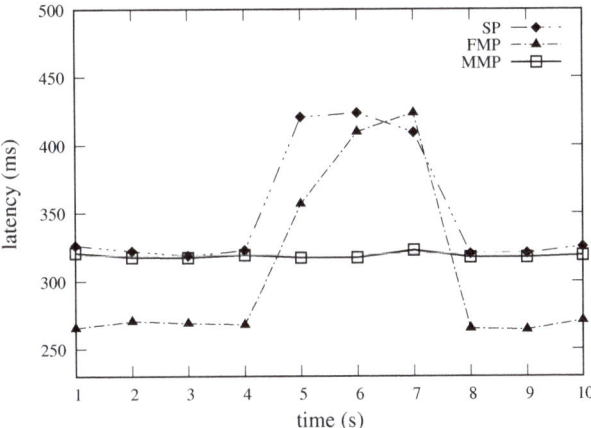

Figure 6. Latency comparison of FMP and MMP when traffic congestion occurs during 4–7 s. Both paths of SP and FMP are affected, whereas the alternative path that MMP found is unaffected.

4.3. Single Path and Concurrent Multi-Path

An evaluation of utilization of multiple available MMP paths is also applied to compare the performance using two and four paths, which are written as a multi-path above. In Figure 7, the node *Src* uses the path $p_0 = \{Src \to A \to B \to D \to F \nearrow Dst\}$ for communication as an SP routing. For two paths, we used p_0 and p_3, and, for four paths, we used p_0 to p_3 as explained below:

- $p_0 = \{Src \to A \to B \to D \to F \nearrow Dst\}$: A low-latency path that will initially be used by *Src* as a single path. An attacker will directly attack this path during communication.
- $p_1 = \{Src \to A \to C \to D \to F \searrow Dst\}$: A low-latency path that will be partially affected by an attack.
- $p_2 = \{Src \to A \to C \to D \to E \to F \nearrow Dst\}$: A high-latency path that will be partially affected by an attack.
- $p_3 = \{Src \to A \to C \to D \to E \to F \searrow Dst\}$: A high-latency path that will mostly be unaffected by an attack.

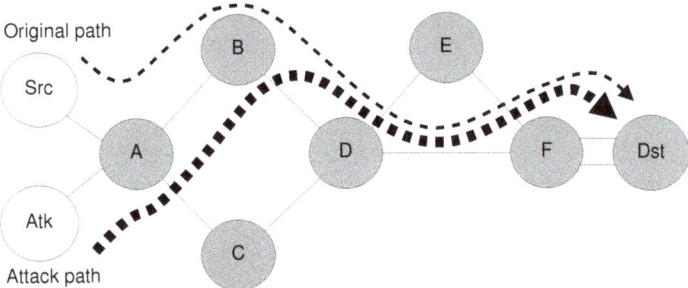

Figure 7. Topology used for MMP evaluation. Dotted curved line shows the user's communication path and the bold dotted line shows the attacker's attack path.

Figure 8 and Table 2 show the results of the comparison when using two paths and four paths for MMP. The paths are selected uniformly at random for the case of '2 paths (random)' and '4 paths (random)'. We sent SCMP packets for 1 s each during a 10 s period (for a total of 10 packets), which consumed the bandwidth for 3 s (from seconds 4 to 7). Because the network status changes owing to the bandwidth consumption of an attacker, selecting stable paths will ensure a much more stable communication with high availability than selecting paths at the same ratio. Here, '4 paths (weighted)' represents selecting 4-times more stable paths than the paths being attacked, and '4 paths (random)' corresponds to randomly selecting all four paths.

The loss rate and average latency of a single path are greater than those obtained using 'MMP: 2 paths (random)' and 'MMP: 4 paths (random)'. Moreover, when using 'MMP: 4 paths (random)', the performance is less affected by an attack with only an approximate 10% increase in the average latency than when using 'MMP: 2 paths (random)' during a bandwidth consumption attack (4–7 s), which shows an increased average latency of approximately 20%. In addition, although the latency showed an improvement, the loss rate shows almost no change because other paths (p_1 and p_2) used for 'MMP: 4 paths (random)' evaluation already encountered a loss rate even when no attack takes place.

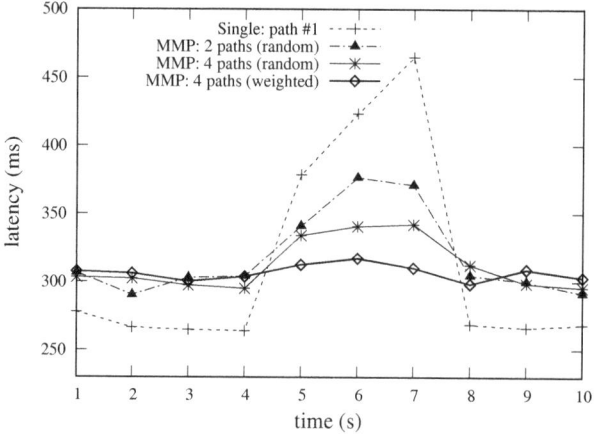

Figure 8. Latency comparison of SP and MMP routing. 'MMP 4 paths' (weighted/random) demonstrates selecting stable paths 4 times more than others and equally selecting all 4 paths by an attack (4–7 s).

Table 2. Experimental results of an SP and MMP for the measurement of loss rates and latency during the attack.

Path	Loss Rate	Avg. Latency
SP: 1 path	7.5%	422.22 ms
MMP: 2 paths (random)	3%	362.77 ms
MMP: 4 paths (random)	3%	330.09 ms
MMP: 4 paths (weighted)	0.5%	313.07 ms

As observed in Figure 8, 'MMP: 4 paths (weighted)' using four times more stable paths mitigate the effect of an attack (during seconds 4 to 7) compared with 'MMP: 4 paths (random)' at the same ratio, because a *weighted* path selection shows an average increased latency of approximately only 4% and a loss rate of 0.5%, whereas a random path selection shows an increased average latency and loss rate of 10% and 2.5%, respectively.

However, there is a limitation on *weighted* multi-path, that is, we assume that the attacker is a passive attacker, and that the user knows which path is being attacked. As a future study, we would like to conduct research on utilizing an optimized path selection algorithm by using the network status of each path and the AS information on the paths to deal with an active attacker who recognizes that the user has changed paths.

4.4. Effectiveness of MMP on Transit-Link DDoS Attack under Large Scale

To measure the effectiveness of MMP under higher latency when using MMP probably increases additional latency by selecting the appropriate paths, we used a larger topology shown in Figure 4 with 6 countries and 18 SCION ASes participating in it. We then used the Google Cloud Platform Computing Engine (GCE) virtual machine as a test client. The following routes $R_{src-dst}$ using the pair of core ASes are used for an evaluation:

- R_{k-s} = {Korea client – Korea core – Switzerland core – Switzerland client},
- R_{k-u} = {Korea client – Korea core – US core – US client},
- R_{k-g} = {Korea client – Korea core – Germany core – Germany client},
- R_{s-u} = {Switzerland client – Switzerland core – US core – US client},
- R_{s-g} = {Switzerland client – Switzerland core – Germany core – Germany client},
- R_{u-g} = {US client – US core – Germany core – Germany client}.

We measured the loss rate of an FMP path and three MMP paths for each $R_{src-dst}$ described above. After the $R_{src-dst}$ is selected, 30,000 SCMP packets with an interval of 0.001 s were sent from one client to another. The test was completed with six $R_{src-dst}$.

Table 3 shows a summary of the test results. From each route $R_{src-dst}$, the FMP path p_0 passes the transit-link directly connected between the selected core source and destination ASes, and MMP paths p_1, p_2, and p_3 are alternative paths not passing the transit-link. The numbers in each table show the loss rates of FMP/MMP paths p_{0-3} under a transit-link DDoS attack. From all routes $R_{src-dst}$, FMP path p_0 that passes the transit-link is affected by 14–32% (average of 19%) significantly by a transit-link DDoS attack, but the other alternative MMP paths show no difference under a DDoS attack and operates properly with MMP routing. Figure 9 shows the average loss rate under bandwidth-based attack. The loss rate of FMP is dramatically increased by the traffic congestion and has 42% when the attack bandwidth is 700 Mbps. However, MMP is stable within 1% loss rate and manages the path to mitigate the congestion.

Table 3. Average loss rate during transit-link DDoS attack. FMP p_0 for each route $R_{src-dst}$ is an FMP path communicating through the transit-link between source and destination. MMP p_1, p_2, and p_3 are alternative paths not passing the transit-link.

Path	R_{k-s}	R_{k-u}	R_{k-g}	R_{s-u}	R_{s-g}	R_{u-g}
FMP: p_0	16%	14%	16%	32%	19%	17%
MMP: p_1	1%	1%	0%	0%	0%	1%
MMP: p_2	1%	1%	1%	0%	0%	1%
MMP: p_3	0%	1%	1%	0%	0%	1%

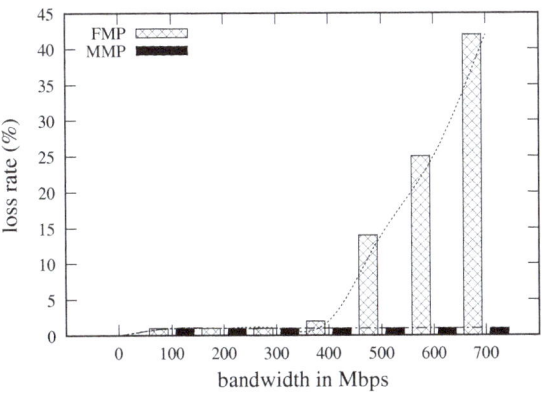

Figure 9. Effect of traffic on FMP. The loss rate is dramatically increased by the traffic congestion on FMP, but MMP manages the path to mitigate the congestion.

5. Research Limitations and Future Work

We measured the latency and the loss rate of SP, FMP, and MMP on a global network using SCION. The results show that MMP routing has a low latency and loss rate compared to FMP routing and more MMP paths show lower latency and loss late. Under the traffic congestion, the loss rate of MMP is stable. However, we can think about the limitation of the operation. If all users of the global network use all possible multi-paths, we could not guarantee the performance of MMP as shown above.

Our suggested future work is to optimize the important factor of MMP operations, such as how many multi-path uses or how many hops of multi-path is the minimum quality for MMP routing. QoS-driven routing is also a possible option for MMP routing policy using QoS metrics, such as latency, loss late, bandwidth, jitter, or other composite metrics. We could make the weight of each multi-path, and some paths could show high latency but stability to communicate.

6. Conclusions

We evaluated multi-path routing on MMP and compared MMP with SP and FMP on a global network. The experimental results show that when using MMP routing, the network performance is affected less by a bandwidth consumption attack than when using SP or FMP routing. In addition, when making a choice among paths under a multi-path situation, selecting more paths with a better network status than paths under an attack causes less effect on the performance of the network under attack.

We verified that the effectiveness of multi-path with real networks deployed globally, especially MMP routing, shows more resilience than FMP used currently. Using the global network, we show that FMP mitigates certain cases of network problems, but MMP gains impressively lower loss rate by comparison.

Author Contributions: Conceptualization, H.A., Y.N., and H.L.; methodology, H.A. and Y.N.; software, Y.N. and A.P.; validation, H.A., H.L., and A.P.; formal analysis, H.A. and H.L.; investigation, H.A.; resources, H.A., Y.N., H.L., and A.P.; data curation, Y.N.; writing—original draft preparation, H.A. and Y.N.; writing—review and editing, H.A., H.L., and A.P.; visualization, H.A. and Y.N.; supervision, H.A. and H.L.; project administration, H.L.; funding acquisition, H.L. All authors have read and agreed to the published version of the manuscript.

Funding: This work is supported in part by the Institute of Information & communications Technology Planning & Evaluation (IITP) grant funded by the Korea government (MSIT) (No.2019-0-01343 Regional strategic Industry convergence security core talent training business, No.2019-0-01697 Development of automated vulnerability discovery technologies for blockchain platform security, and No.IITP-2021-2020-0-01819 ICT creative consilience program).

Conflicts of Interest: The funders had no role in the design of the study; in the collection, analyses, or interpretation of data; in the writing of the manuscript, or in the decision to publish the results.

References

1. An, H.; Lee, H.; Perrig, A. Coordination of anti-spoofing mechanisms in partial deployments. *J. Commun. Netw.* **2016**, *18*, 948–961. [CrossRef]
2. Barrera, D.; Chuat, L.; Perrig, A.; Reischuk, R.M.; Szalachowski, P. The SCION internet architecture. *Commun. ACM* **2017**, *60*, 56–65. [CrossRef]
3. Sun, Y.; Edmundson, A.; Vanbever, L.; Li, O.; Rexford, J.; Chiang, M.; Mittal, P. RAPTOR: Routing Attacks on Privacy in Tor. In Proceedings of the 24th USENIX Security Symposium USENIX Security 15, Washington, DC, USA, 12–14 August 2015; pp. 271–286.
4. Labovitz, C.; Ahuja, A.; Bose, A.; Jahanian, F. Delayed Internet routing convergence. *IEEE/ACM Trans. Netw.* **2001**, *9*, 293–306. [CrossRef]
5. Dahlin, M.; Chandra, B.B.V.; Gao, L.; Nayate, A. End-to-end WAN service availability. *IEEE/ACM Trans. Netw.* **2003**, *11*, 300–313. [CrossRef]
6. Paxson, V. End-to-end routing behavior in the Internet. *IEEE/ACM Trans. Netw.* **1997**, *5*, 601–615. [CrossRef]
7. Paxson, V. End-to-end internet packet dynamics. *IEEE/ACM Trans. Netw.* **1999**, *7*, 277–292. [CrossRef]

8. Rekhter, Y.; Li, T. RFC1771: A Border Gateway Protocol 4 (BGP-4). 1995. Available online: https://datatracker.ietf.org/doc/html/rfc1771 (accessed on 18 May 2021).
9. Medhi, D.; Ramasamy, K. *Network Routing: Algorithms, Protocols, and Architectures*; Morgan Kaufmann: Burlington, NC, Canada, 2017.
10. Andersen, D.; Balakrishnan, H.; Kaashoek, F.; Morris, R. Resilient overlay networks. In Proceedings of the Eighteenth ACM Symposium on Operating Systems Principles, Banff, AB, Canada, 21–24 October 2001; pp. 131–145.
11. Raiciu, C.; Paasch, C.; Barre, S.; Ford, A.; Honda, M.; Duchene, F.; Bonaventure, O.; Handley, M. How Hard Can It Be? Designing and Implementing a Deployable Multipath TCP. In Proceedings of the 9th USENIX Symposium on Networked Systems Design and Implementation, San Jose, CA, USA, 25–27 April 2012; pp. 399–412.
12. Jansen, R.; Vaidya, T.; Sherr, M. Point Break: A Study of Bandwidth Denial-of-Service Attacks against Tor. In Proceedings of the 28th USENIX Security Symposium, Santa Clara, CA, USA, 14–16 August 2019.
13. Perrig, A.; Szalachowski, P.; Reischuk, R.M.; Chuat, L. *SCION: A Secure Internet Architecture*; Springer: Berlin/Heidelberg, Germany, 2017.
14. SCION Internet Architecture. Available online: https://www.scion-architecture.net/ (accessed on 18 May 2021).
15. SCIONLab. Available online: https://www.scionlab.org/ (accessed on 18 May 2021).
16. Kwon, J.; Lee, T.; Hähni, C.; Perrig, A. SVLAN: Secure & scalable network virtualization. In Proceedings of the 2020 Network and Distributed System Security Symposium (NDSS 2020, San Diego, CA, USA, 23–26 February 2020.
17. Studer, A.; Perrig, A. The coremelt attack. In *European Symposium on Research in Computer Security*; Springer: Berlin/Heidelberg, Germany, 2009; pp. 37–52.
18. Kang, M.S.; Lee, S.B.; Gligor, V.D. The crossfire attack. In Proceedings of the 2013 IEEE Symposium on Security and Privacy, Berkeley, CA, USA, 19–22 May 2013; pp. 127–141.
19. Stewart, R.; Ramalho, M.; Xie, Q.; Tuexen, M.; Conrad, P. *Stream Control Transmission Protocol (SCTP) Partial Reliability Extension*; Technical Report; RFC 3758 (Proposed Standard): Virginia Beach, VA, USA, 2004.
20. Chuat, L.; Perrig, A.; Hu, Y.C. Deadline-Aware Multipath Communication: An Optimization Problem. In Proceedings of the 2017 47th Annual IEEE/IFIP International Conference on Dependable Systems and Networks (DSN), Denver, CO, USA, 26–29 June 2017; pp. 487–498.
21. Liu, X.; Mohanraj, S.; Pióro, M.; Medhi, D. Multipath routing from a traffic engineering perspective: How beneficial is it? In Proceedings of the 2014 IEEE 22nd International Conference on Network Protocols, Raleigh, NC, USA, 21–24 October 2014; pp. 143–154.
22. Bellovin, S.M.; Ioannidis, J.; Keromytis, A.D.; Stewart, R.R. On the Use of Stream Control Transmission Protocol (SCTP) with IPSec. 2003. Available online: https://datatracker.ietf.org/doc/html/rfc3554 (accessed on 18 May 2021).
23. Tsirigos, A.; Haas, Z.J. Multipath routing in the presence of frequent topological changes. *IEEE Commun. Mag.* **2001**, *39*, 132–138. [CrossRef]
24. Tsirigos, A.; Haas, Z.J. Analysis of multipath routing-Part I: The effect on the packet delivery ratio. *IEEE Trans. Wirel. Commun.* **2004**, *3*, 138–146. [CrossRef]
25. Papadimitratos, P.; Haas, Z.J. Secure data transmission in mobile ad hoc networks. In Proceedings of the 2nd ACM Workshop on Wireless Security, San Diego, CA, USA, 19 September 2003; pp. 41–50.
26. Pu, C. Jamming-resilient multipath routing protocol for flying ad hoc networks. *IEEE Access* **2018**, *6*, 68472–68486. [CrossRef]
27. Gelenbe, E.; Lent, R.; Nunez, A. Self-aware networks and QoS. *Proc. IEEE* **2004**, *92*, 1478–1489. [CrossRef]
28. Gelenbe, E.; Loukas, G. A self-aware approach to denial of service defence. *Comput. Netw.* **2007**, *51*, 1299–1314. [CrossRef]
29. Sakellari, G.; Gelenbe, E. Demonstrating cognitive packet network resilience to worm attacks. In Proceedings of the 17th ACM Conference on Computer and Communications Security, Chicago, IL, USA, 4–8 October 2010; pp. 636–638.
30. Czachórski, T.; Gelenbe, E.; Kuaban, G.S.; Marek, D. Time-Dependent Performance of a Multi-Hop Software Defined Network. *Appl. Sci.* **2021**, *11*, 2469. [CrossRef]
31. Baldini, G.; Fröhlich, P.; Gelenbe, E.; Hernandez-Ramos, J.L.; Nowak, M.; Nowak, S.; Papadopoulos, S.; Drosou, A.; Tzovaras, D. IoT Network Risk Assessment and Mitigation: The SerIoT Approach. *Secur. Risk Manag.* **2020**, *88*. [CrossRef]
32. Gelenbe, E.; Domanska, J.; Fröhlich, P.; Nowak, M.P.; Nowak, S. Self-aware networks that optimize security, QoS, and energy. *Proc. IEEE* **2020**, *108*, 1150–1167. [CrossRef]
33. Fröhlich, P.; Gelenbe, E.; Nowak, M.P. Smart SDN management of fog services. In Proceedings of the 2020 Global Internet of Things Summit (GIoTS), Dublin, Ireland, 3 June 2020; pp. 1–6.
34. Gelenbe, E. Steps toward self-aware networks. *Commun. ACM* **2009**, *52*, 66–75. [CrossRef]
35. Michel, O.; Keller, E. SDN in wide-area networks: A survey. In Proceedings of the 2017 Fourth International Conference on Software Defined Systems (SDS), Valencia, Spain, 8–11 May 2017; pp. 37–42.
36. Kwon, J.; Seo, D.; Kwon, M.; Lee, H.; Perrig, A.; Kim, H. An incrementally deployable anti-spoofing mechanism for software-defined networks. *Comput. Commun.* **2015**, *64*, 1–20. [CrossRef]
37. Smith, J.M.; Schuchard, M. Routing around congestion: Defeating DDoS attacks and adverse network conditions via reactive BGP routing. In Proceedings of the 2018 IEEE Symposium on Security and Privacy (SP), San Francisco, CA, USA, 20–24 May 2018; pp. 599–617.

38. Katz-Bassett, E.; Scott, C.; Choffnes, D.R.; Cunha, Í.; Valancius, V.; Feamster, N.; Madhyastha, H.V.; Anderson, T.; Krishnamurthy, A. LIFEGUARD: Practical repair of persistent route failures. *ACM SIGCOMM Comput. Commun. Rev.* **2012**, *42*, 395–406. [CrossRef]
39. Ballani, H.; Francis, P.; Zhang, X. A study of prefix hijacking and interception in the Internet. *ACM SIGCOMM Comput. Commun. Rev.* **2007**, *37*, 265–276. [CrossRef]
40. Giuliari, G.; Klenze, T.; Legner, M.; Basin, D.; Perrig, A.; Singla, A. Internet backbones in space. *ACM SIGCOMM Comput. Commun. Rev.* **2020**, *50*, 25–37. [CrossRef]
41. Kushman, N.; Kandula, S.; Katabi, D. Can you hear me now?! it must be BGP. *ACM SIGCOMM Comput. Commun. Rev.* **2007**, *37*, 75–84. [CrossRef]

Review

Autonomous Haulage Systems in the Mining Industry: Cybersecurity, Communication and Safety Issues and Challenges

Tarek Gaber [1,2], Yassine El Jazouli [3], Esraa Eldesouky [1,4] and Ahmed Ali [4,5,*]

1. Faculty of Computers and Informatics, Suez Canal University, Ismailia 41522, Egypt; t.m.a.gaber@salford.ac.uk (T.G.); em.eldesouky@psau.edu.sa (E.E.)
2. School of Science, Engineering, and Environment, University of Salford, Manchester M5 4WT, UK
3. Suncor Energy, 111 5th Ave. SW, Calgary, AB T2P3Y6, Canada; Yassine.el@networksharks.ca
4. Department of Computer Science, College of Computer Engineering and Sciences, Prince Sattam Bin Abdulaziz University, Al-kharj 11942, Saudi Arabia
5. Higher Future Institute for Specialized Technological Studies, Cairo 3044, Egypt
* Correspondence: a.abdalrahman@psau.edu.sa

Abstract: The current advancement of robotics, especially in Cyber-Physical Systems (CPS), leads to a prominent combination between the mining industry and connected-embedded technologies. This progress has arisen in the form of state-of-the-art automated giant vehicles with Autonomous Haulage Systems (AHS) that can transport ore without human intervention. Like CPS, AHS enable autonomous and/or remote control of physical systems (e.g., mining trucks). Thus, similar to CPS, AHS are also susceptible to cyber attacks such as Wi-Fi De-Auth and GPS attacks. With the use of the AHS, several mining activities have been strengthened due to increasing the efficiency of operations. Such activities require ensuring accurate data collection from which precise information about the state of the mine should be generated in a timely and consistent manner. Consequently, the presence of secure and reliable communications is crucial in making AHS mines safer, productive, and sustainable. This paper aims to identify and discuss the relation between safety of AHS in the mining environment and both cybersecurity and communication as well as highlighting their challenges and open issues. We survey the literature that addressed this aim and discuss its pros and cons and then highlight some open issues. We conclude that addressing cybersecurity issues of AHS can ensure the safety of operations in the mining environment as well as providing reliable communication, which will lead to better safety. Additionally, it was found that new communication technologies, such 5G and LTE, could be adopted in AHS-based systems for mining, but further research is needed to considered related cybersecurity issues and attacks.

Keywords: cybersecurity; autonomous haulage systems; operating technology; mining industry; cyber-physical systems; communication; safety

Citation: Gaber, T.; El Jazouli, Y.; Eldesouky, E.; Ali, A. Autonomous Haulage Systems in the Mining Industry: Cybersecurity, Communication and Safety Issues and Challenges. *Electronics* **2021**, *10*, 1357. https://doi.org/10.3390/electronics10111357

Academic Editor: Victor A. Villagrá

Received: 1 May 2021
Accepted: 2 June 2021
Published: 7 June 2021

Publisher's Note: MDPI stays neutral with regard to jurisdictional claims in published maps and institutional affiliations.

Copyright: © 2021 by the authors. Licensee MDPI, Basel, Switzerland. This article is an open access article distributed under the terms and conditions of the Creative Commons Attribution (CC BY) license (https://creativecommons.org/licenses/by/4.0/).

1. Introduction

Since the launch of Industry 4.0, smart machinery and intelligent services have been unveiled, including quality-controlled systems [1]. Industry 4.0 redirected organizations' perspective on technology along with its role in developing a viable business model that increases their profits. Industry 4.0 aims to integrate Operating Technology (OT) and Information Technology (IT) ecosystems to cope with current business requirements, such as information sharing and controlling. In the heart of Industry 4.0, Cyber-Physical Systems (CPS) were shown to be a revolutionary development. CPS are "engineered systems that integrate information technologies, real-time control subsystems, physical components and human operators to influence physical processes by means of cooperative and (semi)automated control functions" [2]. CPS combine real-time communications and computer processes with physical world applications to enable autonomous and/or remote control of physical systems [3].

Mining is an industry that inherits the full advantage of Industry 4.0 by using cutting-edge driverless vehicles called Autonomous Haulage Systems (AHS) or Autonomous Haulage Trucks (AHT). These trucks can carry up to four hundred tons of ore and accurately transport it without human interaction. AHS is the state-of-the-art in the mining industry for autonomous vehicles. Since their first system development endeavor in Chile in 2005, AHS have attracted attention in the last decade from haulage truck manufacturers such as Caterpillar and Komatsu. The successful reputations of existing autonomous mines around the world notably increased the demand for AHS in surface mining during the last few years. Komatsu America has reported that its FrontRunner system has achieved a new milestone record of more than two billion tons of material that was transported autonomously at CODELCO's Gabriela Mistral in Chile since the opening of its first autonomous copper mine in 2008 [4]. Generally, Autonomous Trucks (ATs) have been designed to reduce the vulnerability to the risk of equipment contact with auxiliary equipment or Equipped Manual Vehicles (EMVs).

According to the definition of CPS above, the AHS could be seen as a kind of CPS. The AHS equipment exceedingly relies on wireless communications, including object avoidance/detection systems, Global Positioning Systems (GPS), e.g., GNSS, and artificial intelligence. Inside the AHS intelligence system (i.e., FrontRunner for Komatsu and Command for Caterpillar), all data obtained are compiled so that the software can make a suitable decision. Calculating the maximum speed allowed to a nearby equipment or the estimated time for an AT to break are possible decisions. Thus, determining the exact location of every AT and EMV is mandatory for the trucks to prevent accidents, which increases safety and decreases the maintenance or replacement cost [5]. ATs have demonstrated notable fuel consumption performance as a result of their driving consistency. They can operate on a 24/7 schedule with no idling time as there is no shift change and no breaks required. Manual truck operators can affect 35% of fuel economy whereas ATs can improve fuel usage by 4% with 25–50% reduced idle time [5].

With the use of the AHS, several mining activities have been strengthened due to increasing the efficiency of operations. Such activities require ensuring accurate data collection from which precise information about the state of the mine should be generated in a timely and consistent manner. Consequently, the presence of secure and reliable communications is crucial in making AHS mines safer, productive, and sustainable. The availability and security issues of communication of AHS the mine environment will be discussed below.

The availability of communication is an essential service in AHS systems. To the best of our knowledge, the current AHS mines rely entirely on wireless communication [6], which employs standard 802.11 Wi-Fi technology operating in the unlicensed spectrum. In the meantime, technologies (e.g., Wi-Fi) were deemed unfit for industrial communications due to the lure of Industry 4.0, and the flexibility offered by these technologies made Wi-Fi attractive for AHS solutions. Specifically, ATs require seamless connectivity to the network at all times [7]. The geographical challenges and the nature of CPS (AHS) running in the mine make wireless communications the only possible way to keep track and ATs connected. Yet, these wireless technologies lead to a set of challenges—for example, frequency interference, channel utilization, and signal jamming. Furthermore, some special design concerns should be considered, such as signal propagation and dynamic topology as the mine keeps growing, so as to obtain efficient and stable contact.

With the integration of OT and IT as in the case of AHS, information sharing becomes more susceptible to disclosure, intrusion, and other cybersecurity issues and attacks. As shown above, AHS can be considered as a CPS system, and thus the former can be integrated with OT using mechanisms that are vulnerable to malicious attacks such as Wi-Fi De-Auth attacks [8], GPS attacks [9], and camera attacks [9,10]. Thus, safety remains an area of concern as being targeted by traditional cyber-attacks due to the similarities in infrastructure with the conventional IT environment. The authors of [11] emphasize that a lack of security can lead to equipment damage, loss of production, severe injuries, and fatalities, thus greatly endangering safety.

In this paper, within the mining environment, we aim to identify and discuss the relation between safety of AHS in the mining environment and both cybersecurity and communication. We argue that addressing cybersecurity issues of AHS can ensure the safety of operations in the mining environment, and that providing reliable communication leads to a better safety. Specifically, we surveyed the literature aiming to study the relation between (1) cybersecurity and safety and (2) communication and safety of AHS in the mining environment. In this paper, the term *safety* refers to the control of any hazards in order to avoid human injury and mechanical risk. The survey's main objective is to identify the pros and cons of the published work at this point as well as to highlight some open issues for further investigation regarding this field of study.

This paper is organized as follows. Section 2 summarizes the literature review related to safety accompanied by cybersecurity and communication. In Section 3, future directions and open issues related to cybersecurity, communication, and safety in AHS are discussed. Finally, the paper is concluded in Section 4.

2. Literature Review

In the literature review, we investigate AHS mines from three different angles, which are communication challenges, cybersecurity, and safety. We argue that the AHS environment is a subset of the industrial environment that faces similar challenges to other OT systems with some particularities to the mining environment (e.g., the terrain) and the associated elements with specific challenges that we try to resolve in this study. We also contend that business decision making is significantly affected by information shared between the industrial and IT environments, which invokes the problem of information accessing or sharing with untrusted networks, e.g., the Internet. We study the standard communication technologies that arise in the industrial environment and their drawbacks as well as the possibility to re-enforce them in order to make them more secure. Furthermore, we argue that if communication in the AHS environment is not secure and reliable, the safety of equipment can be impacted since safety is dependent on the availability of communication. Finally, we conclude that safety by association is a security service in the industrial environment, whereas a cyberattack can lead to disastrous consequences.

2.1. Relation between Cybersecurity and Safety in AHS

Security problems have increased and mutated with Industry 4.0; the security issues in the age of Industry 4.0 are discussed in [12]. In addition, the authors clarify how evolving innovations have brought new security risks to the industrial climate. In addition, the convergence of these technologies has provided a gap for new attack surfaces, such that applying unlicensed wireless communication to mining activities moved current hazards and challenges of the underlying technology to the OT ecosystem and introduced a potential for new attacks on the field equipment (AHS trucks).

CPS are dense heterogeneous systems that encompass various sensors and actuators connected to a pool of computing nodes [13]. Hence, a CPS machine works on perceiving and analyzing the surrounding physical environment. Accordingly, it acts appropriately based on the sensed data using intelligence decisions in an autonomous manner. As a result, ATs are classified as CPS networks, making them susceptible to the same type of OT attacks. These attacks might threaten communications, storage, actuators, computing nodes and perceiving sensors [14].

ATs are endpoints (e.g., sensors, actuators) connected to networks and communicating through a command center with different tiers of security. The authors of [15] address how the unawareness manipulation of logical and physical controls of devices is the most devastating effect of taking control of the endpoints, such as field equipment. A successful attack on a CPS (e.g., MITM) could be catastrophic [14]. As a result, such attacks may lead to a loss of quality of services, data integrity/quality, as well as human life.

Since there are no predefined standards that constitute the precise handling of complex industrial environments, it is feasible to use the existing variety of standards and re-direct them for a specific purpose [12]. Yet, the integration of cybersecurity issues in an industry based on ad hoc structures is a naïve approach that can lead to misleading results. This is because a generic attack study may ignore the main security clues of the CPS, which aim to achieve a balance among usability, risk, cost, and convenience [13]. CPS could also be operating in a safe and controlled environment that is secured by other means. Furthermore, as attacks are generalized, defense perimeter modeling and Root of Trust (RoT) mechanisms are often overlooked [13].

Cybersecurity issues have posed a serious problem and complex threats to organizations looking to make the transition to engage in the Industry 4.0 model, according to [16]. The authors established three vulnerability factors in a cyber-physical infrastructure that may be exploited by cyber-attacks: physical, network, and computation. The potential of interfering with wireless network communication in the mining community through a subsequent survey is addressed in [12]. A successful wining attack can have a significant negative impact on service protection and availability (for example, one of the security CIA services). A solution called "security by design" is proposed by the authors of [14,15], which take into consideration several criteria of cybersecurity architecture. Feasibility, robustness, extensibility, as well as authentication, authorization, network enforced policy, and secure analytics are counted as security measures [15].

Another area of contention is the exchange of knowledge over dynamic networks in both industrial and non-industrial atmospheres (i.e., IT) given that the two atmospheres are geographically or logically isolated. In the age of Industry 4.0, data sharing between the OT and IT environments is essential for making the optimal business decisions (i.e., usually higher management is located in a different facility than current operations) [17]. However, this raises the vulnerability of such sensitive infrastructure, which can be exposed to untrustworthy networks and attacks. The dynamic nature of the CPS portion, such as a complex environment, necessitates a unique approach (i.e., considering cybersecurity and safety). Most data exchange methods, according to [17], are incapable of grasping high-dynamic scenarios in which several parties (i.e., vendors) cooperate to achieve a shared purpose, particularly where privacy is factored in. The same authors suggested a solution based on establishing a dynamic trust zone in which decisions are automatically made through identifying flows, evaluating them, and deciding whether to allow the flow or not [17]. Yet, this strategy increases concern about data protection and confidentiality in a multivendor setting. An additional RoT model is introduced in [13], where a heterogeneous and static environment is built to manage confidentiality problems such as key delivery instead of handling each individual variable. This is considered the most effective way to avoid confidentiality issues.

The protection of GPS positions against malicious attacks is one of the main concerns in AHS. Haulage trucks depend on GPS-derived positional coordinates, which are supplemented by a detailed map. In an essential feature in automated AVs, these driverless devices select the shortest routes to reach new locations even without prior knowledge, which gives rise to vulnerability from malware attacks. Ren et al. [9] demonstrated a number of realistic GPS attacks that were presented under two categories: spoofing and jamming. One of the most frequent GPS attacks is to deviate the correct location of the victims to an incorrect position (i.e., spoofing) by fabricating a spurious signal. Nulling, another advanced attacking mechanism, aims to cancel GPS signals by encrypting negative signals that could be used to launch stealthy attacks. Authors in [18] suggested a recent attack strategy that utilizes selected fake locations to direct the AHT into a predefined area, using Google Maps, for example. Unfortunately, the inefficient protection of GPS data can lead to catastrophic truck collisions, which is another safety concern in the AHS.

Furthermore, since the distinction between industrial and IT networks is becoming blurred, a need for a coordinated approach becomes evident. As shown in [19], applying a defense-in-depth strategy to the industrial fields is emphasized since we now have the capability of mounting new threats that were not inherent in OT. These tactics aim to improve the overall system's CIA security triad (i.e., confidentiality, integrity, and availability). This can be accomplished by enabling the implementation of solutions such as RoT introduced in [13], which was previously addressed. Meanwhile, when applying security strategies, antiviruses, patch management technology support, and security compliance, these strategies should take into consideration the distinctions between IT and OT environments. Such strategies are deemed to be effective when we put the environment we are working under into perspective. In our situation, for example, we physically secure the autonomous truck in the beginning, and then the wireless communications on the truck. Afterwards, we ensure secure tower communications and finally the network backbone.

Autopilots rely extensively on computer vision and Artificial Intelligence (AI) techniques since a vehicle perceives visual data very differently than a person does. Cameras are critical in autonomous trucks for a variety of tasks, including lane detection, obstacle detection, parking, and sign recognition [9,10], which raises another security risk. A blurred camera's performance breaks a safety standard, increasing the likelihood of fatal accidents triggered by camera attacks. A typical attack involves the use of a laser matrix to blind cameras at a close range of less than half a meter, for a few seconds, inflicting irreversible damage and thereby ruining the autonomous procedures. Optical features are the camera's weak point, as physical attacks can hinder the existence of a completely secure camera system. Nonetheless, Petit et al. [20] suggest that removable near-infrared cut filters and photochromic lenses provide adequate protection from various angles. A recent study addresses the use of machine learning to detect and mitigate remote attacks via a dedicated anti-hacking device [21]. Notably, these attacks sometimes may not require any physical access to the truck, such as attacks with lasers, and their consequences can be critical. Yet, some of these assaults do not require physical access to the truck and their effects may be serious.

Concerning autonomous mining, Labbe [6] notes that existing AHS standards and literature have always placed a premium on the system's protection aspect, but never on its cybersecurity components. In addition, the proposed study [6] claims that developing a generic threat model for AHS systems is important, which would be applicable to any OEM and mining facility. In that, Labbe [6] introduces an under development solution called MM-ISAC, that would align with safety requirements such as ISO17757:2017 and security standards such as ISO/IEC 27000:2016, ISA99. Although this initiative is still in its early stages, we anticipate that the MM-ISAC will collaborate closely with vendors to refine the system specification across all affected areas, including infrastructure, communications, and cybersecurity, in order to come up with a sound threats model.

Finally, Abdo et al. [22] argue that safety and cybersecurity should be seen in tandem. Currently used risk assessment approaches treat safety and cyber threats as two distinct entities, while in the Industry 4.0 era, a safety risk is also a cybersecurity risk. Consider an AHT and a manual haulage truck; both provide certain safety measures to safeguard and protect the equipment and operators. However, the only safety concern regarding a manual truck is if the operator tampers physically with the truck. On the contrary, an AHT poses the same safety concerns plus a residual risk of it being a CPS connected to an open network (i.e., Wifi). Hence, the cybersecurity risks is a critical factor that might affect the safety of the mine. The bowtie and attack tree analysis are utilized by [22], combining safety and security in risk analysis to generate an exhaustive representation of risk scenarios. Unfortunately, this method would necessitate the collection of qualitative and quantitative data to calculate the probability of safety risks. These data are private and often proprietary to manufacturers as well as not always accessible for analysis, making this study difficult to conduct further studies to confirm or improve the results. Al-Ali et al. [17] propose creating a trust zone to handle sharing such information, but this entails the acceptance of all parties. Table 1 summarizes some proposed solutions and their challenges.

Table 1. Summary of literature on cybersecurity and safety.

Papers	Used Techniques/Technologies	Security and Safety Challenges Addressed	Advantages	Disadvantages
[6]	Combination of existing standards, e.g., ISO 27000, ISA99	Existing solutions are focused on safety aspects of AHS applications only	Utilizing well-established standards in different environments	Proposed solution is still immature and has yet to be evaluated in AHS environment
[12]	Combine pre-existing standards to address the complexity of OT environment	Same forms of attacks that occur in the IT world can be observed in the industrial arena. Lack of standardized approach	Some standards are mature and proven to be effective	Standards do not take into consideration the wireless environment within AHS setups. Adopting some practices could be risky as they are not environment specific
[13,14]	AHTs is a CPS	textbfFocus on specific application rather than standardized specifications that were designed to support general purposed devices	CPS can benefit from a wide range of existing applications in other areas	Manipulating the CPS without systems' owner knowledge, e.g., MITM attack. Loss of integrity that could result in loss of lives
[15]	Security by design	Physical, network, and computation are three fields that may be exploited	Enforce, Authentication, Authorization, Network Enforced Policy, and Secure Analytics as measures to reinforce security	Approaching cybersecurity problems in an industrial ad hoc manner can lead to misleading findings because generalized attack studies lack the specifics of security objectives
[17]	OT–IT data sharing	Address how to access/share data between different environments	Dynamic trust zone where decisions are dynamically made by defining and analyzing flows, then intelligently determine whether the flow is permitted	Raises concerns about data privacy and confidentiality, especially in a multivendor environment
[9]	Superior signals extraction to identify targeted GPS attacks	The GPS data must be protected to avoid GPS-based collisions and ATs deviation from the target positions	Practical techniques can figure out vital characteristics of the network such as signal strength	Secure GPS strategies are not considered in intended AHS and need to be investigated further, especially in terms of spoofing and jamming
[21]	Defense Strategies (removable near-infrared-cut filters, and photochromic lenses)	Camera attacks (a blurred camera's outputs break a safety standard and increase fatal accidents as well)	Photochromic lenses and removable near-infrared cut filters offer sufficient protection from a variety of angles	There are no concrete solutions for camera protection in ATs. Camera attacks can cause inaccurate detections of obstacles, lanes, or traffic signs
[22]	Bowtie analysis and attack tree analysis	Cybersecurity impacts safety of mining equipment	Produce an exhaustive representation of risk scenarios. To measure the risk of safety threats, quantitative and qualitative data are required	Data are usually privately owned by manufacturers and are not always available for analysis, which would pose a challenge to conduct such a study

2.2. Relation between Communications and Safety in AHS

AHS systems have prospered in surface mining with outstanding results, which increased the demand to adopt autonomous mining. Manufacturers have placed the greatest emphasis on safety, which is seen as the essence of mining operations. From collisions to high weather temperatures to difficult ground conditions, robotic autonomy has aided in operating in such harsh environments. This can be done either by improving the efficiency or by allowing robots to operate in areas that humans find inoperable. Marshall et al. [11] discuss how robotics have contributed to mining and other domains regarding the following areas:

- Assist workers in hazardous environments that pose health risks, such as excessive heat, dust, poisonous smoke, or hydrogen sulfide (H2S).
- Fulfill labor shortages.
- Provide an opportunity to increase health and safety.
- Outperform humans in terms of performance.

Reliable communication has a significant role in the prosperity of AHS systems. ATs can communicate via various communications with the command center to collect data from neighbors, control telemetry, and monitor the health and safety of components. Using a secure communication system, the command center can guide trucks as well as manage them to enable tracking the mining operations. It is important to note that the mining environment is the same as any other industrial environment. That is, mining operations would effectively benefit from technological advancement in communications, but they also suffer from the same vulnerabilities with some particularities pertaining to the mining environment. In particular, AHS in mining relies heavily on wireless communications that were considered unfit for industrial operation at some point. However, as stated in [8], with the industrial revolution 4.0 and the integration of CPS and IoT systems in mining (i.e., in the industry in general), communication technologies have become essential for daily operation. Indeed, an autonomous truck must maintain continuous communication with central control. Otherwise, communication failure, even for a single piece of equipment, means the whole fleet will stop running. Technically, this is called "mine shutdown" due to communication loss. According to [6], AHS relies entirely on wireless technology for secure production and supervisory control. As a result, a stable network infrastructure (wired and wireless) is critical to AHS operations.

Since the advent of Industry 4.0, Wi-Fi technology has been an integral part of industrial operation [8]. Wireless Networks (WNs) have also opened new opportunities for business, such as easy deployment with lower cost. Despite the essential role of WNs in linking field devices and mobile assets, they face some difficulties that could affect both credibility and availability of operations. Labbe [6] demonstrates that AHS communications are susceptible to known attacks such as Wi-Fi De-Auth, which is a DoS assault on key operations. The authors in [7] show how existing wireless standards are insufficient to meet the demands of Industry 4.0. Sisinni et al. [23] argue that the advent of IoT and CPS in the industrial environment have caused existing Wi-Fi standards to lose traction as they are not capable of handling dense and large-scale deployments. Signal interference, topology control in the mining environment, and signal jamming are some of the issues inherited from 802.11 standards [8].

In addition, Kiziroglou et al. [24] address the relevance of wireless sensor networks (WSNs) and their capacity to improve safety as well as availability in a mining environment. The authors also highlight current challenges and how the mining industry should take advantage of WSNs. For that reason, WSNs could assist in the following areas of mining operations:

- Localization services, especially for AHS vehicles that require high precision and low latency.
- Data collection and analysis to minimize downtime that is a critical factor in extending the lifetime of an operation along with aiding in optimizing operations and achieving proactive maintenance.
- Health and safety are paramount in mining industry; sensing technology may assist in gathering data from the field to monitor both employee and equipment health, especially in areas where toxic gases are present (e.g., H2S). Furthermore, proximity sensors are designed to prevent and detect obstacles along with dangerous condi-

tions while trucks are driving in an autonomous mode, which is essential in mining operations.

Despite the advantages of WSNs implementation (e.g., low cost, flexible design, and real-time monitoring), they still suffer from some significant drawbacks that restrict their application to specific areas [25]. For instance, WSNs protocols rely on the 802.15.4 standards with lower energy consumption as a primary aim. Thus, the majority of these protocols are designed for low-data-rate proximity applications, which makes them ideal in smaller environments (i.e., mining environments would require a high data rate and significant proximity). Although some 802.14.5 protocols, such as WilressHART, are built to support security in the industrial environment, ensuring confidentiality might be challenging. That is, WSNs sensors consume lower energy that limits their ability to encrypt with more secure algorithms.

Private LTE (pLTE) is a viable alternative solution to traditional 802.11 technology (Wi-Fi) that could provide robust communication and evade WSNs limitations [26]. In [26], the authors demonstrate how pLTE addresses performance attributes. Additionally, they highlight how the global LTE ecosystem enables private enterprises to deploy and operate LTE networks independently of licensed service providers. Furthermore, the provision of an open-access spectrum (e.g., 3.5 GHz in the United States and 5 GHz worldwide [27]) enables organizations to deploy pLTE networks. In addition, pLTE networks guarantee adequate coverage, particularly in remote areas such as mines. It also has the potential to uplink/downlink traffic capacity, especially where video streaming is used. Subsequently, organizations with private LTE have increased control over network traffic, Quality of Service (QoS), and security, and the network can be customized to optimize reliability and latency in challenging environments, such as mining.

Additionally, pLTE would reduce maintenance costs since Long-Term Evolution (LTE) infrastructure does not need as many towers as traditional Wi-Fi due to its higher spectral efficiency. It also could alleviate contention issues associated with other existing networks. In terms of security, pLTE leverages well-established cellular network security infrastructure, e.g., Classic SIM-based and non-SIM options security. Due to the above, pLTE seems to be the savior solution, although it comes with a high price tag and the assumption that there is already an infrastructure ready to be deployed. Furthermore, pLTE solutions might be subject to approval and discretion by local governments, especially when it comes to the licensed spectrum.

Nowadays, several ongoing advancements proceed in the field of autonomous vehicle technologies. Specified standards developed by the IEEE team (i.e., IEEE 802.11p) for vehicular networks are known as Wireless Access for Vehicular Environment (WAVE). Furthermore, Dedicated Short Range Communications (DSRC) is one among these technologies that is deployed for short- to medium-range communications, especially for vehicular networks. The DSRC/WAVE technology has been utilized for distinct vehicular applications including infotainment, resource efficiency, and safety applications [28]. Regarding mining truck autonomy, Abdellah and Paul [29] survey the performance of different routing protocols when used for cooperative collision warning in mines. This study could serve as guidance for the design of new traffic control systems that prioritize safety applications. In addition, faster data packet dissemination is emphasized for cooperative collision notification in underground mining such as deploying 5G technology. In addition, the authors in [29] address the compatibility of vehicular networks (i.e., AHS in our case) with communication standards such as IEEE 802.11x, WiMax, and DSRC/WAVE standards. Table 2 summarizes the most recent technologies, their characteristics and limitations in AHS environments.

Table 2. Summary of literature on cybersecurity and safety.

Papers	Used Techniques/Technologies	Security and Safety Challenges Addressed	Advantages	Disadvantages
[8]	Wi-Fi technology is essential in AHS mining and Industry 4.0	AHS trucks require constant communication with central command and with each other	Presented new advantages for the business, such as low cost and easy deployment	Vulnerabilities such as Wi-Fi De-Auth, which is DoS attacks on critical operations. The authors of [7] explain how the current wireless standards alone are not adequate to address Industry 4.0 requirements
[24]	WSNs to increase safety	Monitor safety of equipment and operators when applicable in mining environment	Collect real-time data in the field and send the data to command centers to monitor persons and equipment in high-risk areas. A proximity sensor is designed to identify and avoid hazards while trucks are in autonomous mode, particularly in mining operations	WSN protocols are based on the 802.15.4 standard with a low energy usage objective. 802.14.5 protocols support security in the industrial environment, e.g., WilressHART, while ensuring confidentiality might be a challenge due to the limited battery life of sensor nodes that affects their ability to encrypt with more secure algorithms
[26]	Use of pLTE as medium of communication in the mine	Lack of topology and coverage issues	Private organizations can (in some countries) deploy and operate LTE networks without relying on licensed service providers. Additionally, pLTE networks allow organizations to guarantee coverage, especially in mines, and the potential to increase the capacity of uplink/downlink traffic for better video streaming	The lack of ASH-based research addressing potential new threats/attacks and their impact on the safety of mines
[29]	DSRC for autonomous mining vehicles	Applying several routing protocols under DSRC/WAVE standards for autonomous vehicles in underground mines	The first to address the usage of DSRC/Wave in the underground mine topology with the Rayleigh fading channel for emergency message dissemination protocols. It also provides a cooperative collision warning in underground mining environment	The lack of research into potential attacks and communication protection in underground mining using DSRC technology

3. Open Issues

Since our primary design goal is functionality, it is not unexpected nor rare to discover security flaws in applications that were designed to perform a particular task. However, lack of security can be detected once these applications are placed in open networks. Hence, we quickly realize the significant degradation effect in functionality because security was not a part of the initial design as mentioned earlier by Kim et al. [14]. Autonomous mines (i.e., mines in general) are not an exception to this phenomenon in which manufacturers strived to develop smart machinery that can be remotely operated and controlled as well as autonomously perform the work. Nevertheless, the cybersecurity aspect of this equipment is being implemented as an afterthought. In such a context, the aim of the cyberattacks may be to sabotage or slow down the ATs due to a competitive company in the market. Through this survey, we lightened how securing communication in an industrial environment is highly dependent on general practices and guidelines, as stated in [6].

According to the literature and discussion above, more real cases should be considered for study to understand the impact of the GPS attacks on AHS in the mining industry. Aspects including the signal strength, direction of arrival, and signal clock can be considered so as to identify fake signals as well prevent GPS attacks. A successful GPS attack can deviate the ATs from the target's true positions. A catastrophic collision of two or more trucks can also occur, resulting in the loss of people and resources. On the other hand, camera attacks are discussed in several studies, such as [9,21,30,31]. These papers have highlighted ideas about protection from camera attacks, without presenting practical solutions or countermeasures to camera attacks. Camera attacks are extremely dangerous since ATs may be unable to detect obstacles or recognize traffic signs in the mine resulting in ATs collisions. Thus, more investigations and solutions are still needed to observe the effect of camera attacks on the safe operations of the AHS environments.

The discussion in Section 2.1 revealed that most of the cybersecurity challenges that exist recently in mining are due to the absence of standardization and oriented solutions. The availability, productivity, and safety of operations are subjected to the security of communications when we project these challenges onto real-world applications, e.g., AHS. Wireless communication technologies such as LTE and 5G would achieve a rapid transformation of the AHS in the mining environment [32]. Thus, incorporating these technologies into the mining operations will give rise to opportunities for new attack vectors in the industrial environment. This leads us to the earlier claim that there is a current necessity for a well-studied procedure or guideline to tackle specific issues in the mine. As an example, what is the ultimate channel utilization and power transmit (Tx) within the mine? How can rouge access points be prohibited from injecting traffic into the mining network in the presence of mine challenges with signal processing? Is the IEEE802.11 wireless protocol a viable mining solution (i.e., especially for AHS) or should different solutions be suggested, e.g., pLTE, 4G, and 5G systems?

Furthermore, there is a strong relationship between cybersecurity and communication technologies in the mining industry. DSRC-WAVE technology is also considered a suitable vehicular communication technology (i.e., vehicle to everything (V2X)) for autonomous vehicles in the mining industry, according to Abdullah et al. [29]. This technology enables a variety of applications for autonomous vehicles, including safety and resource efficiency applications [28]. As a result, the investigation of DSRC communication technology in the mining industry, especially the AHS, is still an open issue. For example, an unauthorized emergency message disseminated among the autonomous trucks can cause disruption in the mining operation. The integrity of the propagated message is not maintained, which could lead to wrong decisions related to the mining operations. Although these communication technologies will outperform current technologies in terms of bandwidth and latency, they would be exposed to different variant of cybersecurity attacks, such the case of jamming attacks [33].

5G, pLTE, and DSRC-WAVE, as any previous wireless communication technologies, are vulnerable to common attacks such as jamming attacks, which produce deliberate interference in order to obstruct genuine users' communication. Such types of attacks (i.e., jamming attacks) pose a serious risk to public safety, and hence mining safety too. Therefore, the mining industry, when developing AHS systems adopting any of these technologies, should develop a comprehensive security strategy for enabling these technologies in automated mines using AHS-based systems.

All of these issues inspired us to pose even more in-depth questions and investigate different directions. For instance, in areas where both AHS and manual mining might be simultaneously used, what are the cybersecurity ramifications of having them both on the same network? Up until the writing of this paper, there were no industry standards or regulations that impose any specifications. Instead, organizations are repurposing other industry standards to fit within the mining environment, as previously mentioned. Overall, this study seeks to bridge the gap between existing standards and mining applications in cybersecurity, specifically network infrastructure. Yet, we believe a comprehensive solution would take into consideration security during the design phase as well as build solutions purposely to accommodate the mining environment and its applications, such as the AHS system.

4. Conclusions

In this paper, a type of Cyber-physical Systems (CPS), i.e., an AHS in the mining environment, was discussed in terms of cybersecurity, reliable communication, and safety. AHS-based trucks have been shown to be very useful in the mining industry. However, the safety of using such trucks has not been thoroughly investigated. The literature was then surveyed to identify and discuss the relation between safety of AHS in the mining environment and both cybersecurity and communication. The relation between cybersecurity and safety in AHS systems was discussed and it was found that compromising cybersecurity would threaten the safety of mining operations leading to loses in people and equipment. In addition, we have shown that the reliability of wireless communication is mandatory for the safety of AHS operations. Through this survey, there are several open issues and challenges that may entail further studies and investigations. It was discussed that new technologies such 5G, pLTE and DSRC-WAVE could be good alternatives for the currently used IEEE 802.11. Although these communication technologies would outperform current technologies in terms of bandwidth and latency, they would be exposed to traditional or new cybersecurity threats or attacks, which still need to be considered while designing AHS systems and studied in the mining environment.

Author Contributions: Conceptualization, T.G. and Y.E.J.; methodology, T.G.; E.E.; and A.A.; validation T.G.; analysis, T.G.; E.E.; and A.A.; investigation, T.G. and Y.E.J.; E.E.; and A.A.; resources, T.G. and Y.E.J.; writing—original draft preparation, T.G. and Y.E.J.; writing—review and editing, T.G.; E.E.; and A.A.; supervision, T.G. All authors have read and agreed to the published version of the manuscript.

Funding: This research received no external funding.

Acknowledgments: We would like to acknowledge with much appreciation Prince Sattam Bin Abdulaziz University for its ongoing support to our research.

Conflicts of Interest: The authors declare no conflict of interest.

Abbreviations

The following abbreviations are used in this manuscript:

CPS	Cyber-physical Systems
AHS	Autonomous Haulage System
OT	Operating Technology
ATs	Autonomous Trucks
EMV	Equipped Manual Vehicle
ICS	Industrial Control Systems
AHT	Autonomous Haulage Trucks
RoT	Root of Trust
IT	Information Technology
AI	Artificial Intelligence
H2S	Hydrogen Sulfide
GPS	Global Positioning System
WNs	Wireless Networks
WSNs	Wireless Sensor Networks
LTE	Long-Term Evolution
DSRC	Dedicated Short Range Communications
WAVE	Wireless Access for the Vehicular Environment
pLTE	private LTE
QoS	Quality of Service

References

1. de Vass, T.; Shee, H.; Miah, S. IoT in Supply Chain Management: Opportunities and Challenges for Businesses in Early Industry 4.0 Context. *Oper. Supply Chain Manag. Int. J.* **2021**, *14*, 148–161. [CrossRef]
2. Carreras Guzman, N.H.; Wied, M.; Kozine, I.; Lundteigen, M.A. Conceptualizing the key features of cyber-physical systems in a multi-layered representation for safety and security analysis. *Syst. Eng.* **2020**, *23*, 189–210. [CrossRef]
3. Lee, E.A.; Seshia, S.A. *Introduction to Embedded Systems: A Cyber-Physical Systems Approach*; Mit Press: Cambridge, MA, USA, 2017.
4. Roth, M. Komatsu's Autonomous Haulage System Sests Record for Surface Material Moved. 2018. Available online: https://www.mining.com/web/frontrunner-autonomous-haulage-system-sets-new-record-latest-industry-milestone/ (accessed on 30 April 2021).
5. Parreira, J. An Interactive Simulation Model to Compare an Autonomous Haulage Truck System with a Manually-Operated System, Autonomous Haulage Truck, Simulation Model. 2013. Available online: https://open.library.ubc.ca/collections/24/items/1.0074111 (accessed on 30 April 2021).
6. Labbe, R. Securing Autonomous Systems, Mining and Metals Information Sharing and Analysis Centre, Canadaian Institute of Mining, AHS, Cybersecurity. 2019. Available online: https://store.cim.org/en/securing-autonomous-systems (accessed on 30 April 2021).
7. Varghese, A.; Tandur, D. Wireless requirements and challenges in Industry 4.0. In Proceedings of the 2014 international conference on contemporary computing and informatics (IC3I), Mysore, India, 27–29 November 2014; pp. 634–638.
8. Li, X.; Li, D.; Wan, J.; Vasilakos, A.V.; Lai, C.F.; Wang, S. A review of industrial wireless networks in the context of industry 4.0. *Wirel. Netw.* **2017**, *23*, 23–41. [CrossRef]
9. Ren, K.; Wang, Q.; Wang, C.; Qin, Z.; Lin, X. The security of autonomous driving: Threats, defenses, and future directions. *Proc. IEEE* **2019**, *108*, 357–372. [CrossRef]
10. Cheng, H.Y.; Jeng, B.S.; Tseng, P.T.; Fan, K.C. Lane detection with moving vehicles in the traffic scenes. *IEEE Trans. Intell. Transp. Syst.* **2006**, *7*, 571–582. [CrossRef]
11. Joshua, A.M.; Adrian, B.; Eduardo, N.; Steven, S. Robotics and the Handbook. In *Springer Handbook of Robotics*; Springer: Berlin/Heidelberg, Germany, 2016; pp. 1–6.
12. Alani, M.M.; Alloghani, M. *Industry 4.0 and Engineering for a Sustainable Future*; Springer: Berlin/Heidelberg, Germany, 2019.
13. Chattopadhyay, A.; Lam, K.Y. Security of autonomous vehicle as a cyber-physical system. In Proceedings of the 2017 7th International Symposium on Embedded Computing and System Design (ISED), Durgapur, India, 18–20 December 2017; pp. 1–6.
14. Kim, S.; Won, Y.; Park, I.H.; Eun, Y.; Park, K.J. Cyber-physical vulnerability analysis of communication-based train control. *IEEE Internet Things J.* **2019**, *6*, 6353–6362. [CrossRef]
15. He, H.; Maple, C.; Watson, T.; Tiwari, A.; Mehnen, J.; Jin, Y.; Gabrys, B. The security challenges in the IoT enabled cyber-physical systems and opportunities for evolutionary computing & other computational intelligen. In Proceedings of the 2016 IEEE Congress on Evolutionary Computation (IEEE CEC), Vancouver, BC, Canada, 24–29 July 2016; pp. 1015–1021.
16. Lezzi, M.; Lazoi, M.; Corallo, A. Cybersecurity for Industry 4.0 in the current literature: A reference framework. *Comput. Ind.* **2018**, *103*, 97–110. [CrossRef]
17. Al-Ali, R.; Heinrich, R.; Hnetynka, P.; Juan-Verdejo, A.; Seifermann, S.; Walter, M. Modeling of dynamic trust contracts for industry 4.0 systems. In Proceedings of the 12th European Conference on Software Architecture: Companion Proceedings, Madrid, Spain, 24–28 September 2018; pp. 1–4.
18. Zeng, K.C.; Liu, S.; Shu, Y.; Wang, D.; Li, H.; Dou, Y.; Wang, G.; Yang, Y. All your GPS are belong to us: Towards stealthy manipulation of road navigation systems. In Proceedings of the 27th USENIX Security Symposium (USENIX Security 18), Baltimore, MD, USA, 15–17 August 2018; pp. 1527–1544.
19. Nccic, I.C. Recommended Practice: Improving Industrial Control System Cybersecurity with Defense-in-Depth Strategies Industrial Control Systems Cyber Emergency Response Team. 2016. Available online: https://ics-cert.us-cert.gov/sites/default/files/recommended_practices/NCCIC_ICS-CERT_Defense_in_Depth_2016_S508C.pdf (accessed on 30 April 2021).
20. Petit, J.; Stottelaar, B.; Feiri, M.; Kargl, F. Remote attacks on automated vehicles sensors: Experiments on camera and lidar. *Black Hat Eur.* **2015**, *11*, 995.
21. Kyrkou, C.; Papachristodoulou, A.; Kloukiniotis, A.; Papandreou, A.; Lalos, A.; Moustakas, K.; Theocharides, T. Towards artificial-intelligence-based cybersecurity for robustifying automated driving systems against camera sensor attacks. In Proceedings of the 2020 IEEE Computer Society Annual Symposium on VLSI (ISVLSI), Limassol, Cyprus, 6–8 July 2020; pp. 476–481.
22. Abdo, H.; Kaouk, M.; Flaus, J.M.; Masse, F. A safety/security risk analysis approach of Industrial Control Systems: A cyber bowtie–combining new version of attack tree with bowtie analysis. *Comput. Secur.* **2018**, *72*, 175–195. [CrossRef]
23. Sisinni, E.; Saifullah, A.; Han, S.; Jennehag, U.; Gidlund, M. Industrial internet of things: Challenges, opportunities, and directions. *IEEE Trans. Ind. Inform.* **2018**, *14*, 4724–4734. [CrossRef]
24. Kiziroglou, M.E.; Boyle, D.E.; Yeatman, E.M.; Cilliers, J.J. Opportunities for sensing systems in mining. *IEEE Trans. Ind. Inform.* **2016**, *13*, 278–286. [CrossRef]
25. Raza, S.; Faheem, M.; Guenes, M. Industrial wireless sensor and actuator networks in industry 4.0: Exploring requirements, protocols, and challenges—A MAC survey. *Int. J. Commun. Syst.* **2019**, *32*, e4074. [CrossRef]
26. Brown, G. Private LTE Networks-Qualcomm. *Qualcomm* **2017**, 1–11. Available online: https://www.qualcomm.com/media/documents/files/private-lte-networks.pdf (accessed on 3 June 2021).

27. Ratasuk, R.; Mangalvedhe, N.; Ghosh, A. LTE in unlicensed spectrum using licensed-assisted access. In Proceedings of the 2014 IEEE Globecom Workshops (GC Wkshps), Austin, TX, USA, 8–12 December 2014; pp. 746–751.
28. Ng, H.H.; Vasudha, R.; Hoang, A.T.; Kwan, C.; Zhou, B.; Cheong, J.; Quek, A. BESAFE: Design and implementation of a DSRC-based test-bed for connected autonomous vehicles. In Proceedings of the 2018 21st International Conference on Intelligent Transportation Systems (ITSC), Maui, HI, USA, 4–7 November 2018; pp. 3742–3748.
29. Chehri, A.; Fortier, P. Autonomous Vehicles in Underground Mines, Where We Are, Where We Are Going? In Proceedings of the 2020 IEEE 91st Vehicular Technology Conference (VTC2020-Spring), Antwerp, Belgium, 25–28 May 2020; pp. 1–5.
30. Khadka, A.; Karypidis, P.; Lytos, A.; Efstathopoulos, G. A benchmarking framework for cyber-attacks on autonomous vehicles. *Transp. Res. Procedia* **2021**, *52*, 323–330. [CrossRef]
31. Stottelaar, B.G. Practical Cyber-Attacks on Autonomous Vehicles. Master's Thesis, University of Twente, Enschede, The Netherlands, 2015.
32. Conway, G. The Evolving Role of Communications for the Digital Mine. *Eng. Min. J.* **2020**, *221*, 54–55.
33. Arjoune, Y.; Faruque, S. Smart jamming attacks in 5G new radio: A review. In Proceedings of the 2020 10th Annual Computing and Communication Workshop and Conference (CCWC), Las Vegas, NV, USA, 6–8 January 2020; pp. 1010–1015.

MDPI
St. Alban-Anlage 66
4052 Basel
Switzerland
www.mdpi.com

Electronics Editorial Office
E-mail: electronics@mdpi.com
www.mdpi.com/journal/electronics

Disclaimer/Publisher's Note: The statements, opinions and data contained in all publications are solely those of the individual author(s) and contributor(s) and not of MDPI and/or the editor(s). MDPI and/or the editor(s) disclaim responsibility for any injury to people or property resulting from any ideas, methods, instructions or products referred to in the content.

www.ingramcontent.com/pod-product-compliance
Lightning Source LLC
LaVergne TN
LVHW070142100526
838202LV00015B/1873